Springer Series in
ADVANCED MICROELECTRONICS 13

Springer-Verlag Berlin Heidelberg GmbH

Springer Series in
ADVANCED MICROELECTRONICS

Series Editors: K. Itoh T. Lee T. Sakurai W. M. C. Sansen D. Schmitt-Landsiedel

The Springer Series in Advanced Microelectronics provides systematic information on all the topics relevant for the design, processing, and manufacturing of microelectronic devices. The books, each prepared by leading researchers or engineers in their fields, cover the basic and advanced aspects of topics such as wafer processing, materials, device design, device technologies, circuit design, VLSI implementation, and subsystem technology. The series forms a bridge between physics and engineering and the volumes will appeal to practicing engineers as well as research scientists.

1 **Cellular Neural Networks**
Chaos, Complexity
and VLSI Processing
By G. Manganaro, P. Arena,
and L. Fortuna

2 **Technology of Integrated Circuits**
By D. Widmann, H. Mader,
and H. Friedrich

3 **Ferroelectric Memories**
By J.F. Scott

4 **Microwave Resonators and Filters for Wireless Communication**
Theory, Design and Application
By M. Makimoto and S. Yamashita

5 **VLSI Memory Chip Design**
By K. Itoh

6 **Smart Power ICs**
Technologies and Applications
Ed. by B. Murari, R. Bertotti,
and G.A. Vignola

7 **Noise in Semiconductor Devices**
Modeling and Simulation
By F. Bonani and G. Ghione

8 **Logic Synthesis for Asynchronous Controllers and Interfaces**
By J. Cortadella, M. Kishinevsky,
A. Kondratyev, L. Lavagno,
and A. Yakovlev

9 **Low Dielectric Constant Materials for IC Applications**
Editors: P.S. Ho, J. Leu, W.W. Lee

10 **Lock-in Thermography**
Basics and Use
for Functional Diagnostics
of Electronic Components
By O. Breitenstein
and M. Langenkamp

11 **High-Frequency Bipolar Transistors**
Physics, Modelling, Applications
By M. Reisch

12 **Current Sense Amplifiers for Embedded SRAM in High-Performance System-on-a-Chip Designs**
By B. Wicht

13 **Silicon Optoelectronic Integrated Circuits**
By H. Zimmermann

Horst Zimmermann

Silicon Optoelectronic Integrated Circuits

With 291 Figures and 19 Tables

Springer

Professor Dr.-Ing. Horst Zimmermann
Institut für Elektrische Mess- und Schaltungstechnik
Technische Universität Wien
Gusshausstraße 25/354, 1040 Wien
Austria
E-mail: horst.zimmermann@ieee.org

Series Editors:

Dr. Kiyoo Itoh
Hitachi Ltd., Central Research Laboratory, 1-280 Higashi-Koigakubo
Kokubunji-shi, Tokyo 185-8601, Japan

Professor Thomas Lee
Stanford University, Department of Electrical Engineering, 420 Via Palou Mall, CIS-205
Stanford, CA 94305-4070, USA

Professor Takayasu Sakurai
Center for Collaborative Research, University of Tokyo, 7-22-1 Roppongi
Minato-ku, Tokyo 106-8558, Japan

Professor Willy M. C. Sansen
Katholieke Universiteit Leuven, ESAT-MICAS, Kasteelpark Arenberg 10
3001 Leuven, Belgium

Professor Doris Schmitt-Landsiedel
Technische Universität München, Lehrstuhl für Technische Elektronik
Theresienstraße 90, Gebäude N3, 80290 München, Germany

ISSN 1437-0387

Library of Congress Cataloging-in-Publication Data: Zimmermann, Horst, 1957- Silicon optoelectronic integrated circuits / Horst Zimmermann. p. cm. – (Springer series in advanced microelectronics ; 13) Includes bibliographical references and index.

DOI 10.1007/978-3-662-09904-9

1. Integrated circuits–Design and construction. 2. Optoelectronic devices. I. Title. II. Series.
TK7874.Z58 2003 621.3815–dc22 2003057315

springeronline.com

Originally published by Springer-Verlag Berlin Heidelberg New York in 2004
MyCopy version of the original edition 2004

Typesetting by the author using a Springer LaTeX macro package
Cover concept by eStudio Calamar Steinen using a background picture from Photo Studio "SONO". Courtesy of Mr. Yukio Sono, 3-18-4 Uchi-Kanda, Chiyoda-ku, Tokyo
Cover design: *design & production* GmbH, Heidelberg

Printed on acid-free paper SPIN: 11735885 57/3111 - 5 4 3 2 1
www.springer.com/mycopy

Preface

Since the book "Integrated Silicon Optoelectronics" appeared in the "Springer Series in Photonics" in 2000, a whole variety of silicon optoelectronic integrated circuits (SOEICs) have been developed and introduced in many journals and conference proceedings. Therefore, a new book on these new SOEICs collecting and selecting the most interesting ones was highly desirable especially because the main part of "Integrated Silicon Optoelectronics" concentrated on integrated photodetectors. This new book, in contrast, describes considerably more circuits implemented as SOEICs together with the photodetectors.

Many design engineers in semiconductor companies, ASIC and design houses have to design SOEICs, OPTO-ASICs, image sensors or even smart pixel sensors. I also feel that the number of Ph.D. students or diploma workers doing research and development in the field of optoelectronic circuits is constantly growing. This book, therefore, is intended as a second bridge (after "Integrated Silicon Optoelectronics") between microelectronics and optoelectronics. Usually, optoelectronics plays a minor role in electrical engineering courses at universities. Physicists are taught optics but not very much semiconductor technology and chip design. This book covers the missing information for engineers and physicists who want to know more about integrated optoelectronic circuits (OEICs) in silicon technologies and about their rapidly emerging applications. The low-cost requirement permanently drives and pushes silicon OEICs in contrast to expensive III/V semiconductor receiver OEICs. This book reflects this trend by stressing CMOS OEICs. BiCMOS OEICs with their better performance, however, are also described in detail, since they are still much cheaper than III/V OEICs.

The importance of OEICs is due to the following advantages of monolithic optoelectronic integrated circuits: (i) good immunity against electromagnetic interference (EMI) because of very short interconnects between photodetectors and amplifiers, (ii) reduced chip area due to the elimination of bondpads, (iii) improved reliability due to the elimination of bondpads and bond wires, (iv) cheaper mass production compared to discrete circuits, wire-bonded circuits, and hybrid integrated circuits, and (v) larger −3dB bandwidth compared to discrete circuits, wire-bonded circuits, and some hybrid integrated circuits due to the avoidance of parasitic bondpad capacitances.

This book describes the basics and theory of photodetecors in compact form. The three chapters on integrated photodectors, thin-film detectors, and SiGe detectors desribe the properties of photodiodes, which were implemented in the circuits discussed later in this book. The chapter on design of integrated circuits covers analytical methods for calculating bandwidth. Methods for calculating input and output resistance as well as electronic noise of transimpedance amplifiers were also added. Furthermore, a transmission-line approach leading to a new π-model for integrated transimpedance amplifiers was included.

In the last and longest chapter, new concepts for DVD and CD-ROM OE-ICs and new results on these still economically more important key devices for optical storage systems were included. The state-of-the-art of fiber receiver OEICs was updated and new market demands like plastic optic fiber receivers and burst-mode optical receivers are covered by the description of these receiver circuits. The key topics systems-on-chip (SoC) and camera-on-chip (CoC) are included. Within CoCs, revolutionary three-dimensional single-chip CMOS cameras are described. Various CMOS and BiCMOS optical sensor chips are introduced. Several innovative smart-sensor circuits are highlighted. Furthermore, speed-enhancement techniques for fiber receivers and large-area, large-capacitance photodiodes are explained. The general trend towards deep-sub-micrometer analog-digital CMOS SoCs is considered with respect to OEICs. Finally, new optical interconnect and free-space receivers based on the optoelectronic phase-locked loop principle are introduced.

Parts of the book have their origin in the lecture "Optoelectronic integrated circuits" started at Vienna University of Technology in 2001 and in the lecture "Optoelectronics" I had given from 1994 to 1999 at Kiel University. This book, however, dives much deeper into the topic. The possibilities of SOEICs are described thoroughly with respect to circuits and some new integrated detectors like the innovative so-called spatially-modulated-light detector and the photonic mixer device (PMD) are added. The newest publications on silicon OEICs up to 2003 are considered.

I would like to thank Prof. Dr.-Ing. P. Seegebrecht from Kiel University, Germany, where I began research and development on silicon OEICs, for the generous possibility to work independently and to acquire the title 'habilitatus'. I am also indebted to Prof. Dr. H. Föll, who offered a waferprober for the characterization of the OEICs. The work of the OEIC group members at Kiel University, A. Ghazi, T. Heide, M. Hohenbild, K. Kieschnick, and G. Volkholz, is highly appreciated. Three students, N. Madeja, F. Sievers, and U. Willecke, carefully performed simulations and measurements. M. Wieseke and F. Wölk helped with the preparation of numerous drawings. Special thanks go to R. Buchner from the Fraunhofer-Institute for Solid-State Technology in Munich and H. Pless from Thesys Microelectronics (now Melexis) in Erfurt, Germany, for their engagement in the fabrication of CMOS OEICs and BiCMOS OEICs, respectively. I gratefully acknowledge the funding of the projects by the German ministry for education, science, research, and technology (BMBF) within the leading project 'Optical Storage'.

At the Institute for Electrical Measurements and Circuit Design (EMST) at Vienna University of Technology, I would like to thank my colleague and head of the institute Gottfried Magerl for his great support towards a quick start of research. I also have to thank my Ph.D. students M. Förtsch, J. Knorr, F. Schlögl, K. Schneider, and R. Swoboda for their careful design and characterization work. M. Hofer, Ch. Sünder, and J. Wissenwasser drew many new figures. I further have to thank G. Langguth and H. Wille from Infineon Technologies in Munich, Germany, J. Sturm from Infineon Technologies in Villach, Austria, as well as A. Martin from Infineon Technologies in Vienna, Austria, for their support and constructive cooperation. I further acknowledge funding from the European Commission in the project INSPIRED.

I extend my sincere thanks to Dr. Ascheron and his team at Springer for the good cooperation and their technical support with the text processor. My deepest gratitude, however, is directed to my wife and my daugthers, Luise and Lina, as well as to my son Frieder, who supported this second book project with their encouragement and patience.

Vienna, August 2003 *Horst Zimmermann*

Contents

List of Symbols XIII

1 Basics and Theory 1
1.1 Basics of Optical Absorption 1
1.1.1 Photons and their Properties 1
1.1.2 Optical Absorption of Important Semiconductor Materials 2
1.1.3 Photogeneration 4
1.2 Semiconductor Equations 5
1.3 Important Models for Photodetectors 8
1.3.1 Carrier Drift 8
1.3.2 Carrier Diffusion 13
1.3.3 Quantum Efficiency and Responsivity 16
1.3.4 Equivalent Circuit of a Photodiode 21

2 Integrated Silicon Photodetectors 25
2.1 Integrated Detectors in Bipolar Technology 25
2.1.1 Bipolar Processes 25
2.1.2 Integrated Detectors in Standard Bipolar Technology 28
2.1.3 Integrated Detectors in Modified Bipolar Technology 31
2.2 CMOS-Integrated Photodetectors 34
2.2.1 One-Well CMOS Processes 35
2.2.2 Twin-Well CMOS Processes 35
2.2.3 Integrated Detectors in Standard CMOS Processes 36
2.2.4 Spatially-Modulated-Light Detector 41
2.2.5 PIN Photodiode 43
2.2.6 Charge-Coupled-Device Image Sensors 58
2.2.7 Active-Pixel Sensors 62
2.2.8 Amorphous-Silicon Detectors 66
2.2.9 CMOS-Integrated Bipolar Phototransistors 68
2.2.10 Photonic Mixer Device 71
2.3 BiCMOS-Integrated Detectors 72
2.3.1 BiCMOS Processes 72
2.3.2 BiCMOS-Integrated Photodiodes 79

3 **Detectors in Thin Crystalline Silicon Films** ... 87
3.1 Photodiodes in Silicon-on-Insulator ... 87
3.2 Silicon on Sapphire ... 92
3.3 Polyimide Bonding ... 93

4 **SiGe Photodetectors** ... 95
4.1 Heteroepitaxial Growth ... 95
4.2 Absorption Coefficient of SiGe Alloys ... 96
4.3 SiGe/Si PIN Hetero-Bipolar-Transistor Integration ... 96

5 **Design of Integrated Optical Receiver Circuits** ... 101
5.1 Circuit Simulators and Transistor Models ... 101
5.2 Layout and Verification Tools ... 106
5.3 Design of OEICs ... 107
5.4 Transimpedance Amplifier ... 110
5.4.1 Frequency Response ... 111
5.4.2 Phase and Group Delay ... 115
5.4.3 Stability and Compensation ... 118
5.4.4 Bandwidth of Transistor Transimpedance Amplifier Circuit ... 118
5.4.5 Bandwidth Limit of Integrated Transimpedance Amplifiers ... 121
5.4.6 New π-Model for Integrated Transimpedance Amplifier ... 123
5.4.7 Shunt-Shunt Feedback ... 133
5.4.8 Input and Output Resistance ... 137
5.4.9 Noise and Sensitivity ... 139

6 **Examples of Optoelectronic Integrated Circuits** ... 147
6.1 Digital CMOS Circuits ... 147
6.1.1 Synchronous Circuits ... 147
6.1.2 Asynchronous Circuits ... 153
6.2 Digital BiCMOS Circuits ... 155
6.3 Laser Driver Circuits ... 155
6.4 Analog Circuits ... 158
6.4.1 Bipolar Circuits ... 159
6.4.2 Two-dimensional CMOS Imagers ... 162
6.4.3 Optical Distance Measurement Circuits ... 172
6.4.4 Three-dimensional CMOS Cameras ... 176
6.4.5 Smart Pixel Sensors ... 179
6.4.6 CMOS Optical Sensors ... 192
6.4.7 BiCMOS Optical Sensors ... 208
6.4.8 CMOS Circuits for Optical Storage Systems ... 212
6.4.9 BiCMOS Circuits for Optical Storage Systems ... 222
6.4.10 Continuous-Mode Fiber Receivers ... 240

6.4.11 Receiver for Wireless Infrared Communication 283
6.4.12 Hybrid Receivers 291
6.4.13 Speed Enhancement Techniques 293
6.4.14 Plastic-Optical-Fiber Receivers 306
6.4.15 Burst-Mode Optical Receivers 310
6.4.16 Deep-sub-μm Receivers 314
6.4.17 Comparison of Fiber Receivers 317
6.4.18 Special Circuits 320
6.5 Summary 324

References 325

Index 347

List of Symbols

Symbol	Description	Unit
A	Area	cm^2
A_0	Low-frequency open loop gain	
$A(\omega)$	Frequency-dependent gain	
c	Speed of light in a medium	cm/s
c_0	Speed of light in vacuum	cm/s
c_{bd}	Small-signal bulk-drain capacitance	F
c_{bs}	Small-signal bulk-source capacitance	F
c_{cs}	Small-signal collector-substrate capacitance	F
c_{gb}	Small-signal gate-bulk capacitance	F
c_{gd}	Small-signal gate-drain capacitance	F
c_{gs}	Small-signal gate-source capacitance	F
C_{ox}	Gate-oxide capacitance per area	F
c_{ws}	Small-signal well-substrate capacitance	F
c_{je}	Small-signal emitter-base capacitance	F
c_{μ}	Small-signal base-collector capacitance	F
C_{e}	Doping concentration of epitaxial layer	cm^{-3}
C_{B}	Capacitance of bondpad	F
C_{BE}	Base-emitter capacitance	F
C_{C}	Base-collector space-charge capacitance	F
C_{D}	Depletion capacitance of photodiode	F
C_{E}	Total base-emitter capacitance of bipolar transistor	F
C_{F}	Feedback capacitance of transimpedance amplifier	F
C_{GS}	Gate-source capacitance	F
C_{I}	Input capacitance of amplifier	F
C_{L}	Load capacitor	F
C_{P}	Capacitance of chip package	F
C_{RF}	Parasitic capacitance of feedback resistor	F
C_{S}	Parasitic capacitance of signal line	F
C_{SE}	Base-emitter space-charge capacitance	F
C_{T}	Input node capacitance	F
d	Distance	m

Symbol	Description	Unit
d_{e}	Thickness of epitaxial layer	µm
d_{I}	Thickness of intrinsic region	µm
d_{p}	Thickness of P-type region	µm
D	Diffusion coefficient	$\mathrm{cm^2/s}$
D_{n}	Diffusion coefficient of electrons	$\mathrm{cm^2/s}$
D_{p}	Diffusion coefficient of holes	$\mathrm{cm^2/s}$
D_{t}	Decision threshold of photocurrent	A
DR	Data rate	Mb/s
E_{C}	Bottom of conduction band	eV
E_{F}	Fermi energy level	eV
E_{g}	Energy bandgap	eV
E_{t}	Energy level of recombination center	eV
E_{V}	Top of valence band	eV
E	Photon energy	eV
$\boldsymbol{E}$	Electric field	V/cm
f	Frequency	Hz
f_{g}	Bandwidth, −3dB frequency	Hz
f_{s}	Sampling frequency	Hz
f_{T}	Transit frequency (gain-bandwidth product)	Hz
f_{GP}	Frequency of gain-peak	Hz
g_{ds}	Transistor output conductance	A/V
G	Photogeneration (e-h-p)	$\mathrm{cm^{-3}/s}$
$G(\omega)$	Frequency response function	V/A
g_{m}	Transconductance	A/V
h	Planck constant	Js
$\hbar$	$h/2\pi$	Js
$h\nu$	Photon energy	eV
i_{L}	Leakage current of photodiode	A
I	Current	A
I_{ph}	Photocurrent	A
I_{th}	Threshold current	A
I_{B}	Base current	A
I_{C}	Collector current	A
I_{D}	Drain current	A
I_{E}	Emitter current	A
I_{S}	Source current	A
$I(x)$	Light intensity distribution	$\mathrm{W/m^2}$
$I_{\mathrm{i,j}}$	Local luminance at pixel (i,j)	$\mathrm{W/m^2}$
j	Current density	$\mathrm{A/cm^2}$
k_{B}	Boltzmann constant	J/K
$k_{\mathrm{B}}T$	Thermal energy	eV
L	Length	µm
L_{n}	Diffusion length of electrons	µm

Symbol	Description	Unit
L_{p}	Diffusion length of holes	µm
L_{B}	Inductance of bond wire	H
L_{G}	Gate length	µm
L_{W}	Inductance of lead wire	H
$\bar{n}$	Refractive index	
$\bar{n}_{\mathrm{s}}$	Refractive index of surroundings	
$\bar{n}_{\mathrm{sc}}$	Refractive index of semiconductor	
$\bar{n}_{\mathrm{ARC}}$	Refractive index of antireflection coating	
n	Density of free electrons	cm^{-3}
n_{i}	Intrinsic carrier density	cm^{-3}
N	Impurity concentration	cm^{-3}
N_{A}	Acceptor concentration	cm^{-3}
N_{D}	Donor concentration	cm^{-3}
N_{t}	Concentration of recombination centers	cm^{-3}
p	Density of free holes	cm^{-3}
Q_{E}	Minority-carrier charge in the base	As
Q_{pix}	Charge on pixel storage capacitor	As
QE	Quantum efficiency	%
P_{opt}	Incident optical power	W
P_{opt}^{av}	Average incident optical power	W
$\bar{P}$	Optical power in a semiconductor	W
q	Magnitude of electronic charge	As
r_{b}	Base series resistance	Ω
r_{o}	Small-signal output resistance	Ω
r_{c}	Small-signal collector series resistance	Ω
r_{ex}	Small-signal emitter series resistance	Ω
r_{d}	Small-signal drain series resistance	Ω
r_{s}	Small-signal source series resistance	Ω
R	Responsivity	A/W
R_{bb}	Responsivity to black-body radiation	A/W
R_{ov}	Oversampling ratio	
R_{D}	Parallel resistance	Ω
R_{F}	Feedback resistance	Ω
R_{S}	Series resistance	Ω
R_{I}	Input resistance of amplifier	Ω
R_{L}	Load resistance	Ω
$\bar{R}$	Reflectivity	
S	Sensitivity of DVD OEICs	mV/µW
t	Time	s
t_{d}	Drift time	s
t_{diff}	Diffusion time	s
t_{f}	Fall time	s
t_{int}	Integration time	s

Symbol	Description	Unit
t_r	Rise time	s
t_{gd}	Group delay	s
T	Absolute temperature	K
T_s	Sampling time	s
U	Voltage	V
U_{BE}	Base-emitter voltage	V
U_D	Built-in voltage	V
U_{DS}	Drain-source voltage	V
U_{Ea}	Early voltage	V
U_{GS}	Gate-source voltage	V
U_T	Thermal voltage $k_B T/q$	V
U_{th}	Thermal generation/recombination rate	$cm^{-3}s^{-1}$
U_{Th}	Threshold voltage	V
V_{Th}	Threshold voltage	V
V_{det}	Detector bias	V
V_o	Output voltage	V
V_{rev}	Reverse voltage	V
v	Carrier velocity	cm/s
v_s	Saturation velocity	cm/s
v_{th}	Thermal velocity	cm/s
W	Width of space-charge region	μm
W_B	Base thickness	μm
W_G	Gate width	μm
Z_F	Feedback impedance of transimpedance amplifier	Ω
x	x direction	μm
y	y direction	μm
α	Absorption coefficient	μm^{-1}
β	Current gain of bipolar transistor	
β_F	Feedback parameter of transimpedance amplifier	Ω^{-1}
η_e	External (total) quantum efficiency	%
η_i	Internal quantum efficiency	%
η_o	Optical quantum efficiency	%
η_{tia}	Efficiency of transimpedance amplifier	
η_{tia}^{bip}	Efficiency of bipolar transimpedance amplifier	
η_{tia}^{MOS}	Efficiency of MOS transimpedance amplifier	
ϵ_0	Permittivity in vacuum	F/cm
ϵ_r	Relative permittivity	
ϵ_s	Semiconductor permittivity	F/cm
$\bar{\epsilon}$	Dielectric function	
σ	Carrier capture cross section	cm^{-2}
τ	Lifetime	s
τ_n	Electron lifetime	s
τ_p	Hole lifetime	s

Symbol	Description	Unit
τ_B	Base transit time	s
$\bar{\kappa}$	Extinction coefficient	
λ	Wavelength in a medium	nm
λ_0	Wavelength in vacuum	nm
λ_c	Wavelength corresponding to E_g	nm
λ_{ch}	Channel length modulation parameter	V^{-1}
ν	Frequency of light	Hz
μ	Mobility	cm^2/Vs
μ_n	Electron mobility	cm^2/Vs
μ_p	Hole mobility	cm^2/Vs
ω	Angular frequency	s^{-1}
ω_c	$2 \cdot \pi \cdot f_g$	s^{-1}
ω_T	$2 \cdot \pi \cdot f_T$	s^{-1}
ω_{GP}	$2 \cdot \pi \cdot f_{GP}$	s^{-1}
ρ	Charge density	As/cm^3
Φ	Photon flux density	$cm^{-2}s^{-1}$
Ψ	Potential	V
Θ	Angle	$^\circ$

1 Basics and Theory

Optical absorption is a fundamental process which is exploited when optical energy is converted into electrical energy. Optoelectronic receivers are based on this energy conversion process. Photodetectors convert optical energy into electrical energy. In this chapter, the most important factors needed for the comprehension of photodetectors will be summarized in a compact form. For a detailed description of the basics of optical absorption, the book [1] can be recommended. Here, emphasis will, of course, be placed on silicon devices. After the collection of the most important optical and optoelectronic definitions, we will summarize the fundamentals of device physics and modeling of solid-state electron devices including photodetectors in a compact form. A detailed review on modeling of solid-state electron devices can be found in [2]. Here, the semiconductor equations with implemented photogeneration and the models for carrier mobility used in device simulators will be listed first. Carrier drift and diffusion as well as their consequences for the speed and the quantum efficiency of photodetectors will be explained. Furthermore, the equivalent circuit of a photodiode will be discussed in order to show further aspects concerning the speed of photoreceivers.

First, however, we will introduce photons and the properties of light. Photogeneration will be defined. Furthermore, optical reflection and its consequences on the efficiency of photodetectors will be described.

1.1 Basics of Optical Absorption

1.1.1 Photons and their Properties

Due to the work of Max Planck and Albert Einstein it is possible to describe light not only by a wave formalism but also by a quantum-mechanical particle formalism. The smallest unit of light is a quantum-mechanical particle called a photon. The photon, consequently, is the smallest unit of an optical signal. Photons are used to characterize electromagnetic radiation in the optical range from the far infrared to the extreme ultraviolet spectrum. The velocity of photons c in a medium with an optical index of refraction $\bar{n}$ is

$$c = \frac{c_0}{\bar{n}}, \tag{1.1}$$

where c_0 is the velocity of light in vacuum. Photons do not possess a quiescent mass and, unfortunately for the construction of purely optical computers, cannot be stored. Photons can be characterized by their frequency ν and by their wavelength λ:

$$\lambda = \frac{c}{\nu}. \tag{1.2}$$

The frequency of a photon is the same in vacuum and in a medium with index of refraction $\bar{n}$ ($\bar{n}=1$ for vacuum, $\bar{n} > 1$ in a medium). The wavelength λ in a medium, therefore, is shorter than the vacuum wavelength λ_0 ($\lambda = \lambda_0/\bar{n}$). As a consequence, the vacuum wavelength is used to characterize light sources like light-emitting diodes (LEDs) or semiconductor lasers, because it is independent of the medium in which the light propagates.

Photons can also be characterized by their energy E (h is Planck's constant):

$$E = h\nu = \frac{hc}{\lambda} = \frac{hc_0}{\lambda_0}. \tag{1.3}$$

According to fundamental absorption, this formula leads to the boundary wavelength, which can be detected with the bandgap $E_g = 1.1\,\text{eV}$ of silicon:

$$\lambda_c = \frac{hc_0}{E_g} = 1110\,\text{nm}. \tag{1.4}$$

Only light with shorter wavelengths than 1110 nm can be detected with silicon photodiodes.

A useful relation is given next which allows a quick calculation of the energy for a certain wavelength and vice versa:

$$E = \frac{1240}{\lambda_0} \tag{1.5}$$

where E is in eV and λ_0 in nm. Let us define the flux density Φ as the number of photons incident per time interval on an area A. The optical power P_{opt} incident on a detector with a light sensitive area A, then, is determined by the photon energy and by the flux density:

$$P_{opt} = E\Phi A = h\nu\Phi A. \tag{1.6}$$

Receivers for optical communication or data transmission are characterized by the average optical power P_{opt}^{av} needed to achieve a certain bit error ratio.

1.1.2 Optical Absorption of Important Semiconductor Materials

The energy of a photon can be transferred to an electron in the valence band of a semiconductor, which is brought to the conduction band, when the photon energy is larger than the bandgap energy E_g. The photon is absorbed during this process and an electron–hole pair is generated. Photons with an energy smaller than E_g, however, cannot be absorbed and the semiconductor is transparent for light with wavelengths longer than $\lambda_c = hc_0/E_g$.

The optical absorption coefficient α is the most important optical constant for photodetectors. The absorption of photons in a photodetector to produce carrier pairs and thus a photocurrent, depends on the absorption coefficient α for the light in the semiconductor used to fabricate the detector. The absorption coefficient determines the penetration depth $1/\alpha$ of the light in the semiconductor material according to Lambert–Beer's law:

$$I(\bar{y}) = I_0 \exp(-\alpha\bar{y}). \tag{1.7}$$

The optical absorption coefficients for the most important semiconductor materials are compared in Fig. 1.1. The absorption coefficients strongly depend on the wavelength of the light. For wavelengths shorter than λ_c, which corresponds to the bandgap energy ($\lambda_c = hc_0/E_g$), the absorption coefficients increase rapidly according to the so-called *fundamental absorption*. The steepness of the onset of absorption depends on the kind of band–band transition. This steepness is large for direct band–band transitions as in GaAs ($E_g^{dir} = 1.42\,eV$ at 300 K), in InP ($E_g^{dir} = 1.35\,eV$ at 300 K), in Ge ($E_g^{dir} = 0.81\,eV$ at 300 K) and in $In_{0.53}Ga_{0.47}As$ ($E_g^{dir} = 0.75\,eV$ at 300 K). For Si ($E_g^{ind} = 1.12\,eV$ at 300 K), for Ge ($E_g^{ind} = 0.67\,eV$ at 300 K) and for the wide bandgap material 6H-SiC ($E_g^{ind} = 3.03\,eV$ at 300 K) the steepness of the onset of absorption is small.

$In_{0.53}Ga_{0.47}As$ and Ge cover the widest wavelength range including the wavelengths 1.3 and 1.54 µm which are used for long distance optical data transmission via optical fibers. The absorption coefficients of GaAs and InP are high in the visible spectrum (≈ 400–700 nm). Silicon detectors are also appropriate for the visible and near infrared spectral range. The absorption coefficient of Si, however, is one to two orders of magnitude lower than that of the direct semiconductors in this spectral range. For Si detectors, therefore, a much thicker absorption zone is needed than for the direct semiconductors. We will, however, see in this work that with silicon photodiodes GHz operation is nevertheless possible. Silicon is the economically most important semiconductor and it is worthwhile to investigate silicon optoelectronic devices and integrated circuits in spite of the nonoptimum optical absorption of silicon.

The absorption coefficients of silicon for wavelengths which are the most important ones in practice are listed in Table 1.1 [3, 4]. In order to compare the quantum efficiencies of photodetectors for different wavelengths, it is advantageous to use the same photon flux for the different wavelengths. The photocurrents of photodetectors are equal for the same fluxes of photons with different energy, i. e. for different light wavelengths, when the quantum efficiency of the photodetector is the same for the different photon energies or wavelengths, respectively. According to the Lambert–Beer law, different

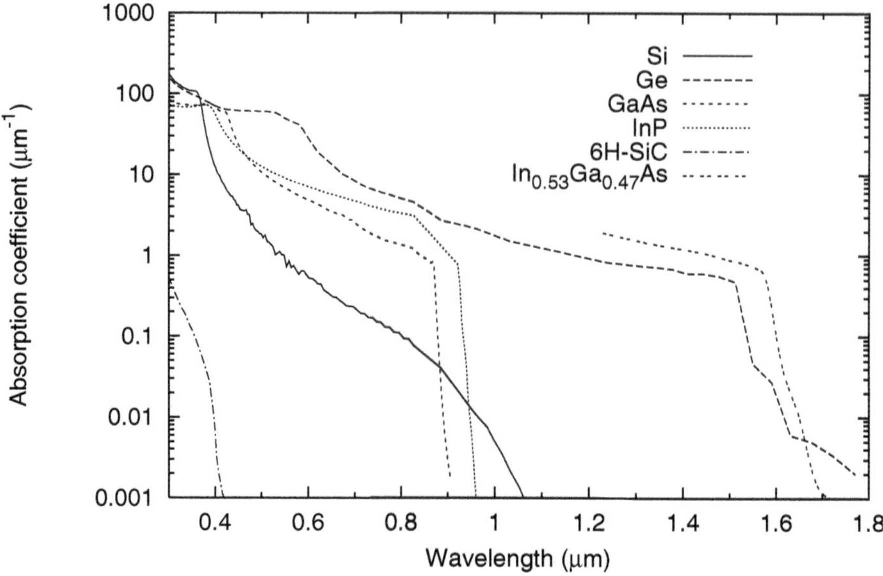

Fig. 1.1. Absorption coefficients of important semiconductor materials versus wavelength

Table 1.1. Absorption coefficients α of silicon and intensity factors I_0 (ehp/cm^3 means electron–hole pairs/cm^3) for several important wavelengths for a constant photon flux density of $\Phi = I_0/\alpha = 1.58 \times 10^{18}$ photons/cm^2

Wavelength (nm)	α (µm^{-1})	I_0 (ehp/cm^3)
850	0.06	9.50×10^{20}
780	0.12	1.89×10^{21}
680	0.24	3.79×10^{21}
635	0.38	6.00×10^{21}
430	5.7	9.00×10^{22}

intensity factors I_0 result for a constant photon flux. As an example, intensity factors are listed for the most important wavelengths in Table 1.1 for a certain arbitrary photon flux density.

1.1.3 Photogeneration

The Lambert–Beer law can be formulated for the optical power $\bar{P}$ analogously to (1.7):

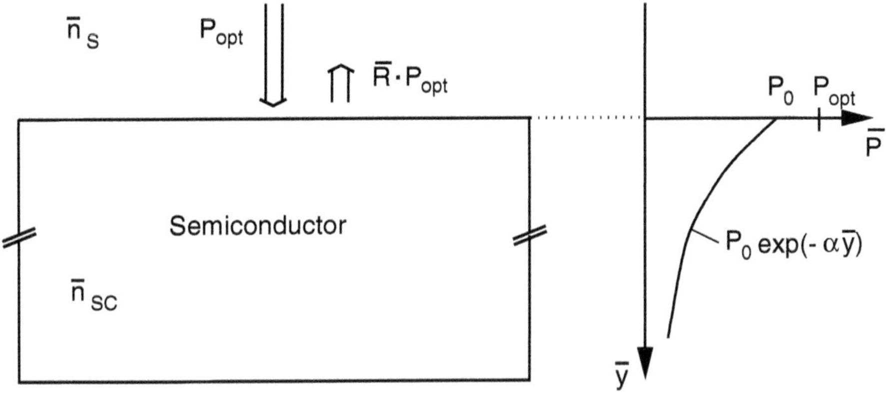

Fig. 1.2. Reflection at a semiconductor surface and decay of the optical power in the semiconductor ($P_0 = (1 - \bar{R})P_{\text{opt}}$)

$$\bar{P}(\bar{y}) = P_0 \exp(-\alpha\bar{y}). \tag{1.8}$$

The optical power at the surface of the semiconductor $\bar{P}(\bar{y} = 0)$ is $P_0 = (1 - \bar{R})$ (see Fig. 1.2). The optical power of the light penetrating into a medium decreases exponentially with the penetration coordinate $\bar{y}$ in the medium (compare Fig. 1.3).

The absorbed light generates electron–hole pairs in a semiconductor due to the internal photoeffect provided that $h\nu > E_g$. Therefore, we can express the generation rate per volume $G(\bar{y})$ as:

$$G(\bar{y}) = \frac{\bar{P}(\bar{y}) - \bar{P}(\bar{y} + \triangle\bar{y})}{\triangle\bar{y}} \frac{1}{Ah\nu}. \tag{1.9}$$

In this equation, A is the area of the cross section for the light incidence and $h\nu$ is the photon energy. For $\triangle\bar{y} \to 0$, we can write $(\bar{P}(\bar{y}) - \bar{P}(\bar{y} + \triangle\bar{y}))/\triangle\bar{y} = -d\bar{P}(\bar{y})/d\bar{y}$. From (1.8), $d\bar{P}(\bar{y})/d\bar{y} = -\alpha\bar{P}(\bar{y})$ then follows and the generation rate is

$$G(\bar{y}) = \frac{\alpha P_0}{Ah\nu} \exp(-\alpha\bar{y}). \tag{1.10}$$

1.2 Semiconductor Equations

The physics of semiconductor devices like diodes, MOSFETs (Metal-Oxide-Silicon Field-Effect Transistors), bipolar transistors, and photodetectors is well known. The equations describing the behavior of these devices in most important cases are the semiconductor equations. These equations have already been implemented in many device simulation programs. The physical models for the parameters used in the semiconductor equations were discussed thoroughly, for instance, in [2].

Device simulation programs have been valuable tools for the development of semiconductor devices for many years. Much time and money can be saved with their help for such a purpose. They are also valuable for the development of photodetectors. We will take the two-dimensional device simulator MEDICI as an example [5]. The drift-diffusion model is implemented in this simulator. MEDICI solves the Poisson equation (1.11), the transport equations (1.13) and (1.14), and the continuity equations (1.16) and (1.17) for electrons and holes. Furthermore, photogeneration is implemented. Due to the photogeneration for the internal photoeffect, electron–hole pairs are created, i. e. the corresponding generation terms for electrons and holes are equal.

The potential Ψ in the device, for which the simulation is performed, is calculated according to the Poisson equation:

$$\triangle\Psi = -\frac{\rho}{\epsilon}\,. \tag{1.11}$$

The quantity ϵ is the product of the relative and absolute dielectric constants: $\epsilon = \epsilon_{\mathrm{r}}\epsilon_0$. The symbol ρ represents the charge density, which can be further broken apart into the product of the elementary charge q times the sum of the hole concentration p (positively charged), of the electron concentration n (negatively charged), of the donor concentration N_{D} (positively charged), and of the acceptor concentration N_{A} (negatively charged):

$$\rho = q(p - n + N_{\mathrm{D}} - N_{\mathrm{A}}). \tag{1.12}$$

The current densities for electrons and holes are the sum of the drift and diffusion current densities:

$$\boldsymbol{j}_{\mathrm{n}} = qn\mu_{\mathrm{n}}\boldsymbol{E} + qD_{\mathrm{n}}\,\mathbf{grad}\,n, \tag{1.13}$$

$$\boldsymbol{j}_{\mathrm{p}} = qp\mu_{\mathrm{p}}\boldsymbol{E} - qD_{\mathrm{p}}\,\mathbf{grad}\,p. \tag{1.14}$$

The total current density results from the electron and hole current densities:

$$\boldsymbol{j} = \boldsymbol{j}_{\mathrm{n}} + \boldsymbol{j}_{\mathrm{p}}. \tag{1.15}$$

The continuity equations with the inclusion of photogeneration $G(x, y)$ due to the penetration of light into the semiconductor can be written as:

$$\frac{\partial n}{\partial t} = \frac{\operatorname{div}\boldsymbol{j}_{\mathrm{n}}}{q} + U_{\mathrm{th}} + G(x,y), \tag{1.16}$$

$$\frac{\partial p}{\partial t} = -\frac{\operatorname{div}\boldsymbol{j}_{\mathrm{p}}}{q} + U_{\mathrm{th}} + G(x,y). \tag{1.17}$$

U_{th} is the thermal generation/recombination term. The photogeneration $G(\bar{y})$ was derived above. The simulator allows us to define nonperpendicular

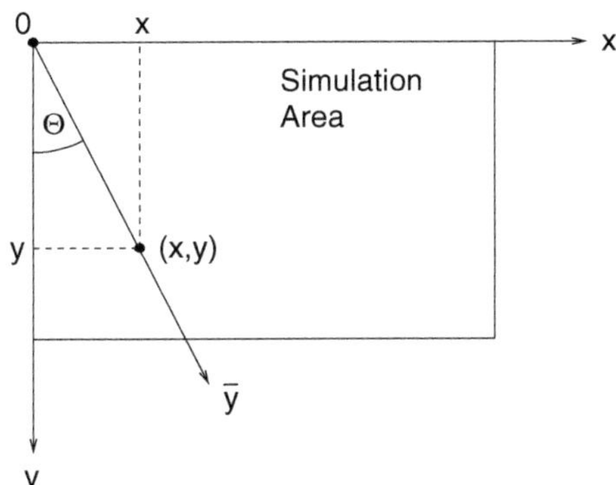

Fig. 1.3. Coordinate transformation from the penetration coordinate $\bar{y}$ to the coordinates (x,y) used in the two-dimensional device simulator for non-perpendicular light incidence

incidence for the light. Then, $G(x,y) = G(\bar{y} = (x^2 + y^2)^{1/2})$ has to be used (compare with Fig. 1.3).

The electric field is determined by (1.11) and obeys:

$$\boldsymbol{E} = -\mathbf{grad}\,\Psi. \tag{1.18}$$

The electric field, for instance, is important for the calculation of the drift velocity $\boldsymbol{v}$ of photogenerated carriers in photodetectors:

$$\boldsymbol{v} = \mu \boldsymbol{E}. \tag{1.19}$$

The mobilities of electrons and holes differ strongly and μ_{n} or μ_{p} has to be used for μ in order to calculate the electron drift velocity or the hole drift velocity, respectively. The mobilities μ_{n} and μ_{p} are only constant for a low electric field. It will be shown below that the drift velocities of electrons and holes saturate for large values of the electric field.

For the calculation of the speed of photodetectors we not only need the mobilities and the electric field but also the width of the space-charge region, where an electric field is present. In the so-called depletion approximation, i. e. with n and p approximately equal to zero in the space-charge region and $\rho = 0$ outside the space-charge region, the distribution of the electric field and the width of the space-charge region W can be calculated analytically for an abrupt PN junction:

$$W = \sqrt{\frac{2\epsilon_{\mathrm{r}}\epsilon_0}{q} \frac{N_{\mathrm{A}} + N_{\mathrm{D}}}{N_{\mathrm{A}} N_{\mathrm{D}}} \left(U_{\mathrm{D}} - U - \frac{2k_{\mathrm{B}}T}{q} \right)}, \tag{1.20}$$

with

$$U_{\mathrm{D}} = \frac{k_{\mathrm{B}}T}{q} \ln \frac{N_{\mathrm{A}} N_{\mathrm{D}}}{n_{\mathrm{i}}^2}. \tag{1.21}$$

Let us assume that one side of the PN junction is doped to a much larger extent than the other, for instance $N_{\mathrm{D}} = 10^{20}\,\mathrm{cm}^{-3} \gg N_{\mathrm{A}} = 10^{16}\,\mathrm{cm}^{-3}$ or

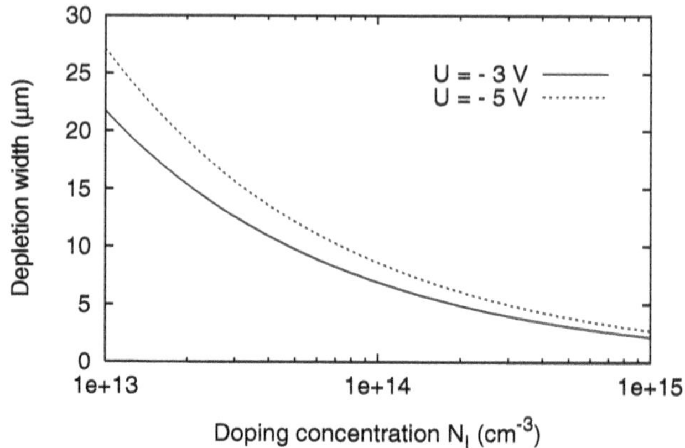

Fig. 1.4. Depletion layer width versus doping concentration of the lower doped side of a PN junction

vice versa. We want to define the lower doping concentration as N_I. Equation (1.20) then can be simplified to

$$W = \sqrt{\frac{2\epsilon_\mathrm{r}\epsilon_0}{qN_\mathrm{I}}\left(U_\mathrm{D} - U - \frac{2k_\mathrm{B}T}{q}\right)}. \tag{1.22}$$

For the reverse voltage U, here, a negative value always has to be used. The space-charge region only spreads into the low doped side of the PN junction due to the simplification. This behavior, however, is a good approximation for most semiconductor diodes and for photodiodes, accordingly. The width of the space-charge region as a function of the doping concentration of the low doped side of the PN junction is shown in Fig. 1.4.

1.3 Important Models for Photodetectors

The most important processes for the characterization of photodetectors with respect to their speed are carrier drift and minority carrier diffusion. The carrier drift in conventional semiconductor materials is a much faster process than the minority carrier diffusion. For the calculation of the frequency response of photodiodes, series resistances and PN junction capacitances are also important. An even more complete equivalent circuit for photodiodes considering wiring capacitances will be discussed. Minority carrier diffusion, series resistances and capacitances reduce the speed of photodiodes.

1.3.1 Carrier Drift

The carrier mobilities in the drift terms (see (1.13) and (1.14)) are the parameters which are responsible for the obtainable speed of photodetectors. The

Table 1.2. Parameter values for the approximation of the electron and hole mobilities in silicon with (1.23) and (1.24)

Parameter	Electrons	Holes
μ_{min} (cm^2/Vs)	17.8	48.0
μ_{max} (cm^2/Vs)	1350	495
N_{ref} (cm^{-3})	1.072×10^{17}	1.606×10^{17}
$\nu_{\mathrm{n,p}}$	−2.3	−2.2
$\chi_{\mathrm{n,p}}$	−3.8	−3.7
$\alpha_{\mathrm{n,p}}$	0.73	0.70

carrier mobilities depend on the doping concentration and on the electric field. An empirical expression is available for the dependence of the mobility on the total impurity concentration N_{total}, which also considers the influence of temperature T in K [6, 7]:

$$\mu_{0\mathrm{n}} = \mu_{\mathrm{n,min}} + \frac{\mu_{\mathrm{n,max}}(T/300)^{\nu_\mathrm{n}} - \mu_{\mathrm{n,min}}}{1 + (T/300)^{\chi_\mathrm{n}}(N_{\mathrm{total}}/N_{\mathrm{ref,n}})^{\alpha_\mathrm{n}}}, \tag{1.23}$$

$$\mu_{0\mathrm{p}} = \mu_{\mathrm{p,min}} + \frac{\mu_{\mathrm{p,max}}(T/300)^{\nu_\mathrm{p}} - \mu_{\mathrm{p,min}}}{1 + (T/300)^{\chi_\mathrm{p}}(N_{\mathrm{total}}/N_{\mathrm{ref,p}})^{\alpha_\mathrm{p}}}. \tag{1.24}$$

N_{total} is the sum of the concentrations of acceptors and donors. The other parameters for silicon are listed in Table 1.2.

The dependence of the carrier mobilities in silicon on the doping concentration according to (1.23) and (1.24) is shown in Fig. 1.5 for $T = 300\,\mathrm{K}$.

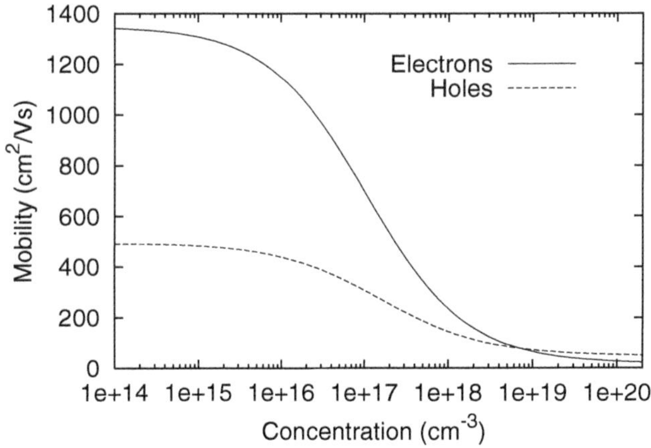

Fig. 1.5. Carrier mobilities in silicon versus total doping concentration [6, 7]

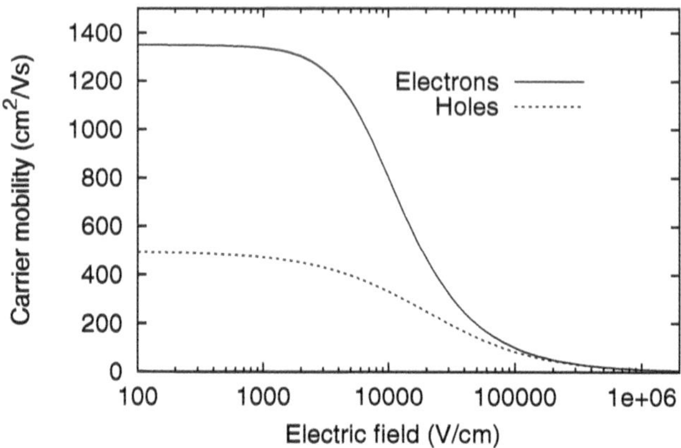

Fig. 1.6. Carrier mobilities versus electric field for weakly doped silicon

According to [6], the dependence of the carrier mobilities in silicon on the electric field can be approximated by

$$\mu_{\mathrm{n}} = \frac{\mu_{0\mathrm{n}}}{1 + (\mu_{0\mathrm{n}} E / v_{\mathrm{n}}^{\mathrm{sat}})^{2}}, \tag{1.25}$$

$$\mu_{\mathrm{p}} = \frac{\mu_{0\mathrm{p}}}{1 + (\mu_{0\mathrm{p}} E / v_{\mathrm{p}}^{\mathrm{sat}})^{1}}. \tag{1.26}$$

The values for the saturation velocities (in cm/s) can be computed from the expression [8] with T in K:

$$v_{\mathrm{n}}^{\mathrm{sat}}(T) = v_{\mathrm{p}}^{\mathrm{sat}}(T) = \frac{2.4 \times 10^{7}}{1 + 0.8 \exp(T/600)}. \tag{1.27}$$

The curves calculated with (1.25) and (1.26) are shown in Fig. 1.6 for a very low doping concentration, i. e. for the 'intrinsic' zone of a PIN photodiode. The carrier mobilities begin to degrade for electric fields in excess of approximately 2000 V/cm.

The drift velocities, therefore, are proportional to the electric field only for smaller values of the electric field (Fig. 1.7). For values of the electric field larger than approximately 10 000 V/cm the electron drift velocity saturates. For holes, the electric field has to exceed a value of approximately 100 000 V/cm in order to obtain the hole saturation velocity.

The dependence of the carrier mobilities on the doping concentration is implemented in MEDICI with the CONMOB model [5]. The model in MEDICI for the dependence of the carrier mobilities, i. e. of the drift velocities on the electric field is called FLDMOB.

One quantity, which should be mentioned here, is the drift time, i. e. the time needed by carriers to drift through the whole width W of the space-charge region or drift zone. The drift time is determined by the drift velocity

v and by the width of the drift zone W. The drift velocity v depends on μ and E. Furthermore, μ depends on the doping concentration. The doping concentration and the bias voltage applied to the photodiode determine the distribution of the electric field. It is, therefore, very difficult to calculate the drift time analytically, because v, μ and E in a real photodetector depend on the space coordinates. Numerical process and device simulations, in general, are necessary in order to compute the transient response of photodetectors.

When we choose the ideal PIN photodiode with a constant electric field for instance (see Fig. 1.8), it is possible to give some analytical expressions for the drift time t_{d} and for the rise and fall times t_{r} and t_{f} of the photocurrent (time difference between the 10 % and 90 % values of the stationary photocurrent) [9]:

$$t_{\mathrm{d}} = t_{\mathrm{r}} = t_{\mathrm{f}} = \frac{d_{\mathrm{I}}}{v(E_0)}. \tag{1.28}$$

Instead of W, the thickness of the intrinsic I-layer of the ideal PIN photodiode d_{I} is used in (1.28). For a large reverse voltage, which results in a large value of the electric field, the drift velocities may be approximated by the saturation velocity v_{s} and

$$t_{\mathrm{d}} = t_{\mathrm{r}} = t_{\mathrm{f}} = \frac{d_{\mathrm{I}}}{v_{\mathrm{s}}} \tag{1.29}$$

results. The rise and fall times limit the maximum data rate DR of optical receivers. The maximum data rate of a photodiode (in the non-return-to-zero transmission mode) can be estimated in a conservative way to be the lower value of DR$= 1/(3t_{\mathrm{r}})$ and DR $= 1/(3t_{\mathrm{f}})$. The mean value of t_{r} and t_{f} may lead to DR$= 2/(3(t_{\mathrm{r}} + t_{\mathrm{f}}))$. In a more aggressive way, instead of the factor 3, the factor 2 can be used, however, in practice the bit-error rate of optical

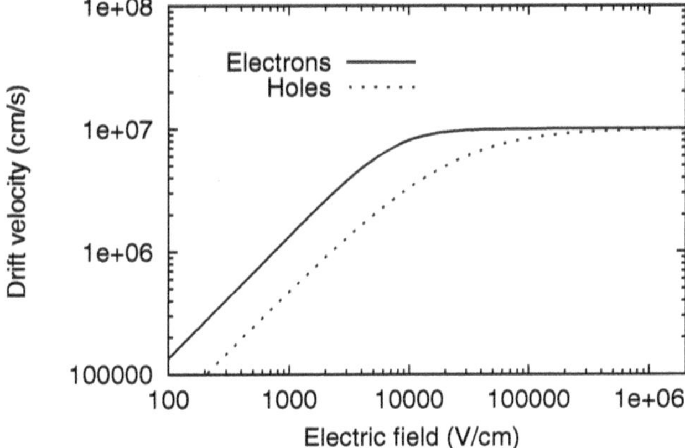

Fig. 1.7. Carrier drift velocities in silicon versus electric field calculated according to (1.19), (1.25) and (1.26) for a low doping concentration

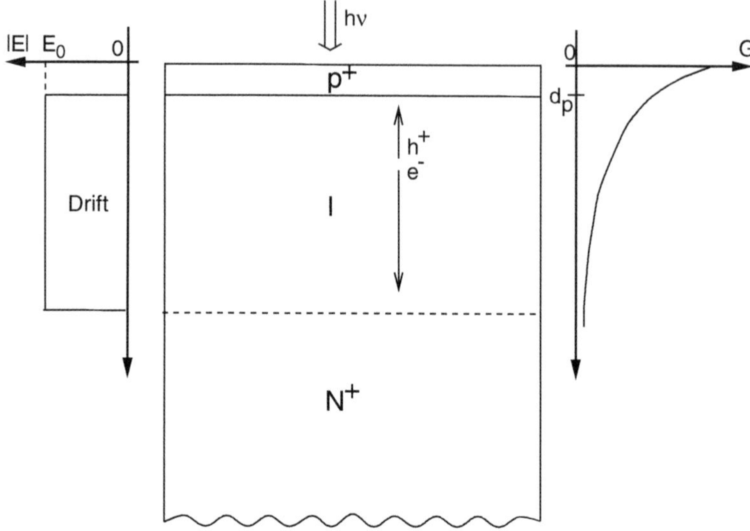

Fig. 1.8. Drift region in an ideal PIN photodiode with an undoped intrinsic I-layer

receivers determines the usable data rate. The relation between the rise time and the -3dB bandwidth should also be mentioned [10, 11]:

$$f_{3\mathrm{dB}} = \frac{2.4}{2\pi t_\mathrm{r}} \approx \frac{0.4 v_\mathrm{s}}{W}. \tag{1.30}$$

This equation was obtained for a time-dependent sinusoidal photogeneration at the surface of the PIN photodiode after integrating the conduction current density and the displacement current density over the drift zone thickness [11]. The amplitude of the short circuit photocurrent density dropped to $1/\sqrt{2}$ of its low-frequency value for $\omega t_\mathrm{d} = 2.4$ leading to (1.30) when assuming that the transit time t_d is equal to the rise time t_r. When the thickness of the intrinsic I-layer is chosen as $W = d_\mathrm{I} = 1/\alpha$, a simple expression is obtained for the -3dB frequency of a PIN photodiode:

$$f_{3\mathrm{dB}} \approx 0.4 \alpha v_\mathrm{s}. \tag{1.31}$$

Another case, for which an analytical solution may be justified, is a PN photodiode with a very large electric field. Then $v \approx v_\mathrm{s}$ and $t_\mathrm{d} = W/v_\mathrm{s}$ may be used.

The electric field in most real photodiodes, however, is much lower than would be necessary in order to justify the assumption of the saturation velocities. The electric field in real PIN photodiodes also is not constant even at the low 'intrinsic' doping level of $10^{13}\,\mathrm{cm}^{-3}$ [12] and, besides, a rather limited validity range of $\alpha d_\mathrm{I} > 3$ for an analytical model, process and device simulations for the computation of the rise and fall times are necessary for real PIN photodiodes. These process and device simulations are also capable of considering the influence of carrier diffusion, which will be discussed next, on the transient behavior of photodetectors.

1.3.2 Carrier Diffusion

The diffusion of minority carriers is important for the calculation of the speed of photodetectors, when electron–hole pairs are generated in the quasineutral regions of the semiconductor where no electric field is present. There can be two such regions in a photodiode. The first region is the heavily doped surface region (P^+ in Fig. 1.9), which, however, in most cases is not very thick. The second and usually more critical region is in the depth of the semiconductor below the edge of the space-charge region SCR (see Fig. 1.9).

The carrier diffusion coefficients D_n and D_p depend on the carrier mobilities according to the Einstein relation

$$D_{n/p} = \mu_{n/p} \frac{k_B T}{q} = \mu_{n/p} U_T. \tag{1.32}$$

The value of the diffusion coefficient of electrons in Si is only approximately $35\,\text{cm}^2/\text{s}$ for low doping concentrations. The value reduces considerably for high doping concentrations. The diffusion coefficient of holes in Si is less than $12.5\,\text{cm}^2/\text{s}$ and also depends on the doping concentration.

The distance over which minority carriers diffuse is called the carrier diffusion length L_n or L_p for electrons or holes, respectively. The carrier

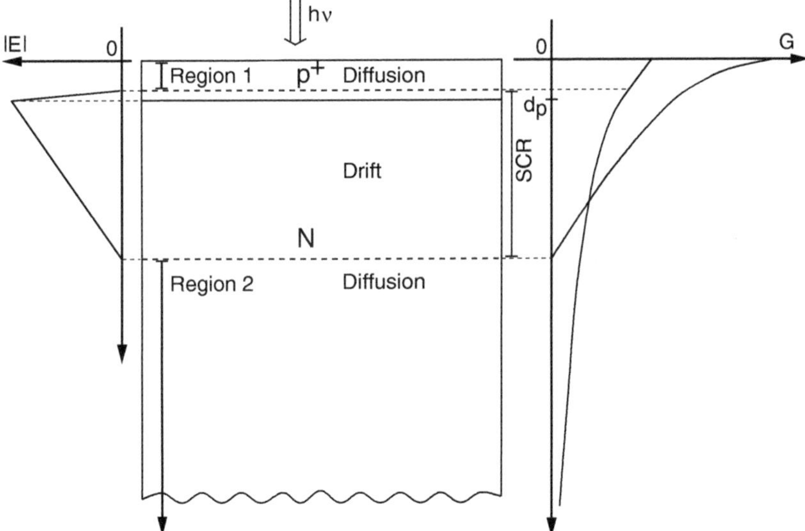

Fig. 1.9. Drift and diffusion regions in a photodiode

diffusion lengths depend on the carrier diffusion coefficients and on the carrier lifetimes τ_n and τ_p:

$$L_\mathrm{n} = \sqrt{D_\mathrm{n}\tau_\mathrm{n}}, \tag{1.33}$$

$$L_\mathrm{p} = \sqrt{D_\mathrm{p}\tau_\mathrm{p}}. \tag{1.34}$$

The carrier lifetimes are due to dopants and unwanted impurities in the semiconductor, which are unavoidably introduced in quite small but noticeable amounts in electronic devices during the fabrication process. These impurities may act as recombination centers. Recombination centers introduce deep energy levels in the bandgap, which enhance the recombination rate of electrons and holes [13–15]. Short carrier lifetimes result in a large recombination rate. The Shockley–Read–Hall (SRH) generation/recombination rate U_th is implemented in MEDICI in this form [5]:

$$U_\mathrm{th} = U_\mathrm{n} = U_\mathrm{p} = \frac{pn - n_\mathrm{i}^2}{\tau_\mathrm{p}[n + n_\mathrm{i}\exp((E_\mathrm{t} - E_\mathrm{i})/k_\mathrm{B}T)] + \tau_\mathrm{n}[p + n_\mathrm{i}\exp((E_\mathrm{i} - E_\mathrm{t})/k_\mathrm{B}T)]}. \tag{1.35}$$

The recombination rate U_th depends on the densities of electrons and holes, on the intrinsic carrier density n_i, on the carrier lifetimes τ_n, τ_p, and on the difference between the energy level of the recombination center E_t and the intrinsic Fermi level E_i. The carrier lifetimes themselves depend on the total impurity density N_total [5]:

$$\tau_\mathrm{n} = \frac{\tau_{0,\mathrm{n}}}{1 + N_\mathrm{total}/N_\mathrm{SRH,n}}, \tag{1.36}$$

$$\tau_\mathrm{p} = \frac{\tau_{0,\mathrm{p}}}{1 + N_\mathrm{total}/N_\mathrm{SRH,p}}. \tag{1.37}$$

In N_total, dopants and recombination centers have to be considered. The carrier lifetimes for low doping concentrations and low densities of recombination centers are $\tau_{0,\mathrm{n}}$ and $\tau_{0,\mathrm{p}}$, respectively. The reference parameters $N_\mathrm{SRH,n}$ and $N_\mathrm{SRH,p}$ for silicon are both equal to $5 \times 10^{16}\,\mathrm{cm}^{-3}$.

Equation (1.35) can be simplified for $\tau_\mathrm{n} = \tau_\mathrm{p}$ and it can be seen that the recombination rate approaches a maximum for $E_\mathrm{t} = E_\mathrm{i}$. The carrier lifetimes are inversely proportional to the recombination rate for low injection conditions, i. e., when the densities of injected carriers $\Delta n = \Delta p$ are much smaller than the density of majority carriers.

The minority carrier lifetime (hole lifetime) in an N-type semiconductor then is

$$\tau_\mathrm{p} = (\sigma_\mathrm{p} v_\mathrm{th} N_\mathrm{t})^{-1}. \tag{1.38}$$

The electron lifetime in a P-type semiconductor similarly is

$$\tau_\mathrm{n} = (\sigma_\mathrm{n} v_\mathrm{th} N_\mathrm{t})^{-1}. \tag{1.39}$$

The carrier capture cross sections σ_n, σ_p, the thermal carrier velocity v_{th} and the density of unwanted impurities N_t determine the minority carrier lifetimes in low doped silicon.

The minority carrier lifetimes in as grown silicon can reach several milliseconds, whereas the fabrication process of photodetectors or integrated circuits can reduce the carrier lifetimes to the order of microseconds. Usually the minority carrier lifetimes are important only when carrier diffusion is the dominating transport mechanism. Drift times are usually of the order of nanoseconds and carrier lifetimes of the order of microseconds do not deteriorate the quantum efficiency of photodetectors when carrier drift is dominating.

Carrier diffusion is the dominant aspect in solar cells, which do not need a very quick transient response to changing light intensities [16]. The spectrum of the sun contains rather long wavelengths, which penetrate several tens of micrometers into silicon. Therefore, many electron–hole pairs are generated in region 2 (see Fig. 1.9), where carrier diffusion occurs. In order to obtain a high quantum efficiency of a solar cell, a long carrier diffusion length and, therefore, a long carrier lifetime is necessary in the semiconductor. The long carrier lifetimes in silicon with a highly developed crystal growth and process technology are the reason why silicon is the dominant material in solar cell production. The interested reader will find the state of the art in solar cell technology in [17–20], for instance.

Other optoelectronic devices, where carrier diffusion is important, are bipolar phototransistors. The diffusion of minority carriers in the base of bipolar transistors limits their switching speed and their transit frequency [21]. We will return to photodiodes here.

We want to estimate the time t_{diff} for the diffusion of electrons through a P^+ region with thickness d_p [22, 23]

$$d_p = \sqrt{2 D_n t_{diff}}. \tag{1.40}$$

This equation has been derived for a time-dependent sinusoidal electron density due to photogeneration in the P^+ layer from the electron diffusion equation [23]. The time response of a low-pass filter with time constant $\tau_r = d_p^2/(2D_n)$ was obtained and this time constant was interpreted as the time t_{diff} that electrons need to diffuse to the space-charge region at the P^+N junction. With (1.32), t_{diff} can be obtained:

$$t_{diff} = \frac{d_p^2 q}{2 \mu_n k_B T}. \tag{1.41}$$

A much slower contribution to the photocurrent usually results from carriers being generated in the second diffusion region (see Fig. 1.9), because the holes have to diffuse a much longer distance than d_p from this region 2 to the space-charge region. The hole diffusion time through 10 μm of silicon is 40 ns whereas the electron diffusion time over the same distance is approximately

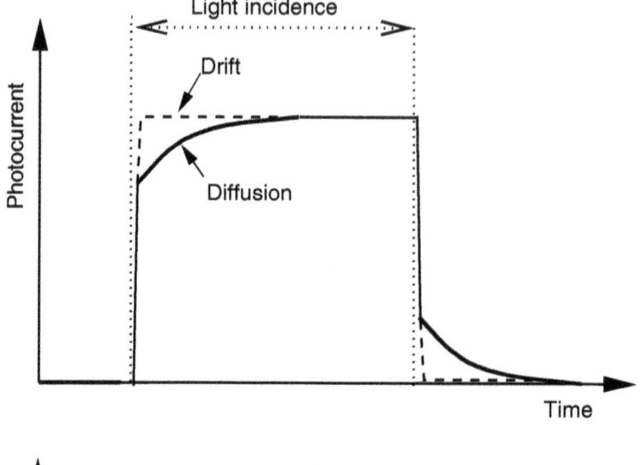

Fig. 1.10. Transient behavior of the photocurrent for carrier drift and diffusion

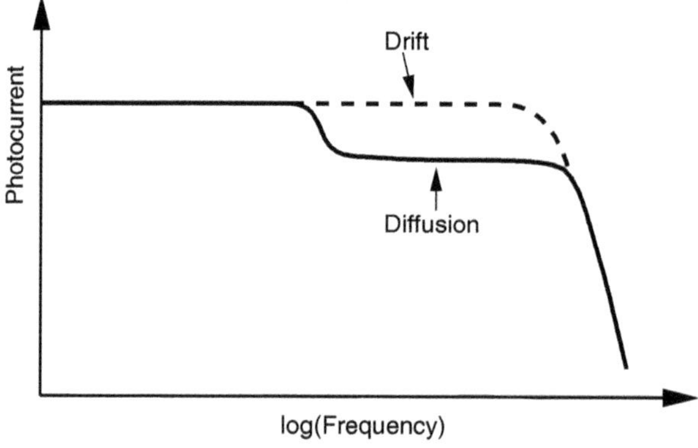

Fig. 1.11. Frequency response of the photocurrent for carrier drift and diffusion

8 ns. The resulting shape of the photocurrent shown in Fig. 1.10 is characteristic of the diffusion of carriers from the second region into the space-charge region. The influence of carrier diffusion on the frequency response of photodetectors is shown in Fig. 1.11.

Summarizing, it can be concluded that carrier diffusion should be avoided in order to obtain fast photodetectors. One possibility is to use light with a short wavelength, i. e. with a large absorption coefficient in Si in order to avoid photogeneration in region two (see Fig. 1.9). Another possibility can be the choice of a semiconductor with larger absorption coefficients. Where neither of these measures is possible, a larger reverse voltage should be applied across the photodiode or the doping concentration in the photodiode should be reduced, in order to obtain a thicker space-charge region.

1.3.3 Quantum Efficiency and Responsivity

The total, external or overall quantum efficiency η is defined as the number of photogenerated electron–hole pairs, which contribute to the photocurrent, divided by the number of the incident photons. The external quantum efficiency

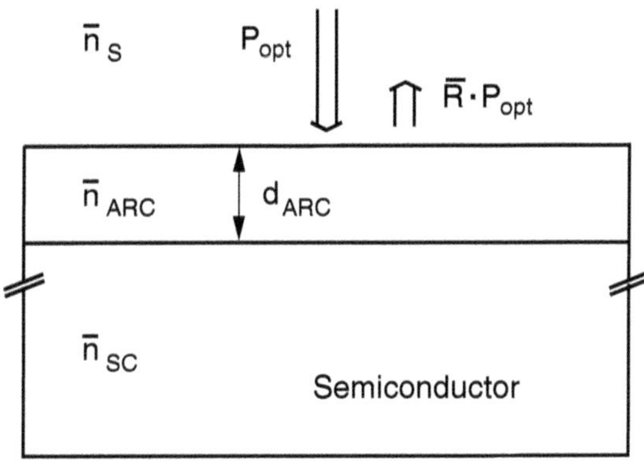

Fig. 1.12. Semiconductor with an antireflection coating

can be determined, when the photocurrent of a photodetector is measured for a known incident optical power.

A fraction of the incident optical power is reflected (see Fig. 1.2) due to the difference in the index of refraction between the surroundings $\bar{n}_s$ (air: $\bar{n}_s = 1.00$) and the semiconductor $\bar{n}_{sc}$ (e. g. Si, $\bar{n}_{sc} \approx 3.5$). The reflectivity $\bar{R}$ depends on the index of refraction $\bar{n}_{sc}$ and on the extinction coefficient $\bar{\kappa}$ of an absorbing medium, for which the dielectric function $\bar{\epsilon} = \bar{\epsilon}_1 + i\bar{\epsilon}_2 = (\bar{n}_{sc} + i\bar{\kappa})^2$ is valid ($\bar{n}_s = 1$) [3].

$$\bar{R} = \frac{(1 - \bar{n}_{sc})^2 + \bar{\kappa}^2}{(1 + \bar{n}_{sc})^2 + \bar{\kappa}^2}. \tag{1.42}$$

The extinction coefficient is sufficient for the description of the absorption. The absorption coefficient α can be expressed as:

$$\alpha = \frac{4\pi\bar{\kappa}}{\lambda_0}. \tag{1.43}$$

The optical quantum efficiency η_o can be defined in order to consider the partial reflection:

$$\eta_o = 1 - \bar{R}. \tag{1.44}$$

The reflected fraction of the optical power can be minimized by introducing an antireflection coating (ARC) with thickness d_{ARC} (see Fig. 1.12):

$$d_{ARC} = \frac{\lambda_0}{4\bar{n}_{ARC}}. \tag{1.45}$$

The index of refraction of the ARC-layer can be calculated:

$$\bar{n}_{ARC} = \sqrt{\bar{n}_s \bar{n}_{sc}}. \tag{1.46}$$

The optimum index of refraction of the ARC-layer is determined by the refractive index $\bar{n}_s$ of the surroundings and by the refractive index $\bar{n}_{sc}$ of the semiconductor. A complete suppression of the partial reflection, however, is not possible in practice. For silicon photodetectors, SiO_2 ($\bar{n}_{sc} = 1.45$) and Si_3N_4 ($\bar{n}_{sc} = 2.0$) ARC-layers are most appropriate.

Because of the partial reflection, it is useful to define the internal quantum efficiency η_i as the number of photogenerated electron–hole pairs, which contribute to the photocurrent, divided by the number of photons which penetrate into the semiconductor.

The external quantum efficiency is the product of the optical quantum efficiency η_o and of the internal quantum efficiency η_i:

$$\eta = \eta_o \eta_i . \tag{1.47}$$

The internal quantum efficiency η_i will be discussed in Sect. 1.3.3 after carrier diffusion and drift have been introduced.

For the development of photoreceiver circuits, and especially of transimpedance amplifiers, it is interesting to know how large the photocurrent is for a specified power of the incident light with a certain wavelength. The responsivity R is a useful quantity for such a purpose:

$$R = \frac{I_{ph}}{P_{opt}} = \frac{q\lambda_0}{hc}\eta = \frac{\lambda_0 \eta}{1.243}\frac{A}{W}, \tag{1.48}$$

where λ_0 is in μm. The responsivity is defined as the photocurrent I_{ph} divided by the incident optical power. R depends on the wavelength, therefore wavelength has to be mentioned if a responsivity value is given.

The dashed line shown in Fig. 1.13 represents the maximum responsivity of an ideal photodetector with $\eta = 1$ or 100 %. The responsivity of real detectors is always lower due to partial reflection of the light at the semiconductor surface and due to partial recombination of photogenerated carriers in the semiconductor or at its surface.

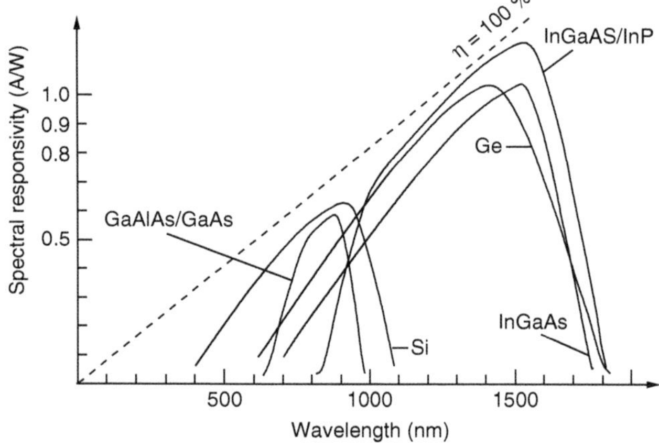

Fig. 1.13. Comparison of the responsivity of real photodetectors with an ideal photodetector with a quantum efficiency $\eta = 1$ (100 %) [24]

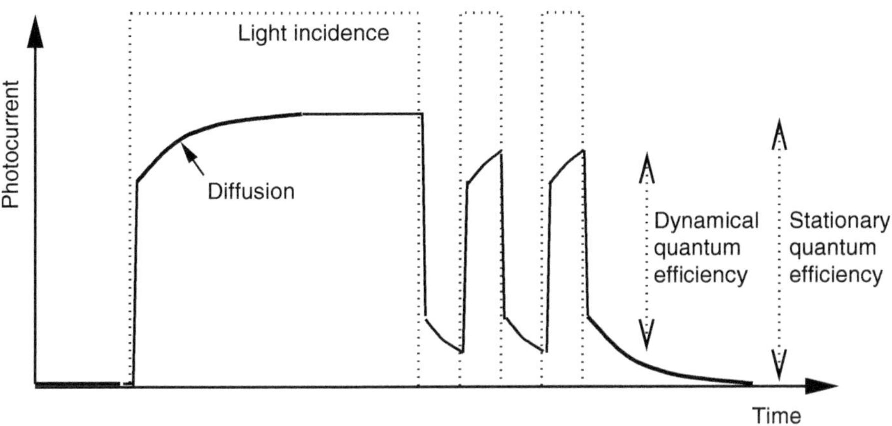

Fig. 1.14. The influence of carrier diffusion on the dynamical quantum efficiency of photodetectors at high data rates

Strictly speaking we have to distinguish between the stationary and the dynamical internal quantum efficiency. In the stationary case, the light intensity and the photocurrent are constant with time. In the dynamical case both change with time. The dynamical quantum efficiency generally is lower than the stationary one.

Let us discuss the stationary case first. As explained above, practically all carriers which are photogenerated in drift regions contribute to the photocurrent. In other words, there is no negative influence of recombination on the internal quantum efficiency in the space-charge regions of photodiodes. The recombination of photogenerated carriers in region 1 and 2 (see Fig. 1.9), however, reduces the internal quantum efficiency. In the highly doped region 1, the carrier lifetime is reduced considerably according to (1.36) and (1.37). This reduces the internal quantum efficiency for short wavelengths considerably, because a large portion of the light is absorbed in region 1.

Light with long wavelengths penetrates deep into silicon and the recombination of photogenerated carriers in region 2 can reduce the internal quantum efficiency. The recombination of photogenerated carriers in region 1 is not very important for long wavelengths due to the large penetration depth and the small portion of photogenerated carriers in region 1.

In the dynamical case, carriers being photogenerated in region 1 and especially in region 2 do not have enough time to diffuse to the space-charge or drift region before the light intensity is reduced again. The diffusion tails of the photocurrent of consecutive light pulses overlap (Fig. 1.14). The photocurrent for a sine-wave light modulation reduces similarly at high frequencies (see Fig. 1.11). It should be mentioned explicitly that the dynamical quantum efficiency depends on the frequency or data rate. The higher both these are, the smaller the dynamical quantum efficiency becomes until the

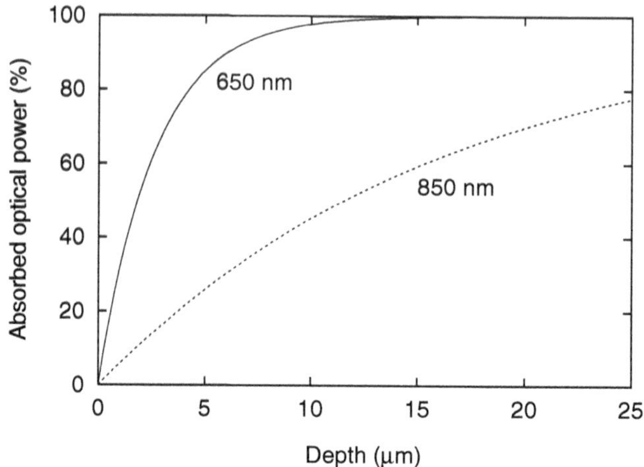

Fig. 1.15. Absorbed optical power as a function of depths in silicon

minimum is reached. This minimum is set by the portion of carriers being generated in the space-charge region, when we assume that the frequency is not extremely high and all drifting carriers still reach the boundary of the space-charge region and contribute to the photocurrent. For this case, we can use the expression

$$\eta_\mathrm{i} = (1 - \exp[-\alpha(d_\mathrm{p} + d_\mathrm{I})])\exp(-\alpha d_\mathrm{p}) \tag{1.49}$$

to describe the dynamical internal quantum efficiency. This expression was derived for ideal PIN photodiodes (compare with Fig. 1.8) with the thickness d_I of the intrinsic region, i.e. the thickness d_I of the drift region. We can also use (1.49) for a PN photodiode shown in Fig. 1.9 to a good approximation because the space-charge region with the thickness d_I does not penetrate far into the highly doped P^+ layer.

Figure 1.15 can be used in a good approximation for the dynamical quantum efficiency of a silicon photodiode with a shallow N^+ or P^+ surface region, if the depth represents the width of the space-charge region and if reflection at the Si surface is neglected, i.e. if an optimized antireflection coating is assumed. Figure 1.15 illustrates that for red light (650 nm) it is much easier to achieve a fast photodiode with a dynamical quantum efficiency close to 100 % than for near infrared light (850 nm), where a much thicker absorption region, i.e. a much wider space-charge region, is required. For 650 nm, an absorption region with a thickness of about 10 μm is sufficient to achieve almost 100 % dynamical quantum efficiency, whereas for 850 nm wavelength only about 45 % result for this absorption region and an absorption region with a thickness of 25 μm results only in a dynamical quantum efficiency of about 80 %.

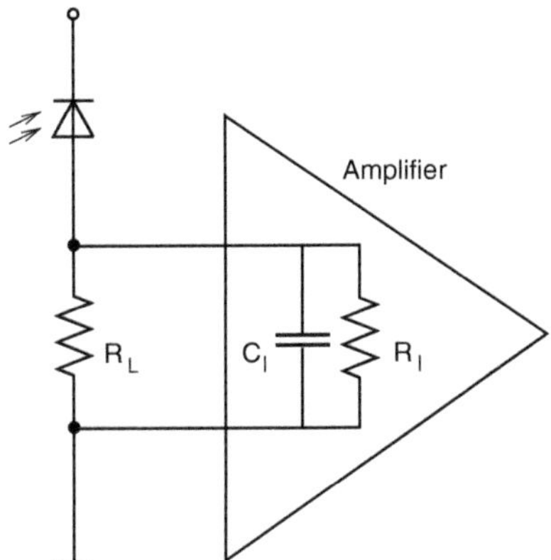

Fig. 1.16. Essential input circuit of a photoreceiver

1.3.4 Equivalent Circuit of a Photodiode

A photoreceiver consists at least of a photodetector, of a load resistor R_L, and of an amplifier with an input capacitance C_I and with an input resistance R_I (Fig. 1.16). The input resistance R_I can be neglected for amplifiers with a JFET or MOS input transistor. R_I, however, is important for amplifiers with a bipolar input transistor.

The complete equivalent input circuit of the photoreceiver considering parasitics due to mounting is shown in Fig. 1.17. The equivalent circuit of a photodiode is obtained without mount parasitics and without R_L, C_I, and R_I.

The current source I_{ph} represents the photocurrent in Fig. 1.17. C_D is the capacitance of the space-charge region of the photodiode. C_B is the capacitance of the bondpad of the photodiode. L_B represents the inductance of the bond wire. The stray capacitance of the detector package is considered in C_P. A typical value for photodiode chips and integrated circuits is $C_B = 0.2$–$0.3\,\text{pF}$. The stray capacitance of package pins of discrete photodiodes and of the output pins of integrated circuits is of the order of 1 pF.

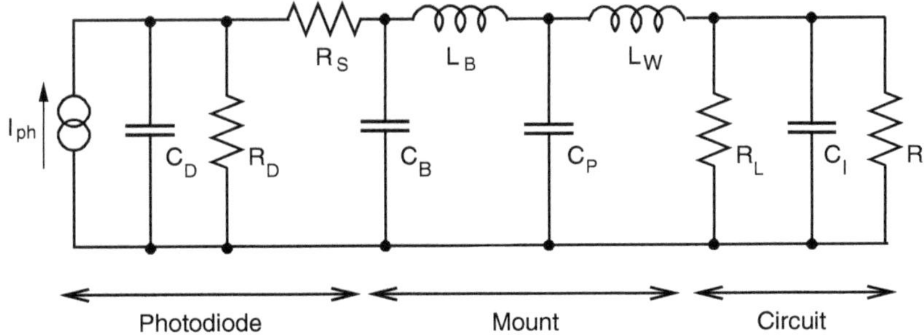

Fig. 1.17. Small-signal equivalent input circuit of a photoreceiver

The lead wires and the wiring on electronic boards introduce the additional inductance L_W. The inductances L_B and especially L_W become important at high frequencies.

The parallel resistor R_D, which models the reverse, leakage, or dark current of a photodiode usually is very large and can be neglected in most cases. The series resistance R_S may not be neglected when the photocurrent has to flow through low doped regions in the photodiode. R_S should be negligible for PIN photodiodes with highly doped P and N regions, when carrier drift dominates, i. e. when the 'intrinsic' zone is fully depleted.

The time constant t_D for a PIN photodiode with a connected amplifier is, in general, when the inductances can be neglected:

$$t_D = (R_L \| R_I)(C_D + C_B + C_P + C_I). \tag{1.50}$$

The -3dB bandwidth f_{3dB} of the input circuit of a photoreceiver is then

$$f_{3dB} = \frac{1}{2\pi(R_L \| R_I)(C_D + C_B + C_P + C_I)}. \tag{1.51}$$

Optoelectronic integrated circuits (OEICs) avoid the bondpad and package capacitances and the inductances of bond and lead wires as well as the inductance of wiring on electronic boards. Optoelectronic integrated circuits, therefore, reach a larger bandwidth $f_{3dB} = 1/(2\pi(R_L \| R_I)(C_D + C_I))$. OEICs with a MOSFET or JFET input stage of the amplifier possess a very large input resistance and the bandwidth can be written as $f_{3dB} = 1/(2\pi R_L(C_D + C_I))$.

For a reverse-biased abrupt PN junction, for which one doping type is present in a concentration several orders of magnitude larger than the other doping type, the capacitance is given by

$$C_D = A\sqrt{\frac{q\epsilon_r\epsilon_0 N_{A/D}}{2}}\frac{1}{\sqrt{U_D - U - (2k_BT/q)}}, \tag{1.52}$$

with

$$U_D = \frac{k_BT}{q}\ln\frac{N_A N_D}{n_i^2}. \tag{1.53}$$

These prerequisites for the doping concentrations are usually fulfilled in photodiodes. For P^+N or N^+P photodiodes, an abrupt PN junction is a good approximation. Equation (1.52), therefore, approximates the capacitance of most photodiodes quite well. For photodiodes with deeply diffused doping regions like N-well to P-substrate photodiodes in CMOS technology, the capacitance can be calculated with the expression for linearly graded junctions [8]. However, deeply diffused doping regions should be avoided in silicon photodiodes at least for short wavelengths in order to keep the quantum efficiency high.

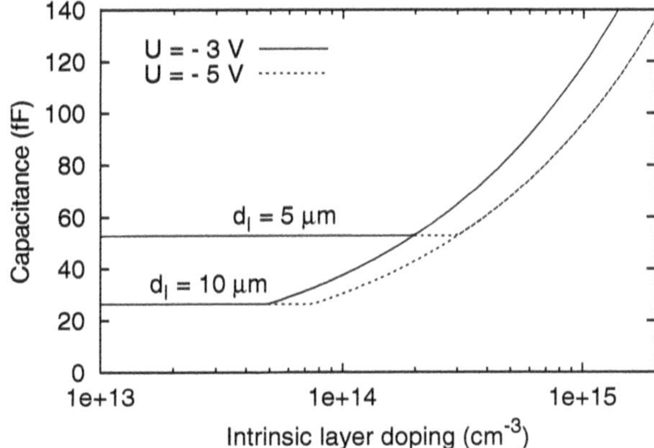

Fig. 1.18. Capacitance of PN and PIN diodes with an area of 2500 μm^2 versus doping concentration. In the constant capacitance regime, the diode can be considered as a PIN diode. For larger doping concentrations, it is not appropriate to call the diode a PIN diode, because the 'intrinsic' zone is not fully depleted

In (1.52), $N_{A/D}$ represents the doping concentration of the low doped side of the PN junction. A negative value has to be used for the reverse bias of the photodiode U.

For a so-called PIN structure, in which high doped P and N regions are at the two sides of a low doped 'intrinsic' zone with thickness d_I, the capacitance is approximately that of a plate capacitor (neglecting the boundary capacitance):

$$C_D^{PIN} = A\frac{\epsilon_r \epsilon_0}{d_I}, \tag{1.54}$$

where A is the area of the P^+-doped (PIN photodiode) or N^+-doped (NIP photodiode) region at the surface of the device.

Figure 1.18 depicts the capacitances according to (1.54) and (1.52) as a function of the doping concentration $N_{A/D}$. A diode area A of 2500 μm^2, thereby, was assumed. When the doping level is reduced starting from $10^{15}\,cm^{-3}$, the capacitance first decreases according to (1.52) until the space-charge region spreads across the whole thickness d_I of the intrinsic zone. Then C_D remains constant according to (1.54), when the doping level is reduced further. For a reverse bias of 3 V and $d_I = 5\,\mu m$, the capacitance finally remains constant at a value of 53 fF, when the doping concentration drops below approximately $2 \times 10^{14}\,cm^{-3}$. For $d_I = 10\,\mu m$, the capacitance remains constant at a value of 26.5 fF, when the doping concentration falls below approximately $5 \times 10^{13}\,cm^{-3}$. For the larger reverse bias of 5 V, constant capacitances are already reached for slightly larger doping levels (see Fig. 1.18).

2 Integrated Silicon Photodetectors

In this chapter the bipolar, CMOS, and BiCMOS process technologies are described. Photodetectors which are produced in these technologies without process modifications and their properties are introduced. Furthermore, the possible improvements of photodetectors resulting from small substrate and process modifications are discussed.

CMOS is the economically most important technology. The section on integrated photodetectors in CMOS technology, therefore, is more comprehensive than the sections on photodetectors in bipolar and BiCMOS technologies. Within the CMOS section, the sophisticated spatially-modulated-light (SML) detector suppressing slow carrier diffusion effects in standard CMOS will be described. Furthermore, the photonic mixer device (PMD) being relevant for future 3D cameras on a chip will be discussed. In addition, the innovative integration of vertical PIN photodiodes will be highlighted, since they allow a considerable improvement of the speed of CMOS OEICs. Furthermore, image sensors using charge-coupled-devices and active pixel image sensors will be described in some detail because of their possible economical importance. Within the BiCMOS section, the exploitation of double photodiodes will be mentioned as another innovation for high-speed OEICs and OPTO-ASICs in standard technology.

2.1 Integrated Detectors in Bipolar Technology

2.1.1 Bipolar Processes

In the standard buried collector (SBC) technology [25], the N^+ collector is implanted in the P-type substrate and an N-type epitaxial layer is grown, which serves as the N^- collector (Fig. 2.1).

The P^+ isolation, the N^+-collector contact, the P-type base, and the N^+ emitter are implanted and annealed subsequently. In modern technologies, polysilicon emitters [26] and polysilicon extrinsic bases [27], oxide isolation [28, 29], or recessed oxide isolation [30, 31] are used. Alternatively, trench isolation [32–34] is also used. Here, however, a simple process flow (Fig. 2.2) will be described, in order to demonstrate more agreement with [35], where a vertical PIN photodiode has been integrated (see Fig. 2.7).

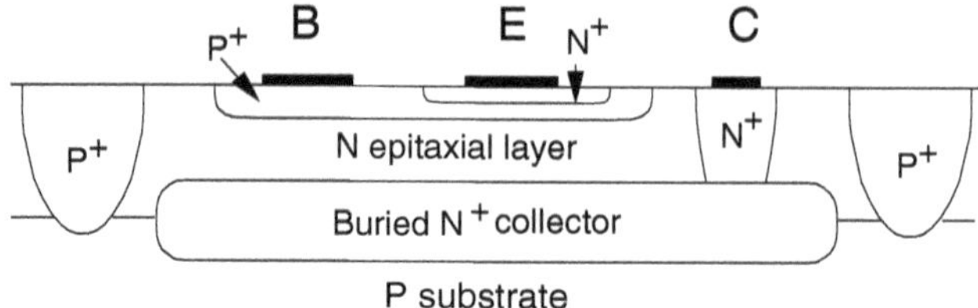

Fig. 2.1. Schematic cross section of an NPN transistor in SBC technology [25]

The P-type substrate with a doping concentration of $\approx 10^{15}\,\text{cm}^{-3}$ is oxidized (Fig. 2.2a). Then the area for the buried collectors is defined by lithography (M1). Arsenic is diffused or, more recently, antimony is implanted (Fig. 2.2b) to form the N^+ subcollector, which is needed to minimize the collector series resistance. Subsequently, the N^- collector layer is grown epitaxially (Fig. 2.2c). An oxidation is performed (Fig. 2.2c) and the areas for device isolation are defined by lithography (M2). Boron is diffused and driven in under oxidizing conditions (Fig. 2.2d). The resulting P^+ regions reach the P substrate and isolate the NPN transistor from neighboring transistors. Then lithography is applied to define the base area (M3). Boron is implanted in this area (Fig. 2.2e) and annealed partially under oxidizing conditions. The emitter area and the collector contact area are defined next (M4) and implanted with phosphorus (Fig. 2.2f) or arsenic. The annealing is again performed partially under oxidizing conditions. The contact holes are defined by lithography (M5) and aluminum is deposited (Fig. 2.2g) and structured by lithography (M6) in order to obtain interconnects (Fig. 2.2h). After oxide and nitride deposition for passivation, a seventh lithography step is needed for the opening of the bondpads. Seven steps of lithography are applied in total for this simple bipolar process with one metal layer. This simple process has often been varied and supplemented [36]. A second metal layer for instance would result in two additional lithography steps. An additional mask before M2 is required for a deep collector contact diffusion (collector plug) in order to eliminate the N^- region and the resulting collector series resistance between N^+ collector contact and N^+ subcollector resulting in a single N^+ path from the collector contact to the subcollector (see Fig. 2.1).

For our purposes, however, this simple process will be sufficient to explain the photodetectors available in bipolar processes without modifications (see Sect. 2.1). It will also be sufficient to understand the complexity of PIN photodiode integration, which will be discussed in Sect. 2.1.3.

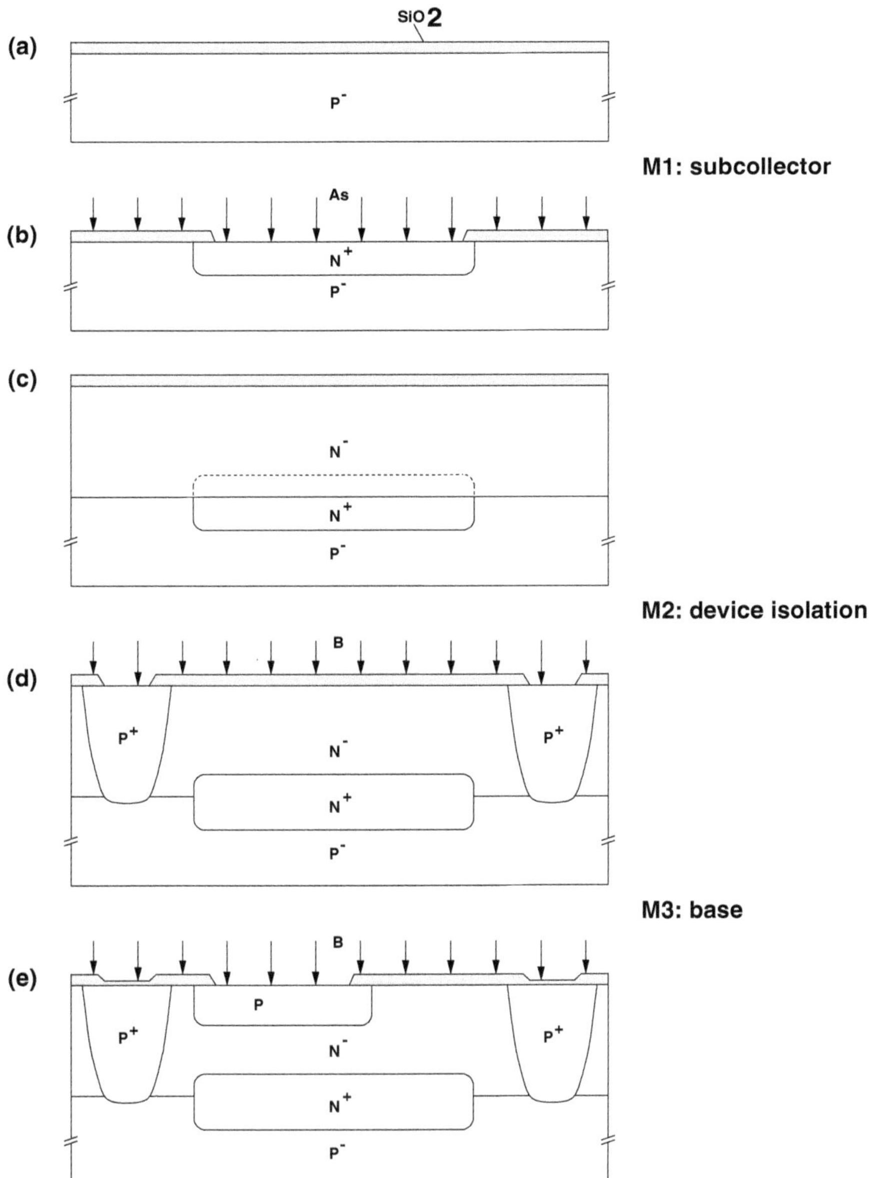

Fig. 2.2a–e. For caption see page 28

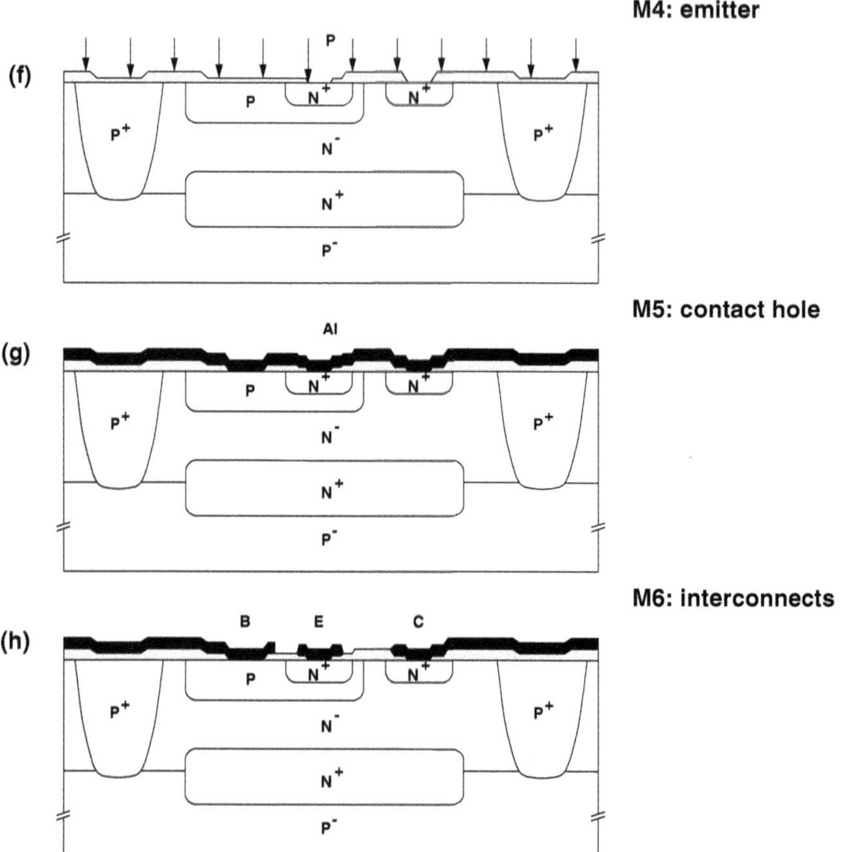

Fig. 2.2a–h. Simplified process flow of SBC bipolar technology. (**a–e**) subcollector, device isolation, and base formation, (**f–h**) emitter, contact hole, and interconnect formation [36]

2.1.2 Integrated Detectors in Standard Bipolar Technology

Photodiodes

An N^+/P substrate photodiode can be obtained in a bipolar OEIC, whereby the N^+ region is formed by the N^+ buried collector and the deep N^+ collector plug (Fig. 2.3).

Such a photodiode was realized in [37]. For an N^+ area of $100 \times 100\,\mu m^2$, a bit rate of 150 Mb/s was reported for $\lambda = 850$ nm. The speed of the photodiode was limited by carrier diffusion in the P substrate with a doping concentration of approximately $10^{15}\,cm^{-3}$. The transient behavior of a smaller photodiode with an N^+ area of $10 \times 10\,\mu m^2$ was much better. A rise time of 0.04 ns and a fall time of 0.4 ns was reported for this smaller photodiode with

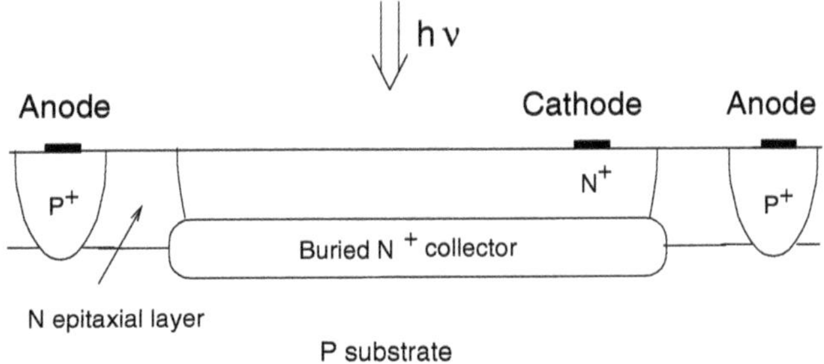

Fig. 2.3. Schematic cross section of a subcollector to substrate photodiode

$\lambda = 820$ nm. These values would allow a data rate DR of about 850 Mb/s, when we use the conservative estimate $DR = 1/(3t_f)$. The reverse bias of the photodiode was approximately 4.2 V. The quantum efficiency of the photodiode was 30 %, which was probably due to optical interferences in the isolation and passivation layers above the photodiode.

The subcollector may be omitted in Fig. 2.3. Then the photodiode is formed by the N^+ contact diffusion and the P substrate. The N^+ penetration depth, however, will be only slightly smaller. The quantum efficiency for blue and green light will be somewhat increased.

Also, without any technological modifications, the buried N^+ collector can serve as the cathode (see Fig. 2.4), the N collector epitaxial layer can serve as the 'intrinsic' layer of a PIN photodiode, and the base implant can serve as the anode in order to integrate PIN photodiodes with a thin 'intrinsic' region in bipolar technologies [38, 39]. The process of [38] was described in [40].

The small thickness of the epitaxial layer of high-speed bipolar processes in the range of about 1 μm causes a low quantum efficiency in the yellow to the infrared spectral region (580–1100 nm). The rise and fall times of the photocurrent for light pulses are very short due to the small epitaxial layer thickness. In [38] a bit rate of 10 Gb/s for the base-collector diode and a responsivity R of merely 48 mA/W for 840 nm were reported. This low re-

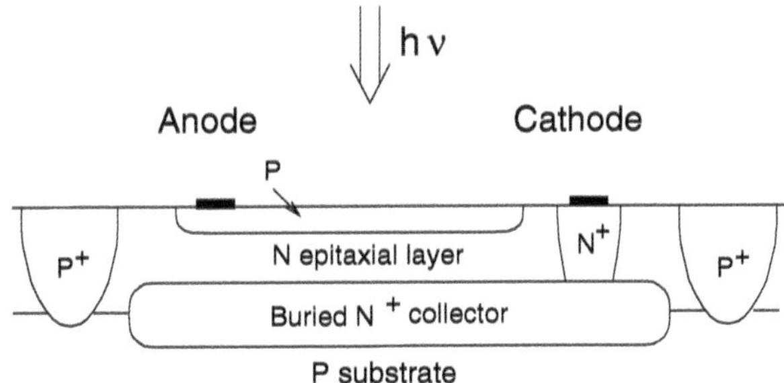

Fig. 2.4. Schematic cross section of a base-collector photodiode

sponsivity at wavelengths from 780 to 850 nm, which are widely used for optical data transmission on short fiber lengths of up to several kilometers, is a major disadvantage of standard bipolar OEICs.

The base-collector diode with a sensitive area of 100 µm^2 fabricated in a 0.8 µm silicon bipolar technology worked up to 3 Gb/s for a wavelength of 850 nm [39]. A sensitivity of 0.045 A/W was reported. A phototransistor with a very small sensitive area of 10 µm^2 reached a data rate of 5 Gb/s. With these monolithic silicon OEICs, a higher data rate was obtained than with a III/V photodetector and a silicon amplifier [41] showing the advantage of monolithic photodetector integration.

Emitter-base photodiodes may be useful for blue and green light when the emitters are implanted. The emitter-base diode, however, does not seem to be very advantageous, when polysilicon emitters are used in the bipolar process. The polysilicon then covers the N^+ emitter and, therefore, the light sensitive area of the photodiode completely, reducing the quantum efficiency by partial light absorption in the polysilicon. In particular, the quantum efficiency for blue and UV light would be very low.

Phototransistors

The NPN transistor with an increased base-collector junction area can, of course, be used as a phototransistor (see Fig. 2.5). Bipolar phototransistors, simply speaking, use the base-collector diode as a photodiode and amplify the photocurrent of this diode. The P type region of this photodiode and the P-type base of the NPN transistor are one P-type region; the cathode of this photodiode and the collector of the NPN transistor consist of one N-type region. The base contact can be omitted. The electron–hole pairs generated in the base-collector space-charge region are separated by the electric field. In this space-charge region the holes are swept into the base and the electrons are swept into the collector. The holes make the base potential positive and the emitter can inject electrons into the base and towards the collector. The photocurrent of the photodiode $I_{\rm pd}$ is amplified by the current gain β of the NPN transistor. This amplified photocurrent is available at the collector electrode ($\beta \cdot I_{\rm pd}$) and at the emitter ($(\beta+1) \cdot I_{\rm pd}$). It should be noted, however, that for long wavelengths, i. e. for large penetration depths of the light, only the fraction of the carriers photogenerated in the base-collector space-charge region will contribute to $I_{\rm pd}$, which is amplified. The other fraction of the photogenerated carriers does not contribute to $I_{\rm pd}$ and is not amplified.

Bipolar phototransistors are known to be much slower than photodiodes. This is due to their current gain β, to the base transit time $\tau_{\rm B}$, for which carrier diffusion may be limiting, to the emitter-base space-charge capacitance $C_{\rm SE}$ and mainly to the rather large base-collector space-charge capacitance $C_{\rm C}$, which results from the necessary light sensitive area. It should be mentioned that the total emitter-base capacitance $C_{\rm E}$ is the sum of the

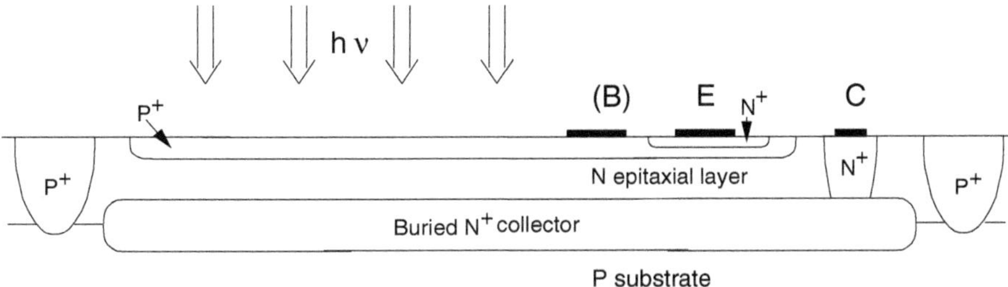

Fig. 2.5. Schematic cross section of a phototransistor in SBC technology

emitter-base diffusion capacitance C_{DE} and the emitter-base space-charge capacitance C_{SE}, i. e. $C_{\mathrm{E}} = C_{\mathrm{DE}} + C_{\mathrm{SE}} = \tau_{\mathrm{B}} g_{\mathrm{m}} + C_{\mathrm{SE}}$. The -3dB frequency of a bipolar phototransistor is [42]:

$$f_{3\mathrm{dB}} = \frac{1}{\beta 2\pi(\tau_{\mathrm{B}} + (k_{\mathrm{B}}T/qI_{\mathrm{E}})(C_{\mathrm{SE}} + C_{\mathrm{C}}))} \tag{2.1}$$

This equation is the same as for a common bipolar transistor [43]. The base-collector capacitance is proportional to the size of the light-sensitive area, which is much larger in a phototransistor than the base-collector junction area of a common bipolar transistor. Therefore, the bipolar phototransistor is much slower than a combination of a PIN photodiode with a small-area common bipolar transistor. Nevertheless, a phototransistor is advantageous at low frequencies and when PIN photodiodes cannot be integrated.

For the application in optical clock distribution, a NPN transistor with a self-adjusting emitter was investigated as a phototransistor receiver for the wavelength $\lambda = 840$ nm [38]. A double-polysilicon process [40] with a 0.5 µm thick epitaxial layer was used. The self-aligned emitter technique was used in order to realize a minimum emitter width of 0.6 µm with a 1.0 µm lithography. The current gain factor β of the transistors of 70 resulted in a relatively high sensitivity of 3.2 A/W, although the thickness of the epitaxial layer was much smaller than the penetration depth $1/\alpha = 16$ µm of the light with this wavelength. A data rate of 1.25 Gb/s was reported for the NPN transistor as a phototransistor with a small size.

2.1.3 Integrated Detectors in Modified Bipolar Technology

UV Sensitive Photodiode

An example of a photodiode integrated in a bipolar OEIC for ultraviolet (UV) detection will also be given. An industrial application for such a sensor system is, for instance, flame detection for combustion monitoring. A spectral range has to be used in which the emitted light of the flame is more intensive than the background radiation, i. e. the black-body radiation, at the temperature of 2000 K. In the UV spectral region from 250 to 400 nm, the optical emission of the flame is much larger than the thermal background radiation from the furnace walls.

Silicon photodiodes usually have a low quantum efficiency in this UV region. The need for cheap sensor systems, however, implied the monolithic integration of the detector together with the electronic circuit in a silicon technology [44]. Additional reasons for the monolithic integration were the insensitivity of OEICs to electromagnetic interference (EMI) and a larger signal-to-noise ratio. The UV sensor was implemented in a bipolar technology [45, 46]. The cross section of the silicon UV sensor is shown in Fig. 2.6. Shallow P^+ and N^+ implants form the depletion region close to the crystal surface, where UV light is absorbed and carriers are generated. The depth of the implanted P^+ and N^+ profiles was not standard and a few additional process steps with low thermal budget and rapid thermal annealing allowed the fabrication of the UV detector and of the interface circuit on a single chip without degrading the electrical performance of the analog components.

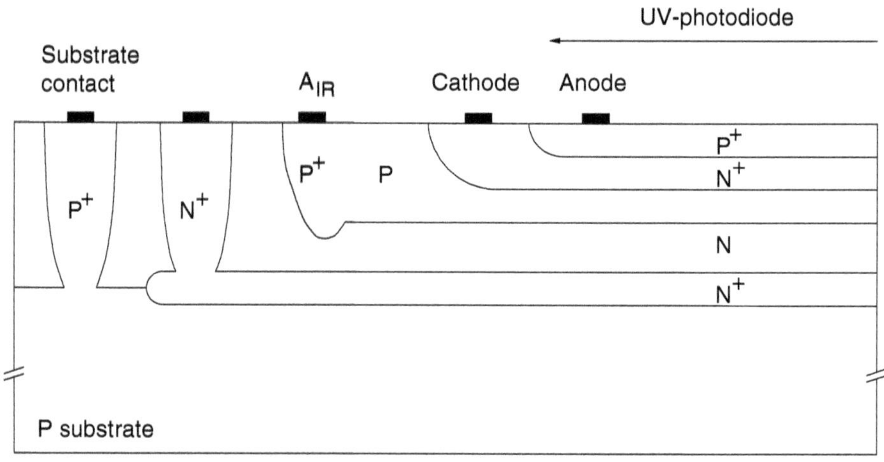

Fig. 2.6. Cross section of a bipolar-integrated UV sensitive photodiode [44]

The P^+/N^+ UV sensitive photodiode was fabricated in the base island of the NPN transistor, which is contacted by A_{IR} (see Fig. 2.6). The diode formed by the base and the added N^+ region was used to reduce contributions of deeper penetrating green, red, and infrared (IR) light to the signal current. The anode current is used as the UV signal current and the photocurrent from longer wavelengths flowing to the cathode and to the A_{IR} anode is not used by the UV sensor system.

The photocurrent of this UV detector ranged from 20 pA to 1 nA in the flame detection application. The UV-OEIC was mounted in a TO5 package with a lens focusing the UV light onto the UV sensitive photodiode with an octagonal area of 1 mm^2. A glass filter was used in addition to perform optical bandpass filtering. The electronic circuit of this UV sensor system will be described in Sect. 6.4.1.

PIN Photodiode

For the wavelengths 780 and 850 nm, a thickness of at least 10 μm is necessary for the so-called intrinsic layer of a PIN photodiode, due to the low optical absorption coefficient of silicon for wavelengths larger than about 700 nm. Such a large I-layer thickness requires the reduction of the epitaxial layer concentration below about 10^{14} cm^{-3} for a voltage of 3 V across the PIN photodiode in order to obtain a spreading of the electric field (drift region) over the whole I-zone [12]. There is, however, the so-called Kirk effect [47] or base push-out effect, which requires a large concentration for the N^- collector of the order of 10^{16} cm^{-3} for high-speed bipolar transistors with transit frequencies in the 10 GHz range or above. According to the Kirk effect, the start and the end of the base-collector space-charge region move closer to the N^+ collector and thereby increase the width of the base drastically, when the concentration of the electrons crossing the base becomes comparable to the doping concentration N_D of the N^- collector. The Kirk effect is a high-injection effect, which was investigated intensively in [48–54]. We will not discuss the details. In short, the Kirk effect becomes active when the collector current density exceeds the critical value $j_{critical}$ [55]:

$$j_{critical} = qN_D v_s. \tag{2.2}$$

To avoid the Kirk effect, the doping level in the N^- collector should be larger than 10^{16} cm^{-3}. A reduction of the doping concentration N_D in the epitaxial layer to below 10^{14} cm^{-3}, as would be necessary for the integration of PIN photodiodes, therefore, causes a dramatic decrease of the critical collector current density $j_{critical}$. In turn, the current gain and the transit frequency of the NPN transistor would drop strongly for $N_D \leq 10^{14}$ cm^{-3}.

In order to combine both of the above mentioned requirements — thick, low-doped I-layer and thin, higher doped N collector — process modifications were suggested in [35].

Figure 2.7 shows the cross section of the OEIC proposed in [35]. The PIN cathode and a P isolation were implanted into the substrate. A three-step epitaxial growth (first 2 μm as a buffer layer in order to avoid autodoping during the epitaxial process, then a 9 μm I-layer, and finally a 4 μm I-layer after the buried collector implant) was performed in order to supply the thick N-type 'intrinsic' layer with a reduced doping level ($< 3 \times 10^{13}$ cm^{-3}) and a thin collector [35]. Assuming that the N^+-buried collector and the N^+-contact implant can be used for the N^+-cathode contact diffusions, that the P^+-isolation implant (compare Figs. 2.1 and 2.2) of the standard process can

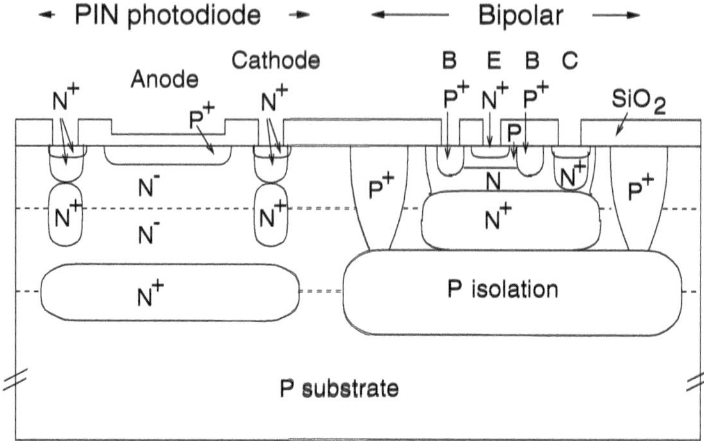

Fig. 2.7. Cross section of a PIN bipolar OEIC [35]

be used for the P isolation contact, and that the base contact implant is used for the PIN anode, the process complexity is increased by three additional lithography steps compared to the original bipolar process (without steps for antireflection coating): (1) for the N^+-buried PIN cathode; (2) for the P isolation in order to isolate the buried collectors from the N^- 'intrinsic' region and to restrict the cathode potential to the photodiode area; and (3) for the N collector implant in order to provide the doping level of about $10^{16}\,\text{cm}^{-3}$. Here, the bias voltage of the photodiode is not limited by the circuit operating voltage U_{CC} or by the breakdown voltage BV_{CE0} of the NPN transistors. Larger voltages, therefore, enable high-speed operation of the PIN photodiode. This possibility, however, was not exploited in [35].

A −3dB bandwidth of 300 MHz for $\lambda = 780$ nm was reported for the PIN photodiode with an area of $0.16\,\text{mm}^2$ at a bias of 3 V. The rise and fall times of the photocurrent were 1.6 ns for these wavelength and bias values. Bandwidth values for higher bias values, unfortunately were not reported in [35]. The responsivity of the PIN photodiode was 0.35 A/W ($\eta = 57\,\%$) for $\lambda = 780$ nm. This responsivity seems to be somewhat low, although an antireflection coating is indicated in Fig. 2.7.

2.2 CMOS-Integrated Photodetectors

The CMOS technology is the economically most important technology for the fabrication of microelectronic circuits. Early processes were one-well processes. Modern processes are twin-well processes. We will briefly discuss one-well processes first and then twin-well processes before a large variety of CMOS-integrated photodetectors will be described.

2.2.1 One-Well CMOS Processes

In [56] an N-well process with a minimum feature size of 1.2 μm is described, which uses epitaxial silicon wafers. The P-type substrate has a doping concentration of $2 \times 10^{18}\,\mathrm{cm}^{-3}$. On this substrate, a 12 μm thick epitaxial P^- layer with a doping concentration of $2 \times 10^{15}\,\mathrm{cm}^{-3}$ is deposited.

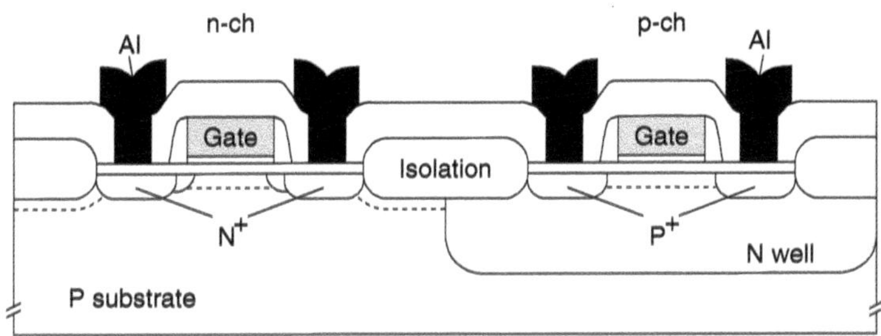

Fig. 2.8. Schematic cross section of a CMOS chip in an N-well technology

The N-channel MOSFET obtains only an anti-punchthrough implant instead of a deep P well (Fig. 2.8). The immunity against latch-up, therefore, is not very good. In order to improve the latch-up immunity of CMOS circuits and to optimize the N- and P-channel transistors independently, twin-well processes were developed.

2.2.2 Twin-Well CMOS Processes

A typical submicrometer CMOS process will be described in the following. It uses twin wells [57], LOCOS (LOCal Oxidation of Silicon) isolation [29], an N^+ polysilicon gate, and lightly doped drain (LDD) N-channel MOSFETs. As illustrated in the final device cross section (Fig. 2.9), a CMOS process consists of six main sections [58] plus passivation:

(1) Formation of the wells by deep N- and P-type diffusions, which allow the independent tailoring of the doping profiles in each well.

(2) LOCOS field isolation and the formation of channel-stopper regions, which isolate neighboring devices from each other electrically.

(3) Ion implantation for the formation of surface- and buried-channel regions for N- and P-MOSFETs, respectively, in order to adjust the threshold voltages.

(4) Gate oxidation and deposition of the N^+ polysilicon layer with subsequent gate definition.

(5) Source/drain junction formation by ion implantation including sidewall spacers for LDD formation.

(6) Deposition of oxide, contact hole opening, and metallization.

(7) Passivation.

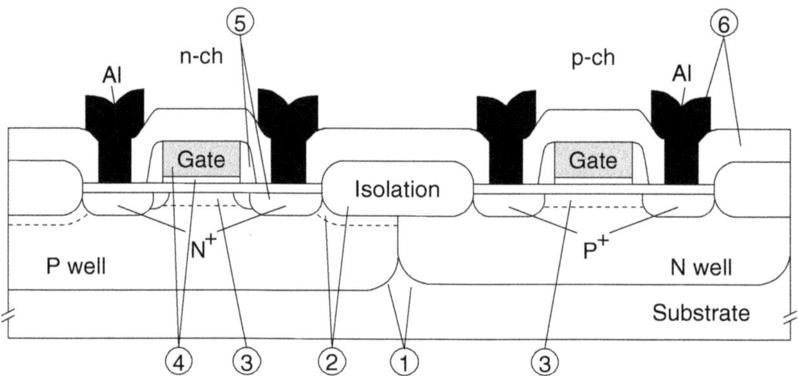

Fig. 2.9. Schematic cross section of a CMOS chip. The common CMOS process can be divided roughly into six main steps [58]

The process flow will be treated in more detail in Sect. 2.2.5. Here several other examples of modern twin-well processes will be mentioned. A 0.6 µm twin-well CMOS process of AT&T uses a 7 µm thick epitaxial layer [59]. The supply voltage is 3.3 V. The wells are drive in to a depth of about 1.8 µm. Further examples of twin-well processes were reported in [60,61]. The first of the two processes is a 0.5 µm process with a specified supply voltage of 3.3 V. The second process is a 0.25 µm process with a specified supply voltage of 2.5 V.

2.2.3 Integrated Detectors in Standard CMOS Processes

PN Photodiodes

The simplest way to build CMOS OEICs is to use the PN junctions available in CMOS processes: source/drain-substrate, source/drain-well, and well-substrate diodes. These PN photodiodes, however, possess regions which are free from electric fields. In these regions, the slow diffusion of photogenerated carriers determines the transient behavior of such PN photodiodes. Published PN CMOS OEICs are characterized by bandwidths of less than 15 MHz [62–64]. Another example is described in [65], where the OEIC was optimized for a dynamic range of six decades in illumination.

The source/drain-substrate and source/drain-well photodiodes are more appropriate for the detection of wavelengths shorter than about 600 nm, whereas the well-substrate photodiode is more appropriate for long wavelengths like 780 or 850 nm. Figure 2.10, for instance, shows an N^+/P-substrate photodiode.

In addition to carrier diffusion, the series resistance of the photodiodes due to lateral anode contacts at the silicon surface together with the relatively large junction capacitance of the photodiode may limit the dynamical PN

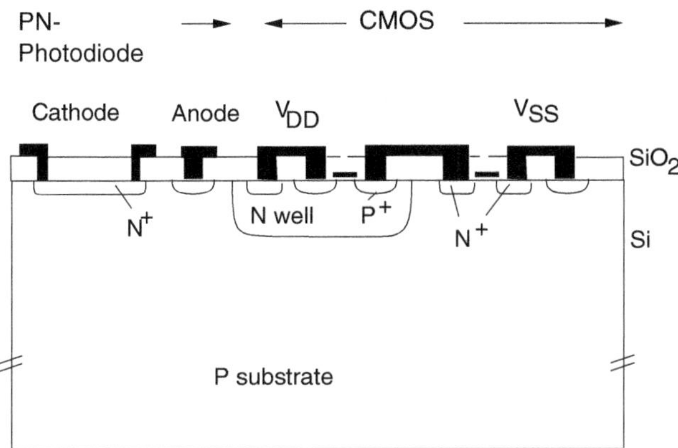

Fig. 2.10. Schematic cross section of a PN photodiode integrated in a one-well CMOS chip

photodiode behavior. In an N-well process, the anode of the N^+/P-substrate photodiode has to be at V_{SS} potential, which may be a restriction for circuit design.

A lateral N^+/P-substrate/P^+ photodiode was used as an optical detector, into which light was coupled by an integrated waveguide [62]. This detector was realized in a 0.8 µm N-well CMOS process. A maximum bandwidth of 10 MHz was achieved with a transimpedance amplifier for a photocurrent of 1 µA with $\lambda = 675$ nm. The speed of the detector was limited by carrier diffusion due to photogeneration outside the diode [63].

The N-well/P-substrate diode in a 2 µm N-well CMOS process was used as a photodiode in [64]. A bandwidth of 1.6 MHz with a wavelength $\lambda = 780$ nm was reported for an unoptimized system. The leakage current density of the photodiode was 15 pA/mm^2 at 5 V. The responsivity for $\lambda = 780$ nm was 0.5 A/W ($\eta = 70$ %). The light was coupled into the photodiode via an integrated waveguide. Therefore, the light transmission of the isolation oxide was not strongly influenced by destructive interference.

In addition to the continuous observation of the photocurrent of a PN photodiode, PN photodiodes can be operated in a storage mode. The capacitance of the photodiode can be used to integrate and store a photogenerated charge and to read it after certain periods (Fig. 2.11).

For such a purpose, the capacitance of the photodiode is initially set to the reverse voltage V_{rev}, when the reset-MOS switch is ON. Then the reset-MOS switch is opened to the OFF state and the photodiode floats for the light integration period. During this time, the capacitance is being discharged with a rate which depends on the light intensity. The remaining charge on the capacitance Q_{pix} is switched by the read-MOSFET to the input of a charge sensitive amplifier (CSA) at the end of the integration period. The smaller this charge being read, the more intensive was the incident light. This discharging

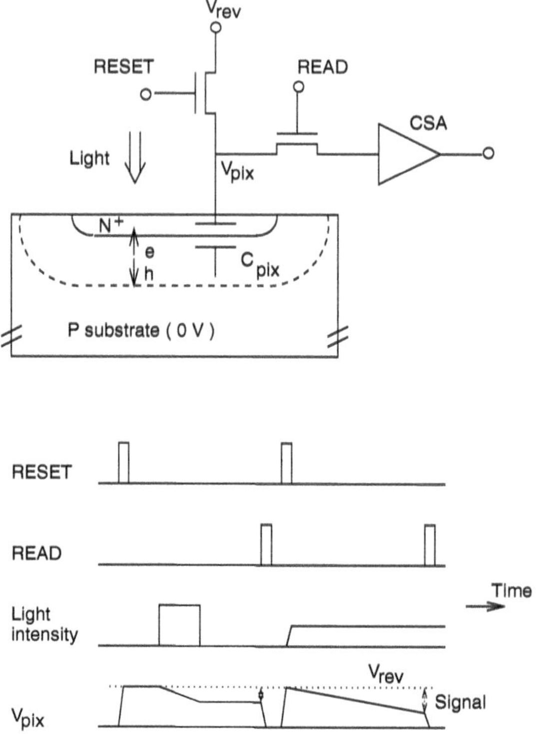

Fig. 2.11. Operation of a PN photodiode in a discharging storage mode

method is advantageous compared to a charging method, because the electric field due to V_{rev} in the photodiode helps to collect the photogenerated charge effectively. With a charging method, the space-charge region would be thinner and charge collection would not be as effective. N^+/P-substrate photodiodes in a 512 one-dimensional array were, for instance, integrated in a 1.2 μm analog CMOS technology and used in the charge storage mode [66].

Double Photodiodes

The double photodiode shown in Fig. 2.12 does not need any process modifications. It consists of the N^+/P-well diode and of the P-well/N^-N^+-substrate diode. The load resistor has to be connected to the common P-well anode in order to collect the photocurrents of both photodiodes.

Figure 2.13 shows the frequency response of a double photodiode (DPD) on a N^-N^+ substrate, which was fabricated with a 1 μm CMOS process. The doping concentration of the N epitaxial layer was approximately $1 \times 10^{15}\,\mathrm{cm}^{-3}$. The integrated 500 Ω resistor was connected to the P-well anode, which was biased at +2 V. The N^+ cathode and the N-substrate cathode were at +5 V. The reverse bias of the double photodiode accordingly was 3 V. The

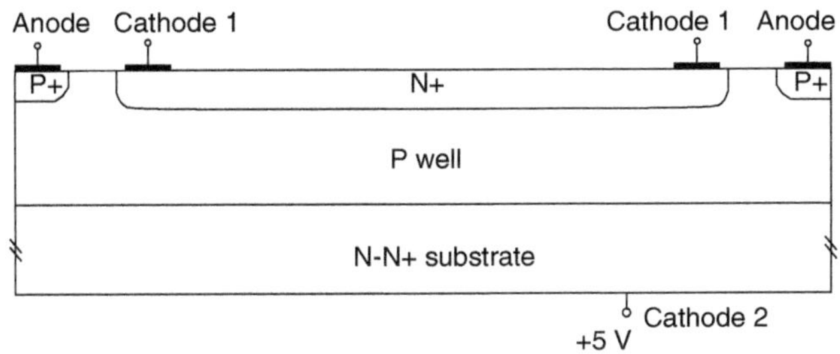

Fig. 2.12. Schematic cross section of a double photodiode

size of the photodiode was approximately $50 \times 50\,\mu m^2$. A -3dB frequency of approximately 90 MHz was measured for this double photodiode with a picoprobe.

In Fig. 2.13 the frequency response of a P^+N photodiode with the same size and fabricated on the same wafer as the double photodiode is shown for comparison. The -3dB frequency of the P^+N photodiode is only 32 MHz and the photocurrent already begins to decrease at a frequency of 2 MHz due to slow diffusion of photogenerated carriers from a depth of more than approximately 2 µm. In the double photodiode two space-charge regions are present. The first space-charge region is formed at the N^+/P-well junction and the second space-charge region is present at the P-well/N-substrate junction at a depth of approximately 4 µm. Furthermore, there is an electric field between these two space-charge regions due to the doping gradient of the N well [67]. The field-free region in the substrate below the P-well/N-substrate

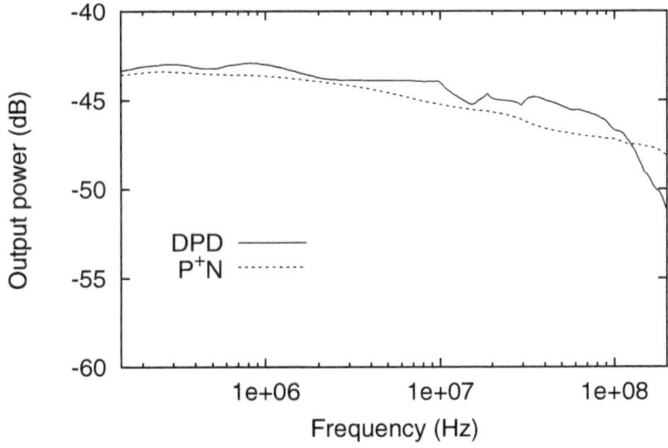

Fig. 2.13. Comparison of the frequency responses of a double photodiode (DPD) and of a P^+N photodiode fabricated in a 1 µm CMOS process and measured with integrated 500 Ω gate polysilicon resistors and a picoprobe for a wavelength of 638 nm

junction where a small portion of carriers is photogenerated with red light, therefore, is thinner in the double photodiode and carrier diffusion is much less important resulting in a much larger −3dB bandwidth.

The rise and fall times of 2.34 ns and 2.48 ns, respectively, for the double photodiode were determined with a picoprobe and a digital oscilloscope (Fig. 2.14). The frequency and transient responses of the double photodiode were limited by the *RC* time constant of its large capacitance of more than 1 pF and the 500 Ω resistor. The capacitance of the picoprobe of 0.1 pF is rather small and can be neglected.

In addition to the application of the double photodiode as a fast photodetector, it is possible to use the DPD as a color detector [68]. The diode with the shallow junction preferentially collects carriers created by shorter wavelength light, while the diode with the deeper junction collects carriers created by longer wavelength light. The shorter wavelengths predominantly generate a photocurrent in the source/drain-well diode, while the longer wavelengths predominantly generate a photocurrent in the well-substrate diode. The quantum efficiency of the source/drain-well diode peaked at about 530 nm and the quantum efficiency of the well-substrate diode showed a maximum at about 710 nm [68]. The ratio of the photocurrents of the source/drain-well diode and the well-substrate diode versus wavelength was a monotonously decreasing function in the range from 450 to 900 nm. Measuring the photocurrents of the two diodes, computing their ratio, and looking up the wavelength for this ratio value from a list stored in an EEPROM for instance allows us, there-

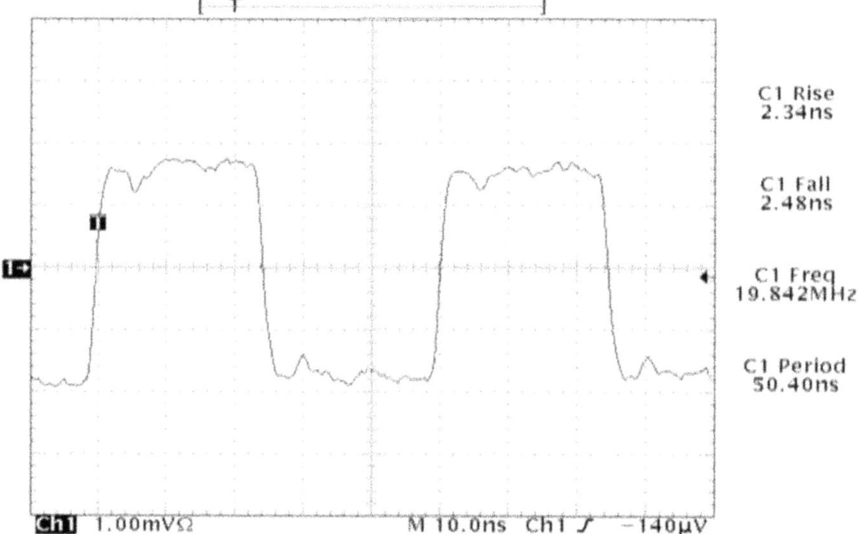

Fig. 2.14. Transient response of a CMOS-integrated double photodiode measured with an integrated 500 Ω polysilicon resistor and a picoprobe for $\lambda = 638$ nm

fore, to determine the wavelength in monochromatic or narrow bandwidth applications.

Interrupted-P-Finger Photodiode

Another example of a high-speed photodiode in a 0.35 µm standard CMOS technology [69] will be given in the following. The photodiode uses the P^+ source/drain implant of the CMOS process for the anode and the N well for the cathode (Fig. 2.15).

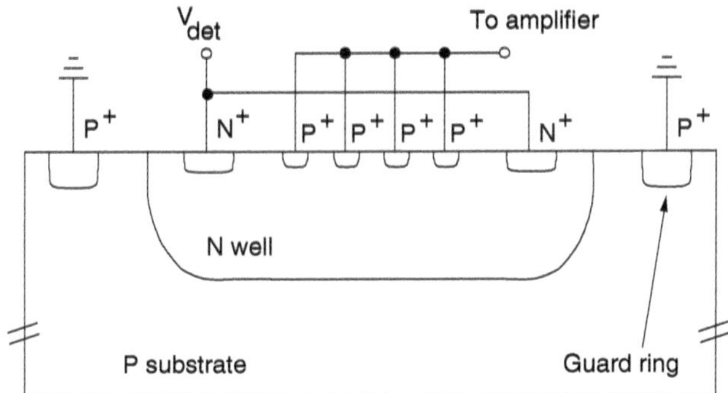

Fig. 2.15. Schematic cross section of an N- well / interrupted - P -finger photodiode [69]

The photoreceiver (see Fig. 6.110) uses the anode current for amplification of the optical signal (see Fig. 2.15). The slow diffusion of carriers generated by light with a wavelength of 850 nm in the P substrate, therefore, does not contribute to the photocurrent. This slow current is collected by the N well and shorted to the power supply V_{det}. To enhance the speed of the photodiode further, an interdigitated network of P^+ fingers is employed instead of a continuous P^+ region for maximizing the depletion regions available for carrier collection, particularly near the surface of the device. These P^+ fingers are connected outside the light sensitive area and form the anode of the photodiode. The N well had a size of $16.5 \times 16.5\,\mu m^2$. With a detector bias of 10 V, a bit rate of 1 GBit/s was reported with a bit error rate of less than 10^{-9}. The responsivity of the photodiode with the reported values of 0.01–0.04 A/W was very low due to the shallow P^+/N-well junction. It should be mentioned that this wide range of 0.01–0.04 A/W probably is due to optical interference in the CMOS isolation and passivation stack.

2.2.4 Spatially-Modulated-Light Detector

A very sophisticated photodetector in standard CMOS technology, the spatially-modulated-light (SML) detector (Fig. 2.16), was proposed to avoid the effect of slow carrier diffusion [70].

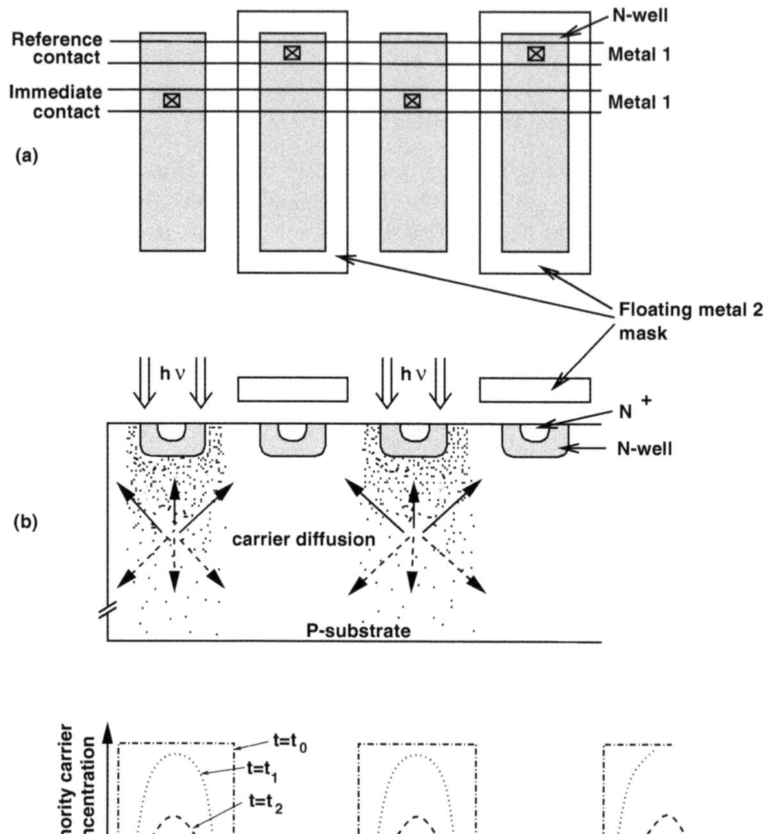

Fig. 2.16. Layout (a), cross section (b), and distribution of photogenerated carriers (c) of SML detector [70, 71]

The basic idea of this light detector is to use a reference diode (deferred detector), which measures the diffusion current caused from carriers in the substrate, and to subtract this diffusion current from the photocurrent in the immediate detector. To achieve a high speed, the immediate and deferred detectors have to be interdigitated as shown in Fig. 2.16. A portion of the diffusing minority charge carriers (electrons) photogenerated below the immediate detectors reaches the N-well of the reference (deferred) detector and causes the reference diffusion current. The deferred detectors are covered by metal, because the incident light has to be prevented from penetrating into the deferred detector. Otherwise, the diffusion current coming from the immediate detectors could not be determined.

Fig. 2.16c shows that immediately after the light incidence the photogenerated carriers are spatially modulated. After an increasing time period more and more carriers diffuse from the immediate detectors to the deferred detectors until they are almost equally distributed after the time t_3 or even better after t_4. With this SML detector the diffusion tail present for simple PN photodiodes can be suppressed to a certain degree. The price which has to be paid, however, is a reduced responsivity due to half of the detector area being covered by metal and because carriers photogenerated below the N-well/p-substrate space-charge region do not contribute to the effective photocurrent. Responsivities of 0.1 A/W (η = 15 %) for 860-nm light and 0.132 A/W for 635-nm light were measured for an implementation in a 0.6 μm CMOS standard technology [70]. A -3 dB bit rate of 500 Mb/s was mentioned for this technology. In a 0.25 μm CMOS technology, an optical receiver with a bit rate of 700 Mb/s was realized (see Sect. 6.4.10).

2.2.5 PIN Photodiode

Lateral PIN Photodiode

A lateral PIN photodiode was fabricated together with NMOS transistors in a 1.0 μm technology using a nominally undoped substrate, which was actually P-type with $N_A = 6 \times 10^{12}\,\mathrm{cm}^{-3}$ [72, 73]. The schematic cross section of the NMOS OEIC is shown in Fig. 2.17. The PIN photodiode was fabricated directly on the high-resistivity substrate. It was stated that the thickness of the depletion layer of the PIN photodiode was approximately equal to the penetration depth of 850 nm light in silicon. The lateral PIN photodiode had a circular form with the cathode in the center surrounded by the anode. The gap between anode and cathode had a width of 10 μm. A dark current of 20 nA at 5 V, a breakdown voltage in excess of 60 V, and a capacitance of 40 fF at 5 V were reported. The external quantum efficiency of the photodiode at 870 nm was 67 % without an antireflection coating.

The −3dB bandwidth of the photodiode was determined by measuring the spectral content of the photodiode response to an optical input approximating a comb of "delta" functions in the time domain. Pulses with a duration of 200 fs and a separation of 13.2 ns from a mode-locked Ti-sapphire 850 nm laser were used as the optical input for the photodiode for this measurement. The output spectrum of the photocurrent was a comb function in the frequency domain with a peak separation of 75 MHz (1/(13.2 ns)) and a pulse height determined by the frequency response of the photodiode. A −3dB bandwidth of the photodiode equal to approximately 1.3 GHz was indicated by this measurement technique at a photodiode bias of 5 V. This large value, however, seems questionable for central light incidence due to the low electric field beneath the wide N^+ cathode.

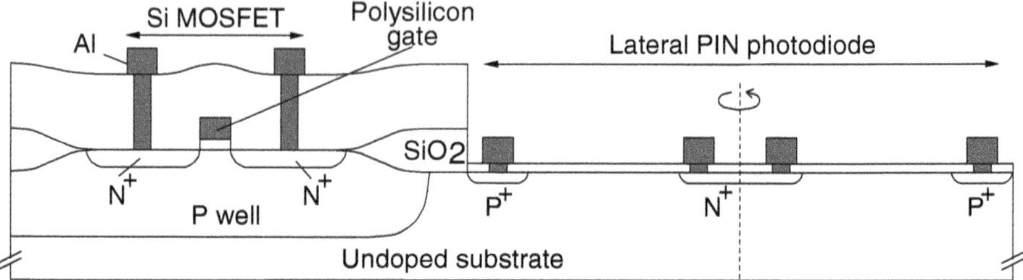

Fig. 2.17. Schematic cross section of an NMOS fiber receiver OEIC with a circular lateral PIN photodiode [72]

Vertical PIN Photodiode

Our goal is the integration of a vertical PIN photodiode in a twin-well CMOS process. We will use Fig. 2.18 to illustrate the CMOS process sequence and the simplicity of PIN photodiode integration by means of the 'third' column on the right of this figure.

The starting material is usually a (100)-oriented silicon substrate — P-type in this case — with an epitaxial layer having a doping concentration of $\approx 10^{15}\,\mathrm{cm}^{-3}$. Steps (a) and (b) show the self-aligned twin-well formation with only one lithographic mask step.

(a) The structured photoresist and nitride layer on a thin oxide layer are used for doping with phosphorus by ion implantation (I^2) into the N-well area selectively.

(b) After removing the photoresist, the wafers are selectively oxidized over the N-well region (LOCOS). The nitride layer is then etched off and boron is implanted to form the P well, whereby the LOCOS oxide prevents the boron from being implanted into the N-well area. The two wells are then driven in by high-temperature annealing to a depth of several micrometers and all oxides are etched off. The final impurity density of the wells at the silicon surface should be higher by at least one order of magnitude than that in the epitaxial layer to ensure independence of the device parameters of tolerances in the doping level of the epitaxial layer. The final impurity density of the wells at the silicon surface is typically of the order of $10^{16}\,\mathrm{cm}^{-3}$.

There are, however, also processes which do not use self-aligned twin-wells. These processes, needing two masks for the twin-well formation, are more appropriate for PIN photodiode integration, because no additional mask, i. e. no process modification is required in order to block out one of the two well implantations in the photodiode area. Although both types of processes end up with two masks for twin-well formation and PIN photodiode integration, for the process using self-aligned twin-wells the additional mask for PIN photodiode integration is a process modification. We will assume a process not using self-aligned twin-wells and start with step (c) for PIN photodiode integration.

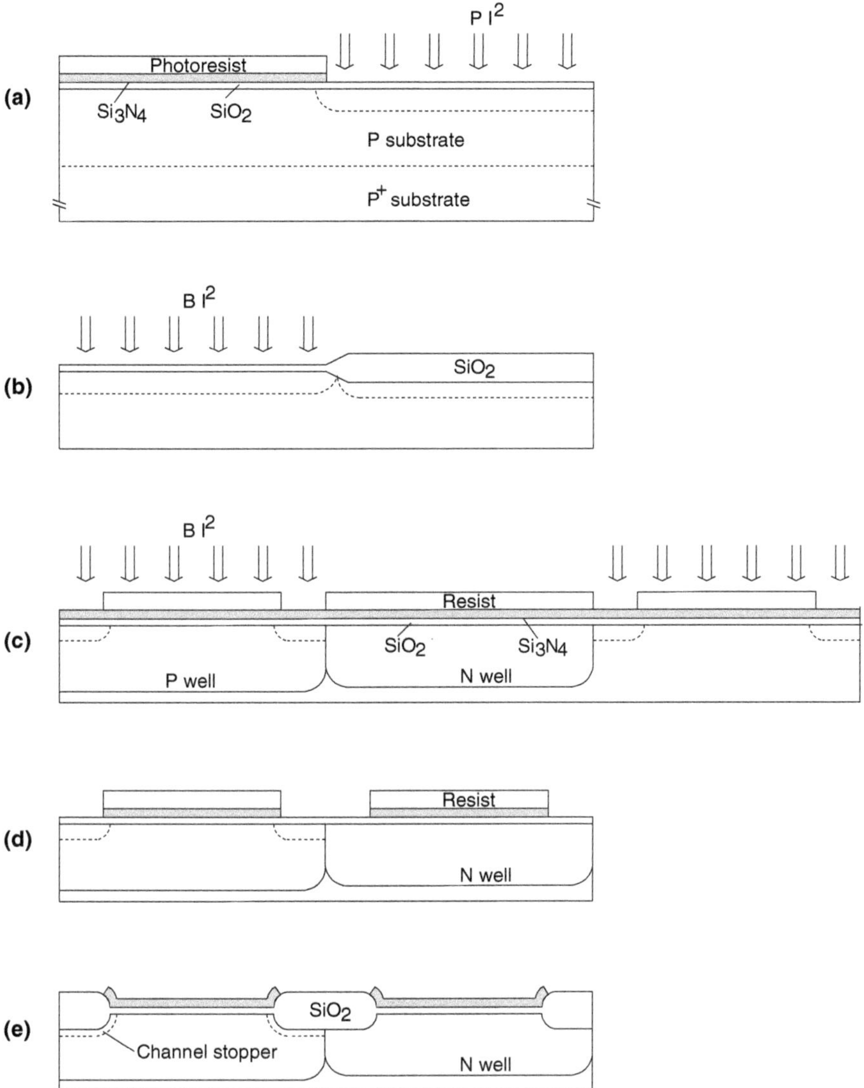

Fig. 2.18a–e. For caption see page 47

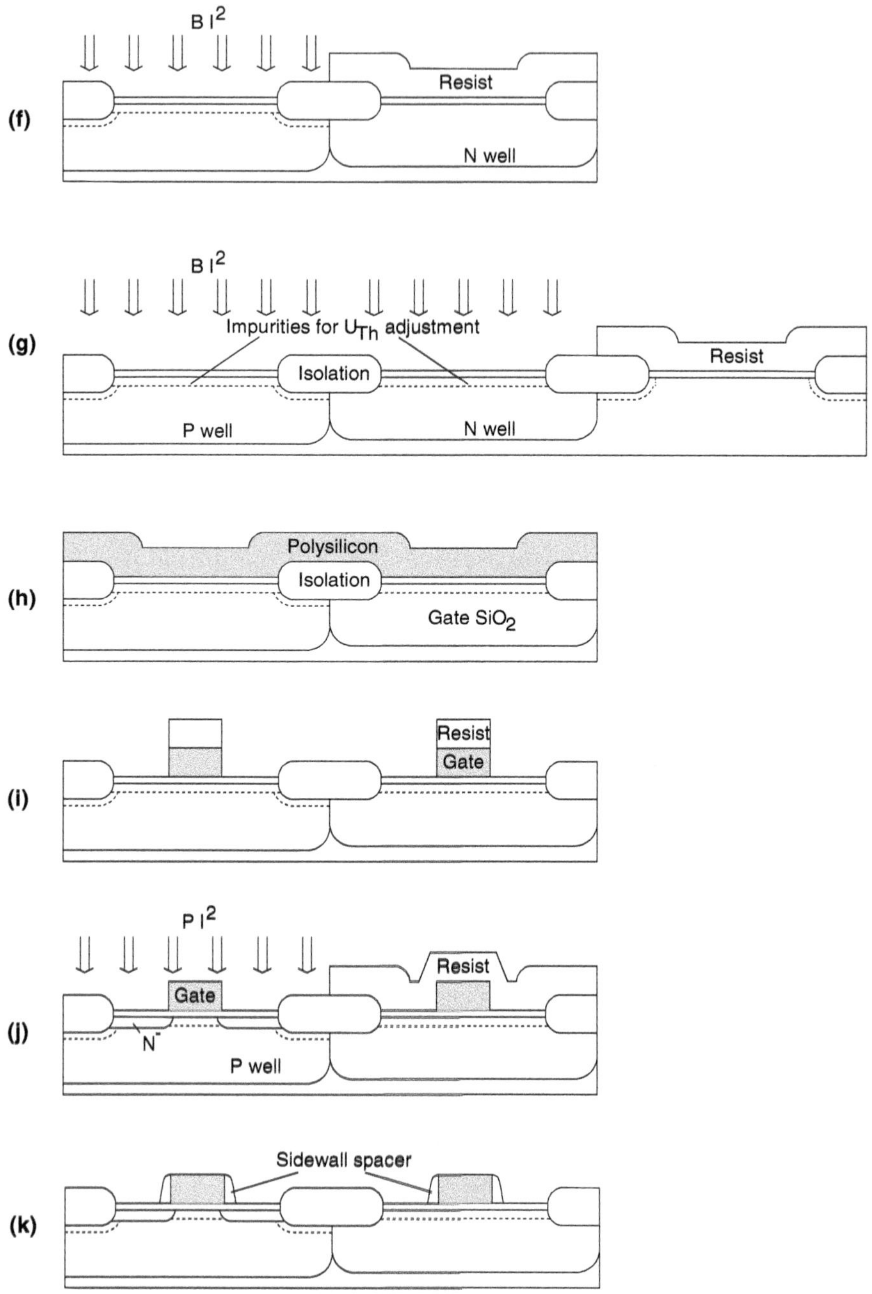

B I2
(f)
Resist
N well
B I2
(g)
Impurities for UTh adjustment
Isolation
Resist
P well
N well
(h)
Polysilicon
Isolation
Gate SiO2
(i)
Resist
Gate
P I2
(j)
Resist
Gate
N-
P well
(k)
Sidewall spacer

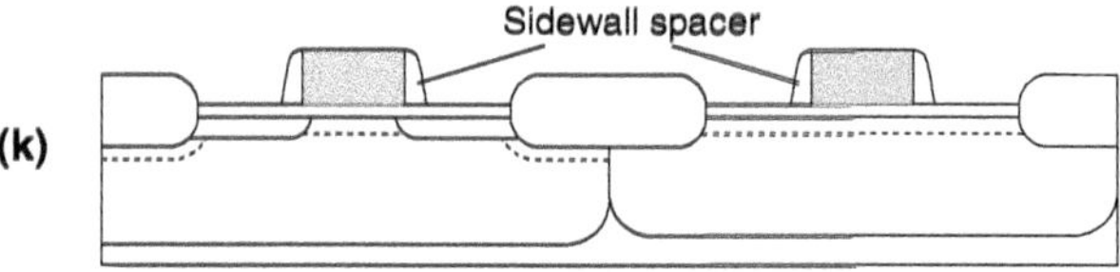

Fig. 2.18f–k. For caption see page 47

(l)

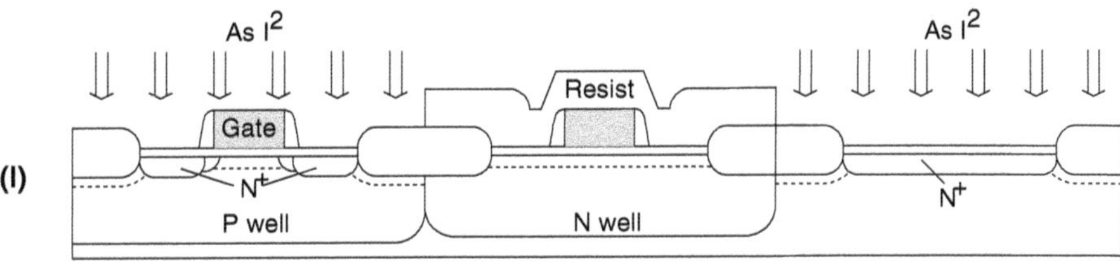

(m)

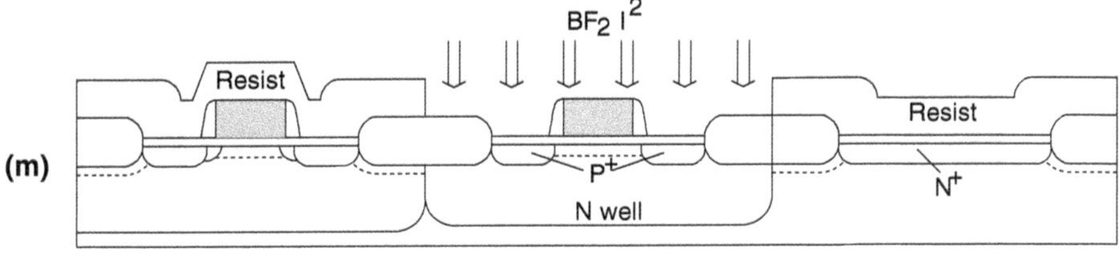

(n)

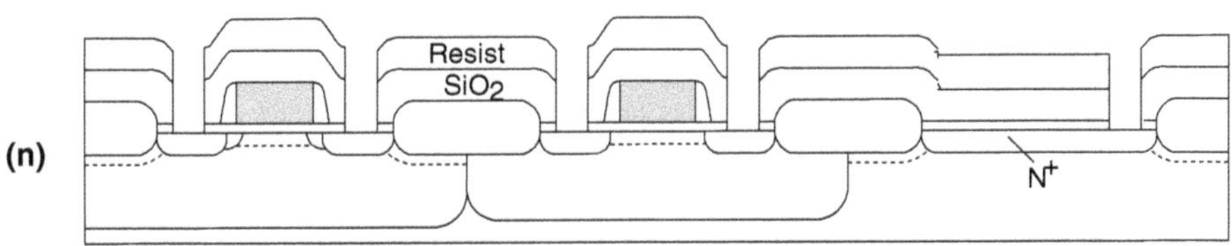

Fig. 2.18a–n. Process flow of a modern CMOS process [58] extended by the right column for PIN photodiode integration: (**a, b**) well formation, (**c–e**) isolation formation with channel stoppers (**f, g**) threshold voltage adjustment, (**h, i**) gate-oxide and gate-electrode formation, (**j**) LDD implantation, (**k–m**) LDD sidewall and source/drain formation, (**n**) contact-hole formation

The steps (c–e) show how LOCOS field isolation with an underlying channel-stopper region is formed.

(c) A nitride layer is deposited on a thin thermal oxide, called *pad oxide*. It is necessary for less mechanical stress concentrated at the LOCOS edge [74]. A photoresist mask is used to implant boron selectively through the oxide/nitride layer into the silicon, where the channel-stopper region is desired. The P-type channel-stopper is needed to increase the impurity density beneath the field isolation in P-well regions. Without this channel stopper, a parasitic N-channel MOSFET between two neighboring N-channel MOSFETs might be normally ON due to a large amount of positive fixed charge induced by subsequent LOCOS oxidation.

(d) The nitride layer is defined by dry etching and the pad oxide is exposed over the isolation region.

(e) The resist is removed and a thick field oxide is thermally grown to a thickness of several hundred nanometers over the isolation region. The nitride protects the remaining *active region* from being oxidized. The channel-stopper implant is driven in during this LOCOS oxidation. The nitride and thin oxide are then removed.

(f) The threshold implantation for the N-MOSFET can be done in two steps making two different threshold voltages available. The first threshold implantation is shown in Fig. 2.18f. A thin oxide is grown prior to the implantation to prevent ion channeling. The doping of the surface channel for the N-MOSFET is performed by boron or BF_2 implantation.

(g) The second threshold implantation for the N-MOSFET is shown in Fig. 2.18g. This boron or BF_2 implantation simultaneously results in the formation of a buried-channel P-MOSFET. The buried-channel P-MOSFET has the advantage of a lower 1/f noise compared to a surface-channel P-MOSFET. The second threshold implantation is carried out without a mask into all active regions. The photodiode region, therefore, has to be protected from being implanted by a resist mask. Let us call this mask a photodiode protection mask. The resist and all thin oxides are then removed to obtain the cross-sectional view before gate oxidation.

(h) The gate oxide is grown thermally in order to obtain a high oxide quality. Subsequently polysilicon for the gate electrode is deposited. The intrinsic polysilicon is then heavily doped with phosphorus in a gas ambient of $POCl_3$ at a temperature between 850 and 950 °C. In a polycide gate process, a silicide layer is then deposited on top of the polysilicon.

(i) The N^+-polysilicon layer is defined by lithography and dry etching. The resist is removed and the gate electrode is exposed to the so-called gate reoxidation or sidewall oxidation to prevent gate-oxide integrity near the gate edges from degradation [75].

The steps (j–m) show how the N^- regions for the LDD N-MOSFET, sidewall spacers, and N^+ as well as P^+ source/drain regions are formed:

(j) Phosphorus — or more recently arsenic — is implanted as the N^- LDD dopant. The implantation energy is low so that the phosphorus or arsenic implantation is masked by the thick field oxide, the gate electrode, and the photoresist mask above the opposite type of well.

(k) After removing the resist, oxide is deposited everywhere on the wafer and anisotropically etched to form sidewall spacers.

(l) Arsenic is implanted for N^+ source/drain formation of the N-MOSFET. A resist mask protects the P-MOSFETs from being implanted. This implant is used for the cathode formation of the PIN photodiode.

(m) BF_2 is implanted selectively into the P-MOSFET source/drain regions using another resist mask.

(n) Two oxide layers are then deposited, from which the first is undoped oxide (TEOS) and the second is boron and/or phosphorus doped for improving the glass flow property, which is needed for planarization. Contact holes are then defined by lithography and opened by dry etching.

Finally, the first aluminum layer for contact and interconnect formation is deposited and defined by lithography and dry etching. Then intermetal dielectric layers and further metal depositions may follow. In industrial practice, passivating oxide, nitride, or oxynitride layers are added by plasma-assisted deposition to protect the chip from impurities and humidity coming from the surroundings. Finally, bondpad areas have to be defined and opened.

Summarizing, we can state that only one additional mask is necessary for PIN photodiode integration on P^-P^+-substrate in order to block out an originally unmasked P-type threshold implantation, when the CMOS process does not use self-adjusting wells. Without this additional mask an N^+/P junction at the surface would result instead of the desired N^+/P^- junction. There would be a high electric field at the N^+/P junction and this would lead to a strongly reduced electric field in the 'intrinsic' P^--zone of the PIN photodiode and to longer rise and fall times.

On an N^-/N^+-substrate, the P-type PIN anode would be at the surface and the unmasked P-threshold implant increases the P^+/N^- junction depth by a small amount. In this case the electric field in the 'intrinsic' zone of the PIN photodiode is not reduced. The quantum efficiency for blue light may be reduced due to the larger junction depth. If this can be tolerated, no additional mask is required for PIN photodiode integration on an N^-/N^+-substrate.

As mentioned above, the doping concentration of the epitaxial layer is usually approximately $10^{15}\,\mathrm{cm}^{-3}$. Considering circuits with integrated photodiodes, a reasonable reverse bias of the photodiodes is approximately 3 V, when a single-supply voltage of 5 V is used. We know from Fig. 1.4 that the width of the space-charge region for a reverse bias of 3 V is only approximately 2 µm for a doping concentration of $10^{15}\,\mathrm{cm}^{-3}$. Due to the larger penetration depth of red and infrared light, slow diffusion of photogenerated carriers results for the epitaxial layer doping level of $10^{15}\,\mathrm{cm}^{-3}$. The doping level of the epitaxial layer has to be reduced in order to avoid this slow carrier diffusion [76]. It should be mentioned that this reduction of the doping concentration in the epitaxial layer is not a process modification. The CMOS manufacturer only has to buy wafers with a lower doping concentration in the epitaxial layer.

For the optical wavelengths of 780 and 850 nm being used in optical data transmission, the thickness of the epitaxial layer of a standard twin-well CMOS process of 7 µm [59], for instance, is too small due to the small absorption coefficients of Si for these wavelengths. In addition to the reduction of the doping concentration in the epitaxial layer, therefore, its layer thickness should be increased.

There are several aspects which must be investigated when the epitaxial layer of a twin-well CMOS process is modified: (i) Latch-up immunity may be reduced. (ii) Reach-through (punch-through) between wells of the some type may occur. (iii) Electrostatic discharge hardness may suffer. Before these aspects are discussed the gain in photodiode performance by the modification of the epitaxial layer, i. e. by the reduction of its doping concentration will be shown in the following.

Results for N-type Substrate

Figure 2.19 shows the structure of a N^-N^+ CMOS-OEIC in the twin-well approach. Here the N^+ substrate serves as the cathode and the P^+ source/drain region as the anode of the integrated PIN photodiode.

PIN-CMOS photodiodes with an area of 2700 μm^2, with a standard doping concentration of $1 \times 10^{15}\,cm^{-3}$ and with reduced doping concentrations in the epitaxial layer down to $2 \times 10^{13}\,cm^{-3}$, and with an integrated polysilicon resistor of 500 Ω, were fabricated in an industrial 1.0 μm CMOS process [78]. For the measurements, a laser with $\lambda = 638.3$ nm was modulated with a commercial ECL generator. The light pulses were coupled into the photodiodes on a wafer prober via a single-mode optical fiber. The rise (t_r) and fall (t_f) times of the photocurrent of the photodiodes were measured with a picoprobe (pp), which possesses a −3dB bandwidth of 3 GHz and an input capacitance of 0.1 pF, and with a 20 GHz digital sampling oscilloscope HP54750/51.

Figure 2.20 contains the waveform of the photocurrent for the photodiode with the standard doping concentration of $1 \times 10^{15}\,cm^{-3}$ in the epitaxial layer. The values $t_r = 5.3$ ns and $t_f = 7.3$ ns were determined from the waveform for the photodiode with the standard doping concentration of $1 \times 10^{15}\,cm^{-3}$ in the epitaxial layer. These large values are due to the diffusion tail of the photocurrent, which is caused by the slow carrier diffusion in the standard epitaxial layer of the photodiode. Values up to $t_r = 15.5$ ns and $t_f = 17.6$ ns were also measured for the photodiode with the nominal doping concentration

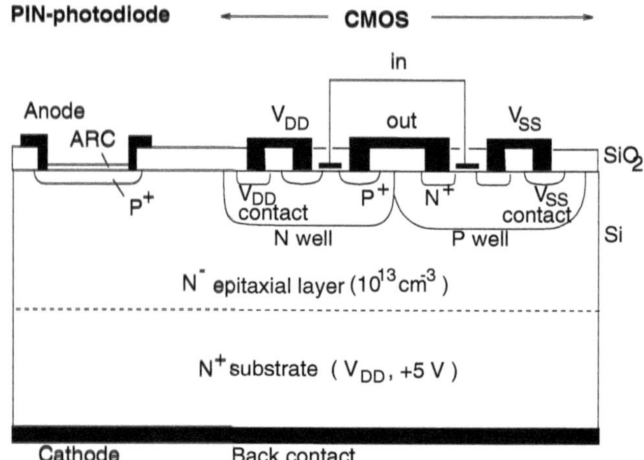

Fig. 2.19. Cross section of a N^-N^+ PIN-CMOS-OEIC [77]

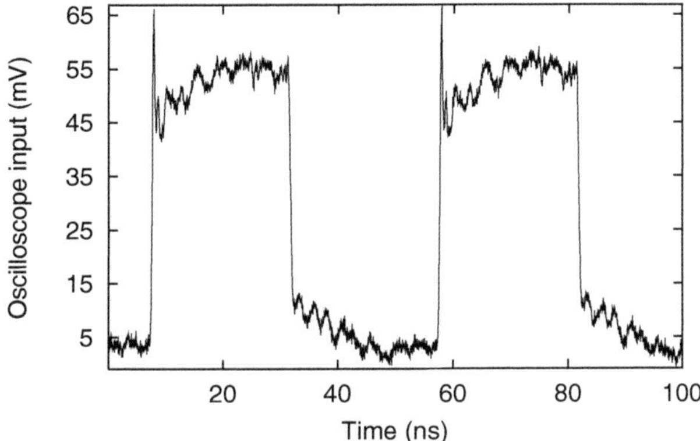

Fig. 2.20. Measured transient response of a CMOS-integrated P^+N photodiode with a standard epitaxial doping concentration of $1 \times 10^{15}\,\mathrm{cm}^{-3}$ for $\lambda = 638$ nm and $|V_{\mathrm{PD}}| = 3.0\,\mathrm{V}$. The overshoot is due to direct laser modulation. The 0 % and 100 % lines for the determination of t_{r} and t_{f} are at 3.5 and 55.5 mV

of $1 \times 10^{15}\,\mathrm{cm}^{-3}$ in the epitaxial layer. The large ranges for t_{r} and t_{f} can be explained like this: The diffusion tail of the photocurrent is very sensitive to the actual doping concentration and thickness of the epitaxial layer. The rise and fall times, therefore, vary strongly with the position of the photodiode on a wafer and from wafer to wafer due to variations of doping concentration and epitaxial layer thickness.

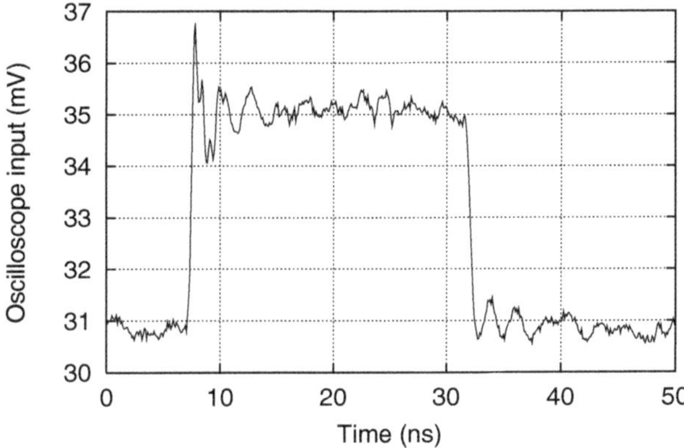

Fig. 2.21. Measured transient response of a CMOS-integrated PIN photodiode with an I-layer doping concentration of $2 \times 10^{13}\,\mathrm{cm}^{-3}$ for $\lambda = 638\,\mathrm{nm}$ and $|V_{\mathrm{PD}}| = 3.0\,\mathrm{V}$. The overshoot in the signal is due to the direct modulation of the laser [79]

For the PIN photodiode with a doping concentration in the epitaxial layer of $2 \times 10^{13}\,\mathrm{cm}^{-3}$, the oscilloscope extracted $t_r^{osc,disp} = 0.37\,\mathrm{ns}$ and $t_f^{osc,disp} = 0.57\,\mathrm{ns}$ from the waveform shown in Fig. 2.21. The evaluation according $(t_r^{PIN})^2 = (t_r^{osc,disp})^2 - (t_r^{laser})^2 - (t_r^{pp})^2 - (t_r^{osc})^2$ with $t_r^{laser} = 0.30\,\mathrm{ns}$, $t_f^{laser} = 0.51\,\mathrm{ns}$, $t_{r/f}^{pp} = 0.1\,\mathrm{ns}$ and $t_{r/f}^{osc} \approx 0.02\,\mathrm{ns}$ results in $t_r^{PIN} = 0.19\,\mathrm{ns}$ and $t_f^{PIN} = 0.24\,\mathrm{ns}$ [77]. With $f_{3dB} = 2.4/(\pi(t_r + t_f))$, the -3dB bandwidth can be estimated to be 1.7 GHz for $|V_{PD}| = 3.0$ V. With the conservative estimate $BR=1/(1.5\ (t_r+t_f))$, a bit rate, BR, of 1.5 Gb/s results for $|V_{PD}| = 3.0$ V. With an antireflection coating (ARC), the quantum efficiency η could be increased from 49 % to 94 %. To our knowledge this is the first time that such a high speed and such a high quantum efficiency have been achieved with an integrated silicon photodiode for a reverse voltage of only 3 V, whereby an only slightly modified standard CMOS process has been used.

Transistor Parameters

In contrast to transistors in bipolar OEICs, the electrical performance of the N- and P-channel MOSFETs is not degraded when the epitaxial layer is modified. This statement was verified by measurements (Table 2.1). The threshold voltages of the NMOS and PMOS transistors are practically independent of the doping concentration in the epitaxial layer, because these transistors are placed inside wells which possess a much higher doping level of several times $10^{16}\,\mathrm{cm}^{-3}$ than the standard epitaxial layer with about $1 \times 10^{15}\,\mathrm{cm}^{-3}$ and because the threshold implants produce an even higher doping level ($\approx 10^{17}\,\mathrm{cm}^{-3}$) than the wells [80].

The reverse, i. e. the leakage current of the drain to well diodes is also listed in Table 2.1 for the NMOS and PMOS transistors [80]. The leakage current for the NMOS transistor in an epitaxial layer with reduced doping concentrations actually seems to be smaller than for the standard concentration.

These results confirm the superiority of the PIN-CMOS integration [76] compared to the PIN-bipolar integration [35]. In contrast to the PIN-bipolar integration, the electrical transistor parameters of the standard twin-well

Table 2.1. Measured threshold voltages U_{Th} and drain leakage currents I_D^r for different doping levels in the epitaxial layer

I-Doping	U_{Th}^{NMOS}	$I_D^{r,NMOS}$	U_{Th}^{PMOS}	$I_D^{r,PMOS}$
(cm^{-3})	(V)	(pA)	(V)	(pA)
Standard	0.79	2.19	−0.62	63.1
1×10^{14}	0.79	0.83	−0.60	66.1
5×10^{13}	0.79	0.81	−0.60	66.1
2×10^{13}	0.78	1.02	−0.60	67.6

CMOS process are completely unaffected for the PIN-CMOS integration and can be used for circuit simulations within the design of OEICs.

Latch-up Effect

At places where N and P wells are in contact, there is the danger of latch-up. Figure 2.22 shows P- and N-type regions acting as parasitic bipolar transistors in a CMOS cross section.

The N^+ source-island of the N-channel MOSFET forms the emitter of the parasitic NPN transistor, the P well forms the base, and the epitaxial substrate forms the collector of the parasitic NPN transistor. Analogously, the emitter of the parasitic PNP transistor consists of the P^+ source island, the base consists of the N well, and the collector of the PNP transistor consists of the P well. The collector of one transistor simultaneously forms the base of the other transistor and vice versa. The NPN and PNP transistors, therefore, form a thyristor structure. The thyristor can turn on, for instance, when the base-emitter-potential of one of the transistors exceeds a value of 0.6 V, approximately, due to a lateral current in one of the wells across R_{wp} or R_{wn} as a result of voltage spikes on the well-interconnect lines. The turn-on of the thyristor leads to a shorting of V_{DD} and V_{SS} and the transistors lose their function in the circuit at least until the next power-off and power-on. They may even be destroyed due to large currents resulting in strong heating.

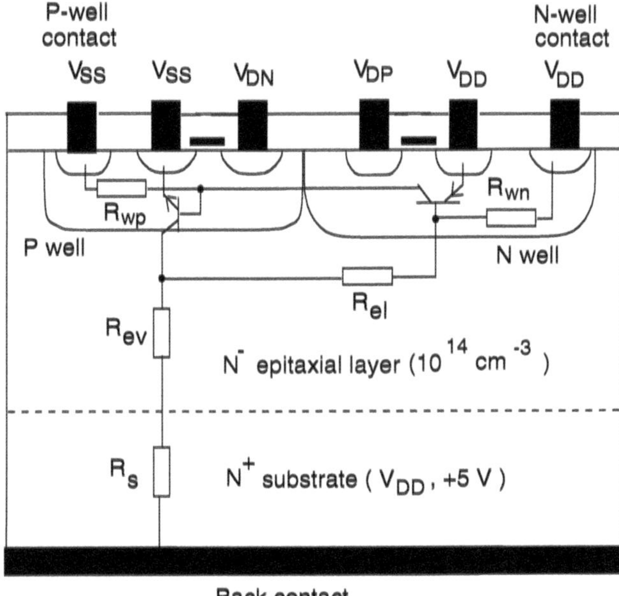

Fig. 2.22. Schematic of the parasitic bipolar transistors being responsible for latch-up in a CMOS circuit [80]

A reduction of the doping concentration in the epitaxial layer for PIN photodiode integration reduces the value of the base Gummel integral of the PNP transistor and increases the PNP current gain accordingly. The value of the NPN transistor's collector series resistance R_{ev} increases when the doping concentration of the epitaxial layer is reduced. These two aspects may lead to a reduced latch-up immunity. The lateral resistor R_{el} (see Fig. 2.22), however, is also increased. This aspect increases the latch-up immunity again, because R_{el} forms a voltage divider together with R_{wn}, which is independent of the doping concentration of the epitaxial layer.

The latch-up immunity can be enhanced by appropriate design measures. When the source of the NMOS transistor is closer to the N well than the drain of the NMOS transistor, the parasitic NPN transistor may partially inject laterally into the N well as a collector instead of injecting only vertically as shown in Fig. 2.22. Therefore, the distance of neighboring N- and P-channel transistors (more accurately between N^+ and P^+ islands in wells of different type which are in contact with each other) must not be smaller than a mimimum N^+–P^+ distance defined in the design rules in order to keep the so-called holding voltage above the V_{DD} voltage value. Using a larger distance increases the latch-up immunity due to a reduction of the bipolar transistor current gain with a wider base. Paying attention to this design rule results in an extinguishing current flow in the thyristor structure after the end of the firing impulse. The minimum N^+–P^+ distance depends on the substrate doping concentration. In order to minimize the N^+–P^+ distance, epitaxial wafers with highly doped substrates are used. Thus the minimum distance depends on the doping concentration of the epitaxial layer. Chapman et al. [81] from Texas Instruments reported a thickness of the epitaxial layer of 5 μm for a 0.8 μm CMOS technology. In order to keep the holding voltage above a value of 5 V, the N^+–P^+ distance had to be larger than 4 μm. For a constant distance of the well contacts, the holding voltage reduced to 1.5 V when the thickness of the epitaxial layer was increased to 7 μm. The holding voltage reduced to approximately 1 V for an epitaxial layer thickness of 9 μm. In order to obtain holding voltage values larger than 5 V for the thicker epitaxial layers, the distance to the well contacts had to be reduced [81].

The influence of the substrate doping concentration on the latch-up behavior of a twin-well CMOS process was investigated in [82]. The result of this work was that with a highly doped substrate (0.01–0.02 Ωcm) and with a 10 μm thick epitaxial layer (1.7–2.5 Ωcm), the emitter firing currents of the NPN and PNP transistors could be increased from 30 mA to 250 mA by implementing an N^+ guard ring around the P emitter in the N well. For a low substrate doping (1.7–2.5 Ωcm), an N^+ guard ring around the P emitter increases the emitter firing current of the NPN transistor from about 7 mA to 235 mA. An additional P^+ guard ring around the N emitter increases the emitter firing current of the PNP transistor from about 9 mA to 60 mA. From the results of [82] we can draw the conclusion that in OEICs with reduced doping concentration in the epitaxial layer or/and increased epitaxial layer thickness, the positive influence of the highly doped substrate on the latch-up immunity is reduced. A second conclusion, however, is that the latch-up immunity can almost be regained with the implementation of guard rings.

Another very effective design measure for the suppression of latch-up is the implementation of guard rings around the wells. These design measures applied as a precaution for OEICs with a reduced doping concentration in the epitaxial layer, however, would require die area and they might result in an inacceptable increase of the die area of highly integrated digital OEICs. For analog OEICs, which contain far fewer transistors, the additional die area for latch-up precaution would be easier to accept. Because of the large economical importance of digital ICs and because of the potential high-volume market for OEICs in optical interconnects, however, the latch-up immunity of PIN-CMOS-OEICs was examined experimentally.

In order to investigate the influence of the doping concentration in the epitaxial layer on latch-up immunity a test structure (see Fig. 2.23) was fabricated together with the CMOS-integrated PIN photodiodes.

The distance of the emitters from the well boundary was 2 μm each. A V_{DD} voltage of 5.5 V was used in order to account for tolerances of power supplies. The hole current injected into the P^+ emitter of the parasitic PNP transistor necessary for triggering the thyristor structure was measured [80]. The injected hole current necessary for triggering latch-up is shown in Fig. 2.24 for several doping concentrations in the epitaxial layer. The injected hole current necessary for triggering latch-up is constant when the doping concentration

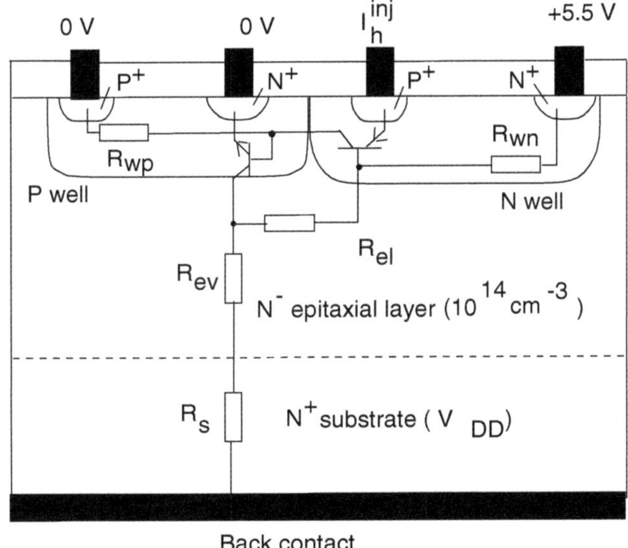

Fig. 2.23. Schematic cross section of the latch-up test structure [80]

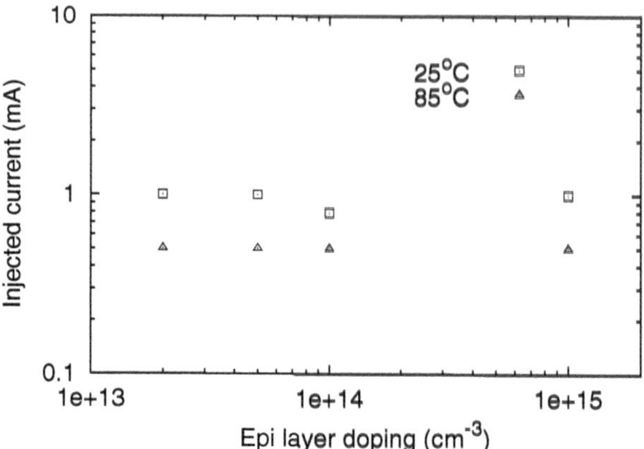

Fig. 2.24. Injected hole current I_h^{inj} necessary for triggering latch-up versus doping concentration in the epitaxial layer [80]

of the epitaxial layer is reduced from $1 \times 10^{15}\,\mathrm{cm}^{-3}$ to $2 \times 10^{13}\,\mathrm{cm}^{-3}$ for room temperature and for an elevated temperature of 85 °C [80]. Latch-up immunity, therefore, is not degraded when PIN photodiodes are integrated. Accordingly, the latch-up aspect does not require a modification of the design rules of the CMOS process used for PIN photodiode integration and die-area-consuming guard rings are not necessary for analog and digital CMOS OEICs.

Reach-through Effect

The reach-through or punch-through effect is important between neighboring analog wells which have a doping type different from that of the substrate. Analog wells are not at the power supply voltage rails like digital wells. Let us assume for simplicity that one of the wells is at the substrate potential (Fig. 2.25).

Then the potential difference between the two wells can be maximum. The consequence is that the space-charge region extends most widely from one well towards the other. When the space-charge region extends to the other well for a large potential difference of the two wells, a current begins to flow between the two wells (Fig. 2.26), which may change the operating point in an analog circuit. Therefore, there is a design rule which sets the minimum distance of analog wells.

When we reduce the doping concentration of the epitaxial layer, the space-charge region extends further for a constant difference voltage. The design rule for the minimum distance of analog wells, therefore, has to be modified [80]. According to (1.22) with the difference voltage of the two wells U_{12} and considering the lateral subdiffusion of the two wells d_{subd} plus a security

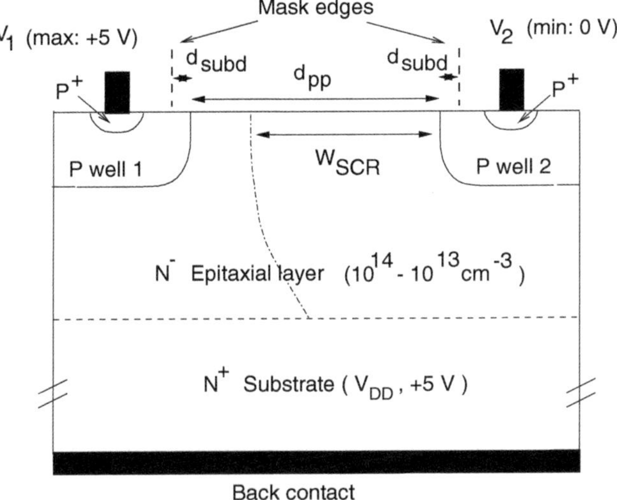

Fig. 2.25. Schematic to illustrate the reach-through effect between analog wells of the same type [83]

distance d_{sec}, the new minimum distance of analog wells $d_{\mathrm{pp}}^{\mathrm{min}}$ [83] results

$$d_{\mathrm{pp}}^{\mathrm{min}} = \sqrt{\frac{2\epsilon\epsilon_0}{qN_{\mathrm{I}}}(U_{\mathrm{D}} + |U_{12}| - \frac{2k_{\mathrm{B}}T}{q})} + 2d_{\mathrm{subd}} + d_{\mathrm{sec}}. \tag{2.3}$$

The difference in this equation compared to the original expression for the design rule is that N_{I} instead of the standard doping concentration in the epitaxial layer is used. For N_{I}, the low tolerance limit of the epitaxial doping

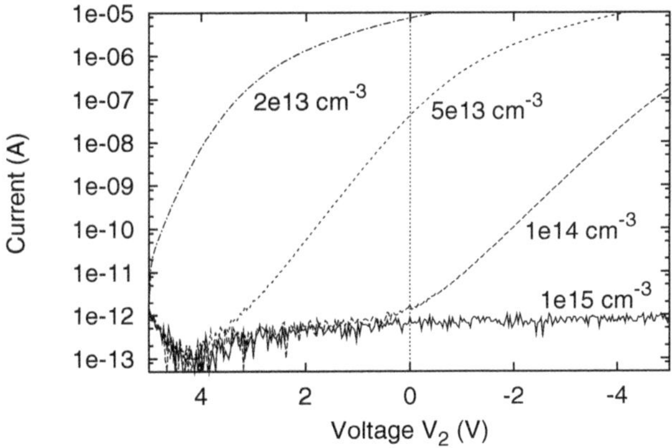

Fig. 2.26. Measured reach-through current for a P-island distance of 10 μm ($V_1 = +5\,\mathrm{V}$; V_2 is varied from +5 to −5 V) [83]

concentration has to be used. $|U_{12}|$ is the maximum potential difference between two neighboring wells which have a doping type opposite to that of the substrate. This maximum potential difference depends on the actual circuit.

The worst case for the maximum potential difference would be, for instance, $V_1 = 5\,\mathrm{V}$ and $V_2 = 0\,\mathrm{V}$. Figure 2.26 shows reach-through currents between a P well and a P^+ island measured in a test structure actually designed for investigation of diffusion of photogenerated carriers. For $N_\mathrm{I} = 10^{15}\,\mathrm{cm}^{-3}$, there is no reach-through current. For $N_\mathrm{I} = 10^{14}\,\mathrm{cm}^{-3}$, there is a very low reach-through current of about $1.5 \times 10^{-12}\,\mathrm{A}$ at $V_2 = 0\,\mathrm{V}$. For $N_\mathrm{I} = 5 \times 10^{13}\,\mathrm{cm}^{-3}$, the reach-through current for $V_2 = 0\,\mathrm{V}$ is about $5 \times 10^{-8}\,\mathrm{A}$, which should be tolerable for most $\mathrm{N}^-\mathrm{N}^+$-CMOS-OEICs [84]. Actually the potential difference between P wells is less than 5 V in most circuits, and measurements confirmed that an $\mathrm{N}^-\mathrm{N}^+$-CMOS-OEIC also works well with a doping level of $N_\mathrm{I} = 2 \times 10^{13}\,\mathrm{cm}^{-3}$. The distance of two analog wells of 10 µm was sufficient for the circuit in [85]. For many analog circuits this value, therefore, does not increase dramatically above the original minimum analog well distance of 6.6 µm.

2.2.6 Charge-Coupled-Device Image Sensors

Charge-Coupled-Device (CCD) image sensors are an economically very important sub-division of integrated optical receivers. CCD image sensors were invented at Bell Labs [86]. Since then, there have been numerous publications on CCD image sensors (for instance [87–99]), because these sensors are key devices for high-volume video cameras and still-photography cameras. There is also a huge number of CCD imaging devices available on the market [100]. Record minimum pixel size was reduced from 40 µm in 1972 to 5 µm in 1995 [101]. The reported record maximum pixel number has increased from less than 2000 to 26 million in the same period [101].

Usually CCD image sensors are simply called CCDs. An advanced CCD image sensor, however, consists of photodetectors, CCDs for readout, and MOSFET source follower output drivers. For the photodetectors, PN photodiodes can be used. The CCD is a much more complicated device than the photodiode. Therefore, the CCD will be explained in some detail. The CCD is basically an array of closely spaced MOS capacitors (see Fig. 2.27). A quantity of charge can be stored below one gate when an appropriate biasing of this and the neighboring gates is chosen (see Fig. 2.27a). The electrons representing the stored charge are confined in the potential well below the electrode with the higher voltage. When the third electrode to the right of this electrode is pulsed to a higher voltage the electrons begin to transfer to its deeper potential well (see Fig. 2.27b). The voltages 5, 10, and 15 V, are chosen arbitrarily; meanwhile they have lower values (e. g. 5, 8, and 10 V [102]; 2.0 and −8.0 V [103]; 1.8 V [104]; 1.1 V [96]) in advanced CCDs.

Figure 2.28 shows the schematic cross section of a three phase N-channel CCD together with its basic input and output structures. The six (in real

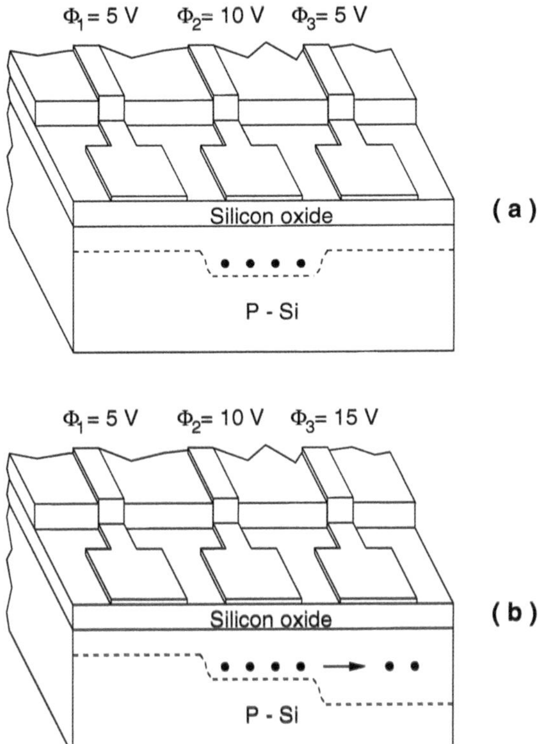

Fig. 2.27. Charge-coupled device. (**a**) Charge storage; (**b**) charge transfer [105]

devices of course many more) MOS capacitors with their gates connected to the clock lines Φ_1, Φ_2, and Φ_3 form the main body of the CCD. The input diode (ID) and input gate (IG) inject charge packets into the main CCD body. The output gate (OG) and output diode (OD) extract the transferred charge packets from the main CCD body. Before the charge transfer, the input and output diodes are at high positive voltage in order to deeply deplete the surfaces under IG and OG. ID and IG, therefore, cannot supply electrons into the main CCD array. The potential wells under the CCD array, therefore, can be assumed to be empty. Then voltages $\Phi_1 > \Phi_2$, Φ_3 are applied. The potential well below Φ_1 is deeper than the others. When the voltage of ID is lowered electrons flow to the potential well under the first Φ_1 electrode through IG which is at a potential higher than 5 V but lower than 10 V. At the end of the injection, the surface potentials under the IG and Φ_1 electrodes are the same as the input diode potential and the electrons are now stored under the IG and first Φ_1 electrode. Now ID can be returned to a higher voltage. Then, the voltage at the Φ_2 electrode is increased and the voltage at the Φ_1 electrode is reduced in order to transfer the charge to the second electrode. Now, the voltage at the Φ_3 electrode is increased and the voltage

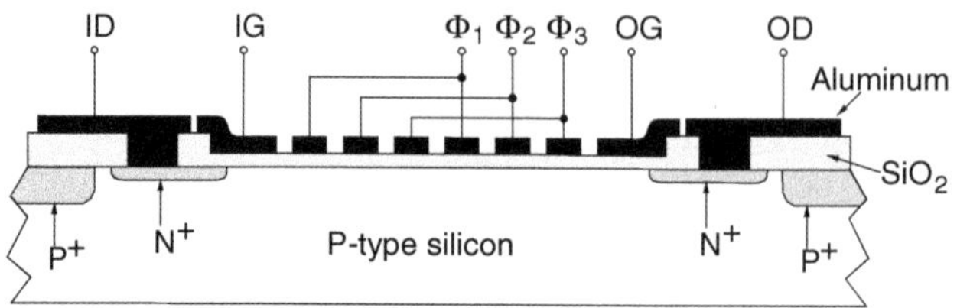

Fig. 2.28. Schematic cross section of a CCD [37]

at the Φ_2 electrode is reduced in order to transfer the charge to the third electrode. This pulse sequence is repeated until the charge packet is stored under the second Φ_3 (or in a real device the last Φ_3) electrode. OG and OD are at high potentials and when the voltage at the Φ_3 electrode is reduced, the charge is pushed to the output diode, thereby giving an output signal proportional to the size of the charge packet at the output terminal.

In a CCD image sensor, the input diode is the photodiode. Therefore, ID is larger in a CCD image sensor than shown in Fig. 2.28 and not covered by metal, of course. During the image integration time i. e. when the ID voltage is high, the electrons photogenerated in the space-charge region are collected and stored under the N^+ island of ID. The photogenerated holes drift to the substrate. Electrons photogenerated in the substrate below the space-charge region diffuse slowly and may recombine with holes or diffuse to neighboring pixels giving rise to cross-talk. The collected charge is proportional to the integration time and to the light intensity as long as the well is not filled. During the charge transfer in the CCD, almost no charge is lost and the output signal is proportional to the light intensity detected by the ID photodiode.

The minimum of the potential well is at the Si/SiO_2 interface, which, therefore, is a critical part of the described MOS structure. The Si/SiO_2 interface may capture electrons in localized states, preventing their proper readout. To improve the situation, buried-channel structures are used, where a thin top layer of the substrate is N-type doped [106]. The resulting potential distribution has a minimum in the silicon somewhat below the Si/SiO_2 interface, which represents a 'cleaner' place for storing electrons. A transfer inefficiency, defined as the relative amount of electrons left behind after transfer from one stage to the next, as low as 10^{-5} per stage is typical for buried-channel CCD structures [107].

In an advanced CCD image sensor, there are arrays of photodiodes and arrays of input gates (Fig. 2.29). The advantage of this interline transfer CCD (IT-CCD) arrangement [108–110] is that the charge transfer from the detector to the storage cells can be made very quickly since only one transfer stage is involved. In older CCD image sensors the CCD main bodies in image registers themselves were used as the photodetectors and the complete image was transferred to a storage register requiring 3N charge transfers [111]. The electrodes were made from polysilicon resulting in a low sensitivity for the blue spectral range. These two disadvantages led to the IT-CCD structure. The larger noise of the IT-CCD with photodiodes [101, 112] was less important. Each third MOS capacitor can have its own photodiode and input gate. One line of a picture with N pixels can be read out through the CCD output register in a serial mode to a single video output (Fig. 2.29). A complete CCD image sensor consists of a two-dimensional matrix ($N \times M$) of photodiodes and input gates as well as M CCD main bodies. The active areas are separated by channel stops consisting of narrow regions with heavy doping in order to prevent blooming, i. e. the spilling over of electrons from full wells into neighboring channels.

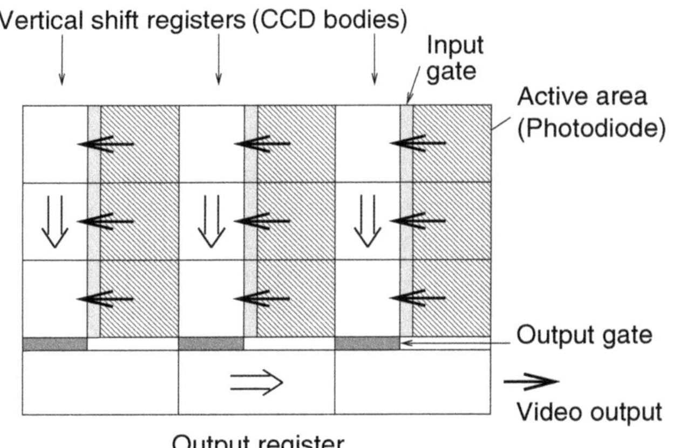

Fig. 2.29. Schematic of a interline-transfer CCD (IT-CCD)

For the sake of completeness it should be mentioned that two-phase CCDs are also possible. These two-phase CCDs require an additional doping structure for generating an internal lateral electric field [96, 104, 111].

Single-chip color image sensors are obtained using three photodiodes in each pixel. These three photodiodes carry a red, a green, or a blue color filter, which are printed on the insulator on the CCD image sensor surface. The color filters are arranged in a mosaic pattern for achieving a better isotropy in resolution than using stripes of the three color filters [111].

A 2/3-inch 1280×1600 CCD image sensor for still photography was fabricated with a minimum feature size of 0.5 µm [98]. Its chip size was 90 mm^2 with a pixel size of 5.1×5.1 µm^2. Supply voltages of 3.3 and 12 V were necessary. The maximum charge storage capacity per pixel was 40 000 electrons. A sensitivity of 390 mV/(Lux s) was reported for green light. The output driver consisted of a triple source follower. The output sensitivity was larger than 19 µV/electron with a saturation output voltage of 0.9 V and an output noise of 10 electrons root-mean-square (RMS). This CCD image sensor was capable of taking six images per second in full resolution [98].

The fill factor of CCD imagers is considerably less than 1. Therefore microlenses were integrated in order to increase the sensitivity [113, 114]. A clock frequency of 37.125 MHz was used in a 2 million pixel CCD image sensor for a HDTV camera [115]. A similar clock frequency of 40 MHz was reported in [103]. Maximum output data rates of 250 Mb/s for CCD arrays [116] were reported. The clock rate of CCD arrays finally is limited as the transfer inefficiency increases at higher clock frequencies, depending on the actual device structure [101]. A clock rate of 325 MHz has been reached [117]. The larger the pixel number the smaller the frame rate and vice versa. A high frame rate of 10^6 frames/s was obtained for a rather small 360×360 pixel CCD imager. The readout frequency of CCDs, therefore, is rather limited. It is, however, sufficient for video applications. CCD image sensors are rather well developed and the reduction in cell size is coming to an end. The limit is set by diffraction occurring at the aperture above the photodiode in IT-CCD image sensors [118]. When the width of the aperture in the tungsten photoshield becomes smaller than about 1 μm, the response of the photodiode begins to drop [118].

CCD image sensors were integrated in CMOS technology. For instance a CMOS/buried-N-channel CCD compatible process has been described [119]. Another CCD/CMOS process was suggested in [120]. The efforts to integrate CMOS with high-pixel-count CCD imagers to increase CCD functionality, however, has not been fruitful due to cost [121]. The aspect which makes it difficult to integrate CCD devices in CMOS technology is the need for closely spaced, thin polysilicon electrodes for charge transfer being incompatible with the self-adjusting gate-source/drain processing scheme. Furthermore, the well doping profiles and threshold implants are not optimized for CCD devices. Silicided gates in advanced CMOS processes increase the difficulties.

A low-pixel-count 128×128 pixel CCD image sensor in a 2 μm CMOS technology with 128 charge-sensitive amplifiers and 128 analog-to-digital converters has been reported [122]. Only one 5 V power supply was necessary. The maximum picture rate was 25 pictures per second.

2.2.7 Active-Pixel Sensors

Nowadays, there is a trend towards a camera-on-a-chip (CoC), which consists of optical sensor elements and highly integrated signal-processing circuitry with a low power consumption. CCDs, therefore, have to compete with a severe rival, the active-pixel sensor (APS), for many imaging applications. It has a performance competitive with CCDs with respect to output noise, dynamic range, and sensitivity but with vastly increased functionality due to its CMOS compatibility and with the potential for lower system cost as well as for a camera-on-a-chip [121].

The APS was described first by Noble in 1968 [123]. Advanced APSs are now emerging from the most advanced image-sensor research and development laboratories in Japan [124–127]. Laboratories in the United States

[128–134] and in Europe [135, 136] are trying to catch up. The charge is shifted from pixel to pixel in a CCD in order to read out the image. An APS (see Fig. 2.30), by contrast, acts similarly to a random access memory (RAM), wherein each pixel contains its own selection and readout transistors, which serve as amplifiers or buffers. The steadily decreasing feature sizes of VLSI/ULSI technologies make it possible to integrate source follower MOSFETs or amplifiers together with each photodetector for all pixels in a large two-dimensional array.

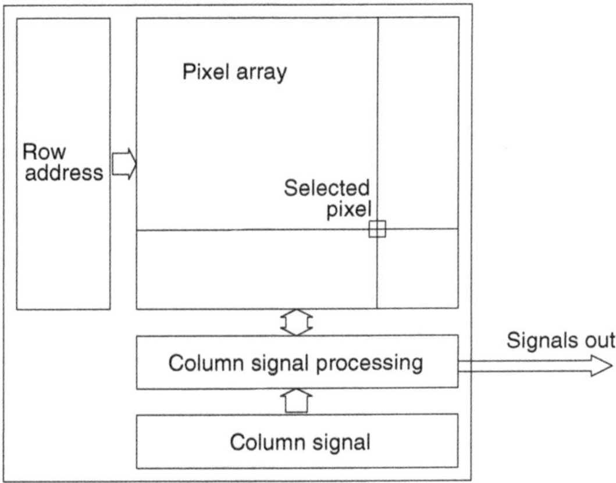

Fig. 2.30. Block diagram of an active pixel sensor imager with random access to each pixel [137, 138]

The advantages of an APS compared to a CCD sensor accordingly are:

(i) random access
(ii) nondestructive readout
(iii) easy integratability with on-chip electronics.

Although CCDs can be integrated in CMOS processes, undesirable compromises in the performance of CMOS or CCD devices have to be made [121]. Important weaknesses of the CCDs are that (i) the charge-transfer efficiency (CTE) has to come very close to 100 % for large pixel numbers, i. e. that the charge-transfer inefficiency has to come close to 0 %, that (ii) large pixel sizes at high data rates are difficult to implement, and that (iii) readout rates are limited.

There are two types of active-pixel sensors: (i) photo-gate APS and (ii) photodiode-type APS. A photo-gate APS requires one charge transfer from the photo-gate into the storage region FD (Fig. 2.31). The photo-gate APS in its function, therefore, resembles a sample&hold circuit. The pixel is selected using transistor S in order to read out the signal. The storage region or output node is reset using transistor R. After reset, the signal charge is transferred from under the photo-gate into the storage node FD. The change in the source-follower voltage between the reset level and the final level is the output signal from the pixel. In more detail, the operation of the sensor is as follows: The photogenerated charge is integrated under the pixel photo-gate PG, which is biased at +5 V. The transfer gate TX and the gate of the reset transistor R are biased at +2.5 V and the selection transistor S is also biased off with 0 V at its gate for charge integration. After the signal integration, all pixels or only the selected ones in the row can be read out simultaneously onto column lines. For this purpose, the pixels are first addressed by the row selection transistor S biased at +5 V, which activates the source-follower output transistor in each pixel. Now, the reset gate R can be pulsed briefly to +5 V in order to reset the floating diffusion output node FD. Then, the photo-gate PG is pulsed low to 0 V, whereby TX is held at +2.5 V, to transfer the integrated signal charge under the photo-gate to the floating diffusion output node FD. The gate potential of the source follower changes and the output voltage of the pixel changes depending on the amount of integrated charge. It should be mentioned explicitly that without resetting the storage node FD before pixel selection, the readout is nondestructive, i. e. the pixel can be read out much more often than once.

A 128 × 128 array with the pixel element shown in Fig. 2.31 was integrated in a 2 μm P-well CMOS process. A 100 % compatibility of the image sensor with the CMOS process was reported [139]. A total area of $6.8 \times 6.8\,\mathrm{mm}^2$ was necessary for this array of pixels with the size of $40 \times 40\,\mu\mathrm{m}^2$. The larger part of this pixel size was occupied by the transistors and the active, i. e. light sensitive, area of each pixel was 26 %. The output conversion factor was 4.0 μV/electron. A noise floor of 42 electrons-rms and a dynamic range of 71 dB were reported [139].

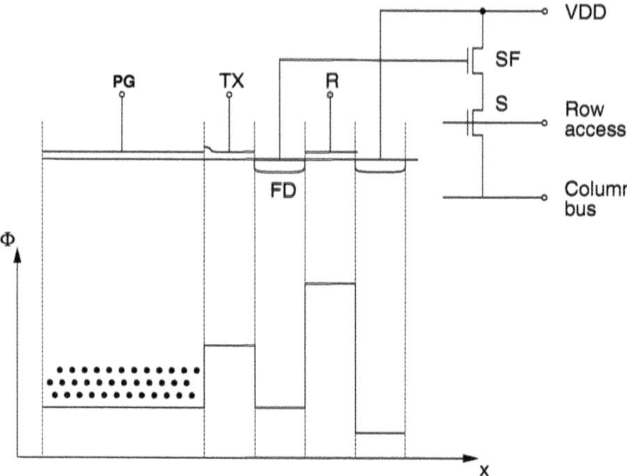

Fig. 2.31. Pixel circuit of a photo-gate active-pixel image sensor [139]

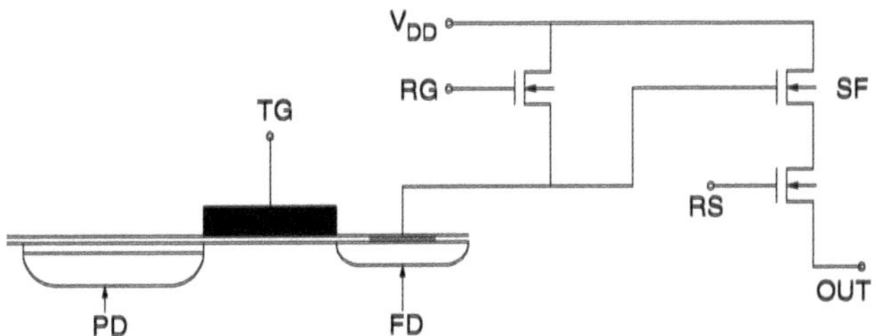

Fig. 2.32. Cross section of photodiode and transfer gate with the schematic of the remaining transistors in the 4-transistor pixel cell of a photodiode-type APS [138]

An example of a photodiode-type APS pixel unit with a 4-transistor pixel cell architecture is shown in Fig. 2.32. Here the charge is integrated with the photodiode PD. For readout, this charge can be transferred by the transfer gate TG to the storage node FD. The rest of the circuit is similar to the schematic in Fig. 2.31. A larger conversion gain of 13 μV/electron was measured for a 256 × 256 pixel imager array with a pixel size of 7.8 μm and a fill factor of 35 % in a 0.6 μm CMOS technology [138]. A dynamic range of 66 dB was obtained. The above mentioned inferiority of photodiode-type CCD imagers to MOS-detector-type CCDs with respect to noise is avoided here by providing each photodiode with its own amplifier and reset transistor [101].

A photodiode-type APS actually does not require a transfer gate. The photogenerated charge can be stored in the photodiode itself [140]. Such an APS schematic is shown in Fig. 2.33.

It is worthwhile to compare the photodiode-type APS and the photo-gate APS. The photo-gate APS has a lower quantum efficiency due to photons being absorbed in the polysilicon gate. It also needs a larger pixel size due to an additional transfer switch and it is less compatible with standard CMOS processes because it requires a thin polysilicon layer with a thickness of only

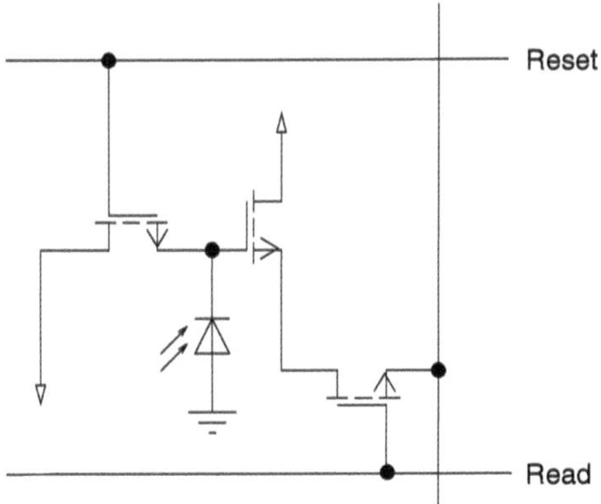

Fig. 2.33. Active pixel with three MOSFETs [124]

about 50 nm. The advantages of the photo-gate APS are a lower noise and a higher conversion gain due to a smaller floating node capacitance [140]. A 256×256 CMOS active pixel sensor was previously combined with circuitry for motion detection [137]. The area of each pixel was $20 \times 20\,\mu m^2$ with a fill factor (photo-gate area/total pixel area) of 25 %. This APS was realized in a 0.9 μm CMOS process. A conversion gain of 7 μV/electron was measured. The pixel output transistor was found to be the primary APS noise source. A typical output referred noise level of 29 electrons rms and a dynamic range of 74 dB was reported.

2.2.8 Amorphous-Silicon Detectors

Hydrogenated amorphous silicon (a-Si:H) has been employed for a variety of imaging devices for low cost applications in large area sensor technology [141–145]. Page-sized two-dimensional a-Si:H photodetector arrays have been realized [145]. Usually, glass substrates are used for the low-temperature fabrication of a-Si:H imaging devices. A huge number of dangling bonds is present in amorphous silicon. In order to obtain stable devices, the a-Si is hydrogenated, i. e. the dangling bonds are saturated and in such a way passivated with hydrogen atoms.

The bandgap of a-Si:H is approximately 1.72–1.78 eV slightly depending on the deposition process. The bandgap of a-Si:H is much larger than that of crystalline silicon and the spectral response of a-Si:H suits the sensitivity of the human eye much better than that of crystalline silicon. The sensitivity of a-Si:H ranges from blue to red. Infrared light with $\lambda > 780\,nm$ cannot be detected with a-Si:H devices. The absorption coefficient of a-Si:H depends on the hydrogen content and on the deposition process. A typical curve for the absorption coefficient is shown in Fig. 2.34 [146]. The absorption coefficient of a-Si:H is much larger than that of crystalline Si in the visible spectrum. An a-Si:H layer with a thickness of 1 μm is sufficient for completely absorbing light with $\lambda < 0.7\,\mu m$ [147].

Amorphous Silicon Image Sensors

Amorphous silicon image sensors can be integrated vertically on microelectronic circuits due to the low-temperature deposition of a-Si:H (below 300 °C). In contrast to the low fill factor of active-pixel sensors, vertically integrated a-Si:H image sensors provide a fill factor close to 100 %. The crystalline integrated circuit typically consists of identical sub-circuits underneath each pixel detector and peripheral circuitry at the boundary of the light sensitive area. An insulating layer separates the optical detector from the crystalline chip.

Another advantage of a-Si:H image sensors is their large dynamic range of 60 dB, which can be further extended considerably by electronic measures.

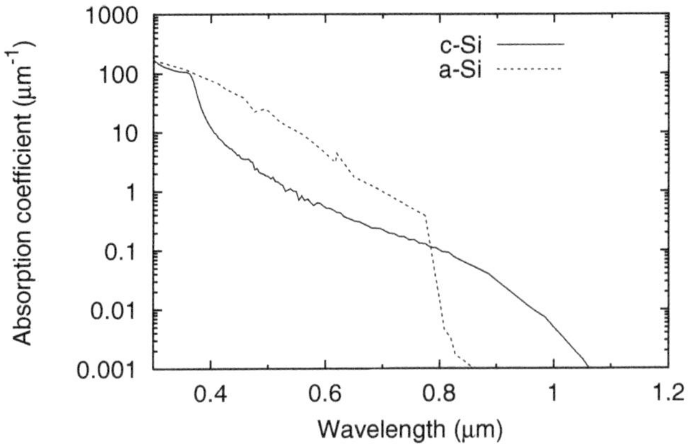

Fig. 2.34. Absorption coefficient of a-Si:H [146]

The concept of *Thin Film on ASIC* (TFA) [148, 149] adds much flexibility to a-Si:H image sensors. The TFA concept allows to design and optimize both the thin-film detector and the application specific integrated circuit (ASIC) independently. Figure 2.35 shows the layer sequence of a TFA sensor.

The a-Si:H thin film system of a TFA sensor does not need to be patterned and can be fabricated in a plasma-enhanced chemical vapor deposition (PECVD) cluster tool without temporarily being taken out of the vacuum for lithography. A very high yield of almost 100 %, therefore, is possible for the thin film system. Furthermore, fabrication costs are very low and barely exceed those of an ordinary ASIC. Without lithography, the pixel is simply defined by the size of its rear electrode. The continuous thin-film layer permits lateral balance currents between adjacent pixel detectors, resulting in cross-talk and in a reduced local contrast. A self-structuring technology for the suppression of this effect was suggested in [150].

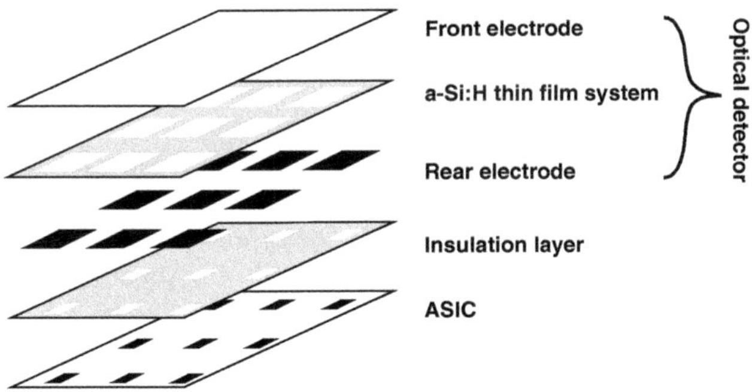

Fig. 2.35. Layer sequence of a TFA sensor [148]

A pixel of the least complex a-Si:H image sensor, which provides minimum pixel sizes and maximum resolution, consists of an a-Si:H photodiode and a PMOS transfer transistor. The function principle is based on a simple charge transfer from the pixel to the readout line [151]. The photocurrent discharges the capacitance of the reverse-biased photodiode during the integration time, resulting in a signal charge proportional to the illumination intensity and to the integration time. This operation mode is called charge storage mode. The amount of storable charge can be increased by an additional MOS capacitance, which can be integrated into each pixel on the ASIC in order to achieve a higher signal-to-noise ratio. In [149] this type of image sensor was called a varactor analog image detector (VALID). VALID is addressed linewise and read out columnwise. A random access sensor is also feasible employing a modified peripheral circuitry. 128×128 pixels with an area of $16 \times 16\,\mu m^2$ each were integrated in VALID using a standard $0.7\,\mu m$ CMOS ASIC process. A saturation illumination of 2500 Lux after an integration time of 20 ms was achieved with an integration capacitance of 250 fF. The dynamic range amounted to 60 dB. A good linearity and no visible image lag was observed. The circuits of more complex TFA image sensors for even higher dynamic ranges, which are essential in the field of artificial vision, will be discussed in Sect. 6.4.2.

2.2.9 CMOS-Integrated Bipolar Phototransistors

Photodetectors with a built-in amplification are most desirable for weak optical signals, especially in a comparatively simple CMOS process because of the high degree of standardization, its low cost, and the widespread interest in CMOS circuits.

The simplest way to integrate a bipolar phototransistor in a CMOS process without any process modifications is to use the N well as the floating base of a PNP transistor, the P^+ source/drain diffusion as the emitter and the P^-P^+ substrate as the collector (Fig. 2.36). This vertical bipolar phototransistor (VBPT) then can be used as an emitter follower only, because the collector is at substrate potential, i. e. at 0 V. The complementary NPN phototransistor on an N^-N^+ substrate material is also possible, of course. The simple structure shown in Fig. 2.36, however, has the additional disadvantage of a low sensitivity for short and medium wavelengths due to the large N-well/P-substrate junction depth.

The lateral bipolar transistor [153] being also available without process modification, however, can be freely connected. A typical lateral bipolar transistor usually suffers from a lower current gain and lower collector efficiency than the vertical bipolar device. In order to achieve a higher responsivity as an identically sized vertical bipolar phototransistor, modifications have been made to the typical lateral bipolar transistor [154]. The structure of this improved lateral bipolar phototransistor (LBPT) is shown in Fig. 2.37.

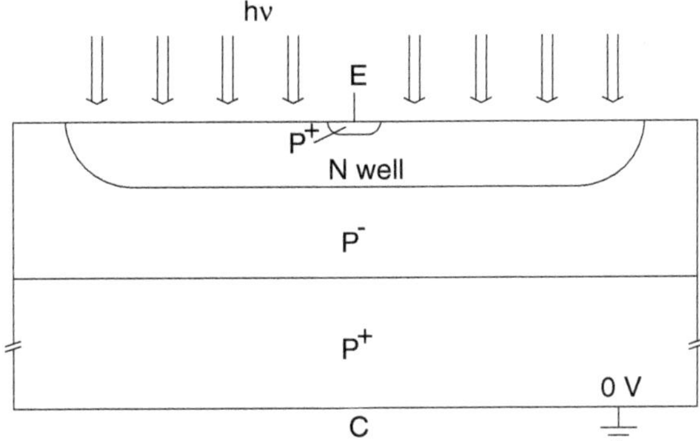

Fig. 2.36. Cross section of the simplest CMOS-integrated PNP phototransistor [152]

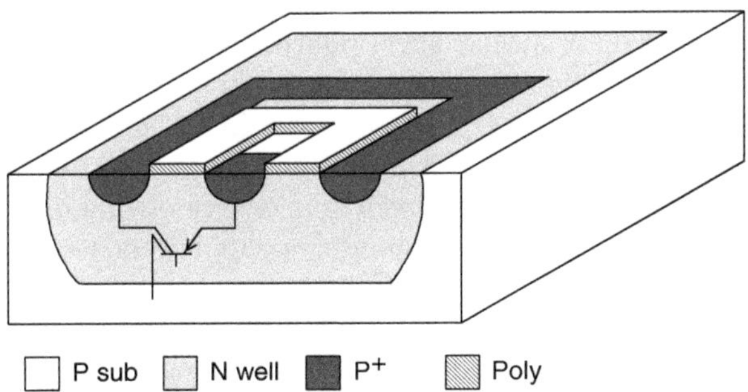

Fig. 2.37. Structure of a CMOS-integrated lateral PNP phototransistor with increased current gain [154]

The lateral PNP phototransistor in a rectangular structure is formed with a P^+ emitter surrounded by a P^+ collector situated in an N-well base. The rectangular structure with the emitter in the center allows the optimum exploitation of the emitter efficiency. Because the base region is more lightly doped than the collector, the collector-base depletion region extends almost completely into the base. This space-charge region must not reach the emitter in order to avoid reach-through between emitter and collector. The base, therefore, has to be made wide enough. The current gain β, therefore, is rather low since β decreases when the base width W_B increases [154]:

$$\beta \approx \frac{1}{(W_\mathrm{B}^2/2\tau_\mathrm{B}D_\mathrm{N}) + (D_\mathrm{P}W_\mathrm{B}N_\mathrm{A}/D_\mathrm{N}L_\mathrm{P}N_\mathrm{D})}. \tag{2.4}$$

In this equation, τ_B is the minority carrier lifetime in the base, D_N and D_P are the carrier diffusion coefficients in the base, L_P is the diffusion length of holes in the emitter, and N_A and N_D are the acceptor and donor densities in the emitter and base, respectively. In [154], therefore, a trick [155] has been used in order to increase the current gain β. For this purpose, a polysilicon ring around the emitter is implemented (see Fig. 2.37) in order to set the base width. By connecting the polysilicon gate to the most positive potential in the circuit, majority carriers in the base accumulate directly under the gate, retarding the widening of the collector-base depletion region and setting the base width to that of the minimum polysilicon gate width.

It should be mentioned that the lateral phototransistor of Fig. 2.37 also contains a parasitic vertical PNP transistor. The effect is a reduction of the collector efficiency of up to 40 % [153, 155], i. e. up to 40 % of the collector current is lost to the substrate. In order to achieve a 100 % exploitation of the amplification of the LBPT, the photodetector current, therefore, has to be drawn from the emitter.

The base of the LBPT is left floating. Base bias current is provided when incident photons create electron–hole pairs. The photodetector current taken from the emitter is $(\beta+1)$ times the base current, which would correspond to the photocurrent of a similar sized photodiode.

LBPT and VBPT devices with a size of $60 \times 60\,\mu\text{m}^2$ were fabricated in a standard 2 μm digital CMOS process. An LED light source with a peak wavelength of 660 nm was used for the illumination of the devices and its illumination intensity was varied over four orders of magnitude. LBPT and VBPT output currents were approximately proportional to the light intensity. The LBPT output current was constantly about nine times greater than the VBPT output current. The output current of the LBPT was about 20 μA for an illumination intensity of $1\,\text{mW/cm}^2$. For these values we can calculate a photoamplification of about 1040 compared to a photodiode of the same size. For an illumination intensity of $P' = 1\,\text{mW/cm}^2$ at 660 nm we obtain a photon flux density $\Phi = P'\lambda/(hc) = 3.33 \times 10^{15}\,\text{cm}^{-2}\text{s}^{-1}$. The photocurrent density of a photodiode $j_{\text{ph}}^{\text{PD}} = \Phi q$, therefore, is $5.33 \times 10^{-4}\,\text{A/cm}^2$. For the area of $60 \times 60\,\mu\text{m}^2$, $I_{\text{ph}}^{\text{PD}} = 19.2\,\text{nA}$ results. If we divide the measured LBPT photocurrent of 20 μA by this value, we obtain a photoamplification of 1040 for the LBPT.

LBPT devices also were fabricated in 1.2 and 0.8 μm CMOS processes. The active area of these devices scaled directly with the processes, i. e. the LBPT cell dimensions in the 1.2 μm process were 36 μm per side and 24 μm in the 0.8 μm process. The highest LBPT output current was observed for the CMOS process with the 1.2 μm design rules. However, no explanation was given for this finding.

The transient behavior of a photodetector determines how fast an array of photodetectors can be scanned in an image sensor. Due to charge storage, the fall time of a phototransistor is usually greater than the rise time [156]. The rise time, however, fortunately limits the scan speed of an image sensor. The fall time does not pose a problem as long as it is shorter than the time needed to scan the array. The worst-case scenario for the rise time is when a brightly illuminated cell is next to a dark cell. Step response of the LBPT with

$W_B = 2\,\mu m$ was measured by switching between dark and brightly illuminated cells in an array. In such a way, a rise time of 3.1 μs was determined for the LBPT. In order to be complete, the dark current of the LBPT of less than 10 pA has to be mentioned.

2.2.10 Photonic Mixer Device

An innovative optoelectronic device the so-called photonic mixer device (PMD) was suggested for optical distance measurement using a time-of-flight method with a pulsed laser signal [157]. This PMD is in principle a MOSFET with a split gate (Fig. 2.38). The light penetrates partially through the two polysilicon gates. The differential electrical modulation signal (u_m) derived from the laser modulation signal is applied to the gates. In dependence on the polarity of u_m, the photocurrent is directed to the readout electrode aK or bK. The PMD therefore operates like a seesaw for photocharges [157].

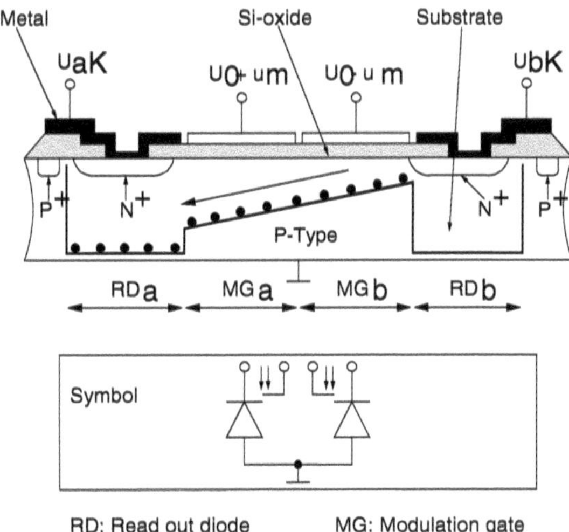

Fig. 2.38. Cross section and symbol of photonic mixer device [157]

In the case of square-wave modulation the correlated measured light $p_{om}(t)$ (see Fig. 2.39) is usually superimposed by a much brighter background light $p_{onc}(t)$. The intensity change of the background light, however, is slow in practice and $p_{om}(t)$ is correlated with the modulation signal.

In this example (Fig. 2.38) the total modulated photocurrent drifts to the left readout electrode during the complete half period and the correlation function $\Delta U_{abm} = U_{aK} - U_{bK}$ reaches its maximum. Usually there is a phase shift (for not too large distances) between received photo signal and u_m and the magnitude of the correlation function is proportional to the distance to

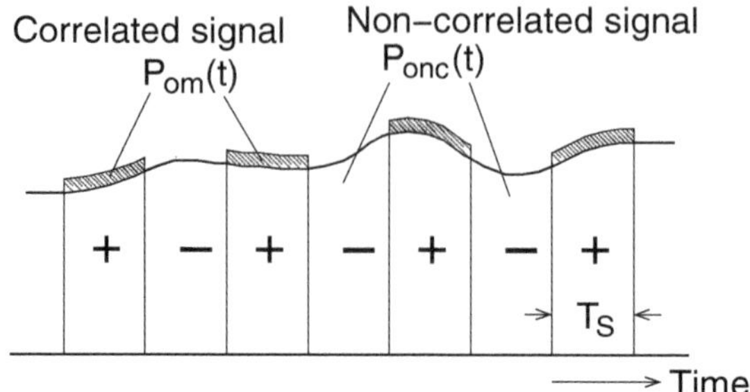

Fig. 2.39. Correlated balanced sampling [158]

be determined. It has to be mentioned that the output signals of the PMD are actually currents, which have to be integrated on two capacitors to obtain the voltage signals.

In a distance measurement range from 80 to 230 cm, a standard deviation of 2 mm over 1 hour of observation time was reported [158]. Newest work verified a distance error less than 1 mm corresponding to a time uncertainty (jitter) of 6.6 ps [158]. Modulation frequencies up to several 10s of MHz were applied.

2.3 BiCMOS-Integrated Detectors

2.3.1 BiCMOS Processes

BiCMOS stands for Bipolar&CMOS. It combines both bipolar transistors and NMOS as well as PMOS transistors in the same chip. In digital BiCMOS circuits it is possible to exploit the advantages of bipolar transistors and of MOS transistors. The capabilities of BiCMOS circuits, therefore, exceed those that are obtained when only one of the device types is used. For example, CMOS allows low-power, high-density digital ICs, which, however, are slower than bipolar ECL-based ICs due to the poor driver capability of MOS transistors, especially when large capacitive loads are encountered. Such loads occur at long interconnects on chip, at high fan-out circuits, as well as at chip outputs. The driver capability of bipolar transistors can be used to advantage in BiCMOS circuits to charge such heavy loads rapidly. Bipolar digital circuits implemented with ECL gates can also operate with small logic swings and high noise immunity. ECL gates, however, exhibit high power consumption, poor density, and limited circuit options. BiCMOS offers the benefits of both bipolar and CMOS circuits by appropriately trading off the characteristics of each technology.

BiCMOS can also offer analog and digital functions on the same chip. For instance, better device matching, lower offset voltages of operational amplifiers, and enhanced bandwidths compared to analog CMOS circuits can

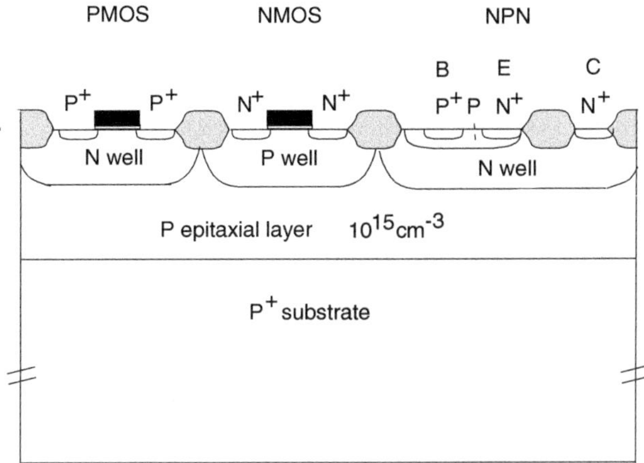

Fig. 2.40. Cross section of a triple-diffused BiCMOS structure [81, 160]

be obtained with analog bipolar sub-circuits in BiCMOS ICs or with BiCMOS circuits. From CMOS, the high input resistance, i. e. the zero input bias current, can be used advantageously in BiCMOS circuits. BiCMOS circuits, therefore, pushed mixed analog/digital (mixed signal) applications tremendously.

The interested reader is recommended to read [159] to study the advantages of BiCMOS in more detail. The benefits of BiCMOS, however, are attained at the expense of a more difficult technology development and at the expense of more complex chip-manufacturing tasks leading to longer chip-fabrication times and higher costs.

BiCMOS processes can be derived from bipolar and CMOS processes. The simplest low-cost BiCMOS process is obtained adding one mask for the P base of the NPN device to a well established CMOS process (Fig. 2.40). The added mask is used to selectively produce lightly doped P regions with a doping level of about $10^{17}\,\mathrm{cm}^{-3}$. The N-well fabrication step also forms the collector regions of the NPN transistors. The heavy NMOS source/drain implant is employed to produce the emitter regions and collector contacts of the bipolar transistors. The PMOS source/drain implant is used to create the base contact, since this serves to lower the base series resistance R_B. The starting substrate is either a P-type wafer or a P-type epitaxial layer on a P^+ wafer. Such simple processes are called *triple-diffused (3-D)* BiCMOS processes, because they produce triple-diffused (C, B, E) bipolar transistor structures [159, 161]. In order to optimize both the CMOS and bipolar device characteristics independently, the doping profiles in the CMOS N well and in the bipolar N well have to be made different. The base-collector breakdown voltage and therefore the collector-emitter breakdown voltage of the NPN transistor may be rather small with the CMOS N well as the collector. In order to increase these breakdown voltages the doping concentration has to be reduced making a separate collector doping, i. e. a bipolar N well, necessary. The resulting cross section is shown in Fig. 2.41.

The main drawback of 3-D bipolar transistors is a large collector series resistance R_C. In order to improve the properties of the NPN transistors (e. g. current gain β and transit frequency f_T), polysilicon emitters allowing a thinner base were added to a 3-D BiCMOS process in [162], of course increasing the process complexity.

The most widely used high-performance BiCMOS technology is based on a twin-well CMOS process. Figure 2.42 shows the cross section of such a high-performance BiCMOS structure [160, 163]. N^+ buried layers are used to reduce the collector series resistance. These N^+ buried layers and the P buried layers also reduce the susceptibility to latch-up in the CMOS part of the BiCMOS ICs.

Four masking steps were added to an advanced 0.8 μm 12-mask CMOS process described in [81] to integrate high-performance bipolar transistors. The starting material is a lightly doped (≈10 Ωcm) silicon wafer with a <100> orientation. At the beginning of the process, a mask has to be added to pattern the N^+ buried layers (see Fig. 2.43a). Next, a selective boron implant is performed, which places an additional P-type dopant between the N^+ buried layers (see Fig. 2.43b), to reduce the minimum buried layer distance. This boron implant also creates the P-type buried layers. This self-aligned approach requiring only one mask to create both the N^+- and P-buried layers is widely used [159]. After the removal of all oxides, a thin (1.0–1.5 μm), near-intrinsic (less than $10^{15}\,\mathrm{cm}^{-3}$) epitaxial layer is then deposited (see Fig. 2.43c). The steps for twin-well formation are conducted next (compare Fig. 2.18a and b). A single-mask process can be used to form both the N well and the P well. A dual-phosphorus implant (deep and shallow) is used

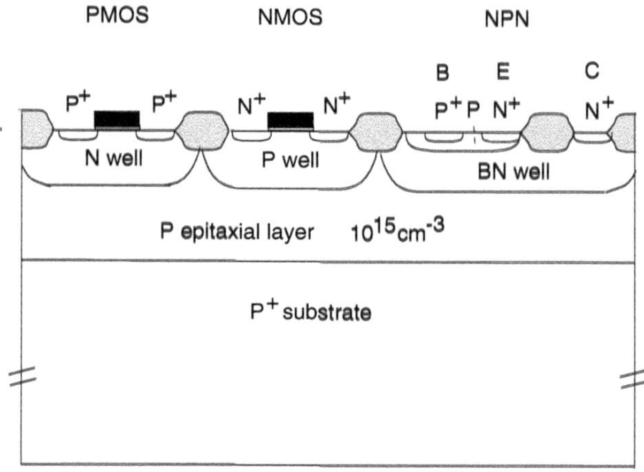

Fig. 2.41. Cross section of a 3-D BiCMOS structure with separately optimized CMOS and bipolar N wells

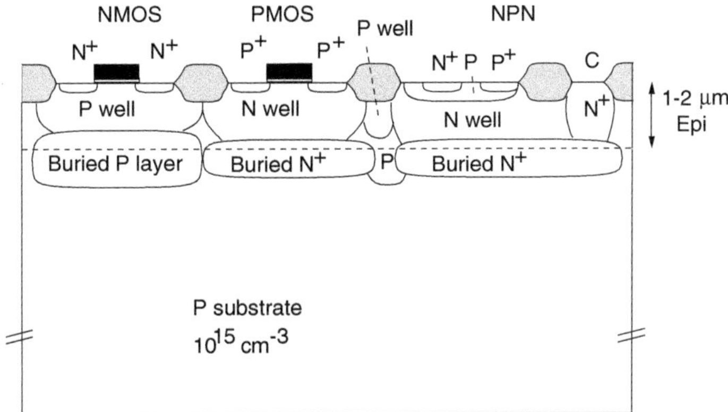

Fig. 2.42. Optimized high-performance BiCMOS structure

to form the N well (Fig. 2.43d). A boron implant is used to form the P well (Fig. 2.43e). Following the implantations that introduce the well dopants, the well drive-in is performed. Then the active regions are defined (see Fig. 2.43f and compare with Fig. 2.18d). The boron channel stop is implanted next and the field oxide is grown.

At this stage, the second mask is added for the formation of the deep N^+-collector contact (Fig. 2.43g). A deep phosphorus implant and a subsequent drive-in/anneal step are applied to form the deep-collector-contact structure. The third added mask (Fig. 2.43h) is used for the patterning of the P base, which is implanted by boron and annealed subsequently.

The fourth added mask is needed to open the areas for the polysilicon emitters (Fig. 2.43i). This process is rather complex and the interested reader will find a thorough description in [159]. The polysilicon on top of the emitters is patterned together with the polysilicon gate definition (Fig. 2.43j). From this stage on, processing (Figs. 2.43k–n) is identical to the CMOS process, from which the BiCMOS process was derived.

In order to achieve transit frequencies of the order of 10 GHz for the NPN transistors, the epitaxial N collector has a thickness of only about 1–2 µm [163–168]. In a 0.8 µm BiCMOS process, an epitaxial layer with a thickness of only 0.8 µm was used [169]. In this process, different from Fig. 2.43a, As instead of Sb was used for the buried collector and due to the thin epitaxial layer, autodoping caused by the N^+ As buried collectors during epitaxy had to be avoided. For this purpose, the pressure in the epitaxial dichlorosilane process was reduced to 25 Torr and a special temperature ramp starting at 850 °C and increasing the temperature to 1150 °C within 10 min resulted after 1 min at 1150 °C in an intrinsic — i. e. intentionally undoped — epitaxial layer thickness of 0.15 µm. The main epitaxial cycle was carried out at 860 °C for approximately 18 min where the epitaxial layer was grown to a thickness of

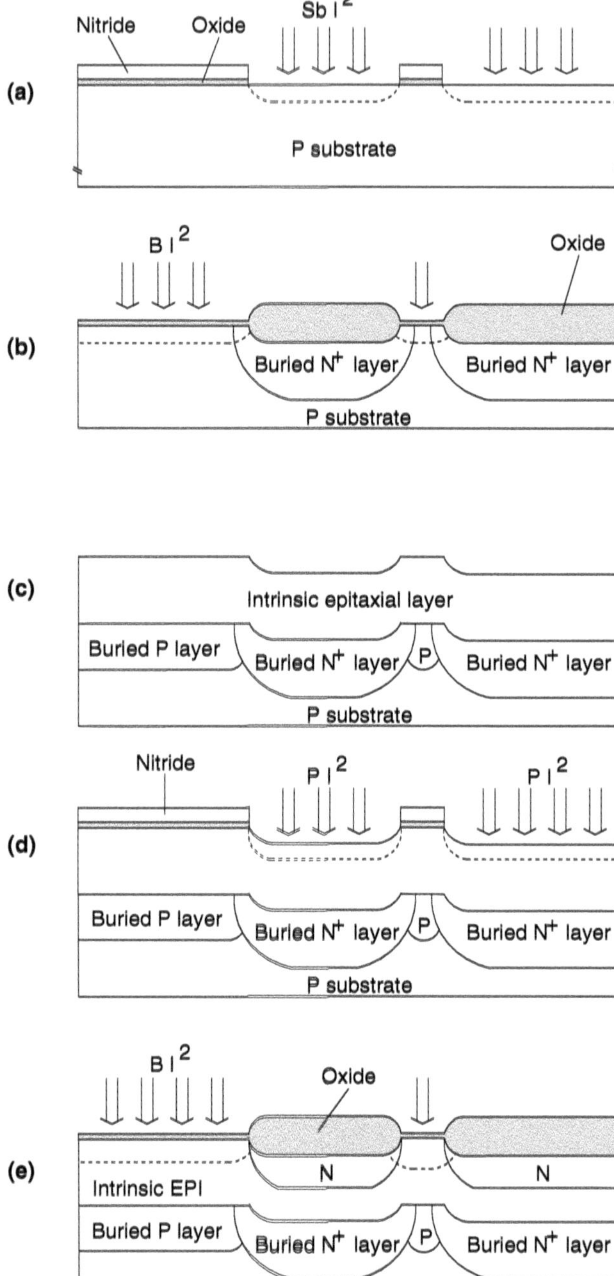

Fig. 2.43a–e. For caption see page 78

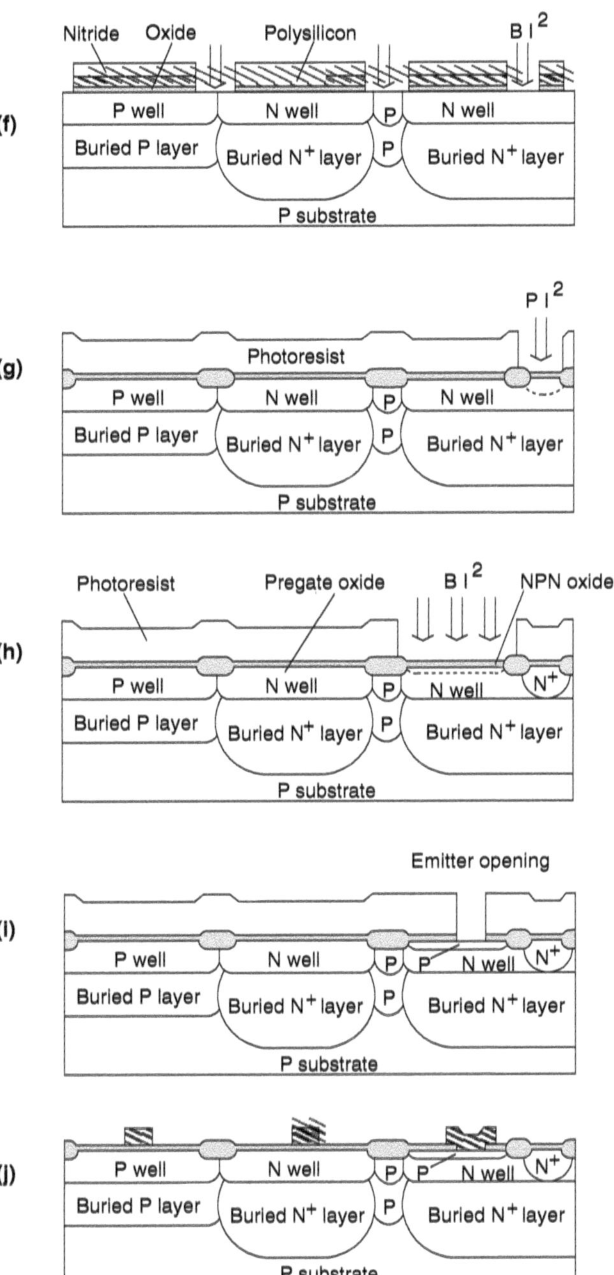

Fig. 2.43f–j. For caption see page 78

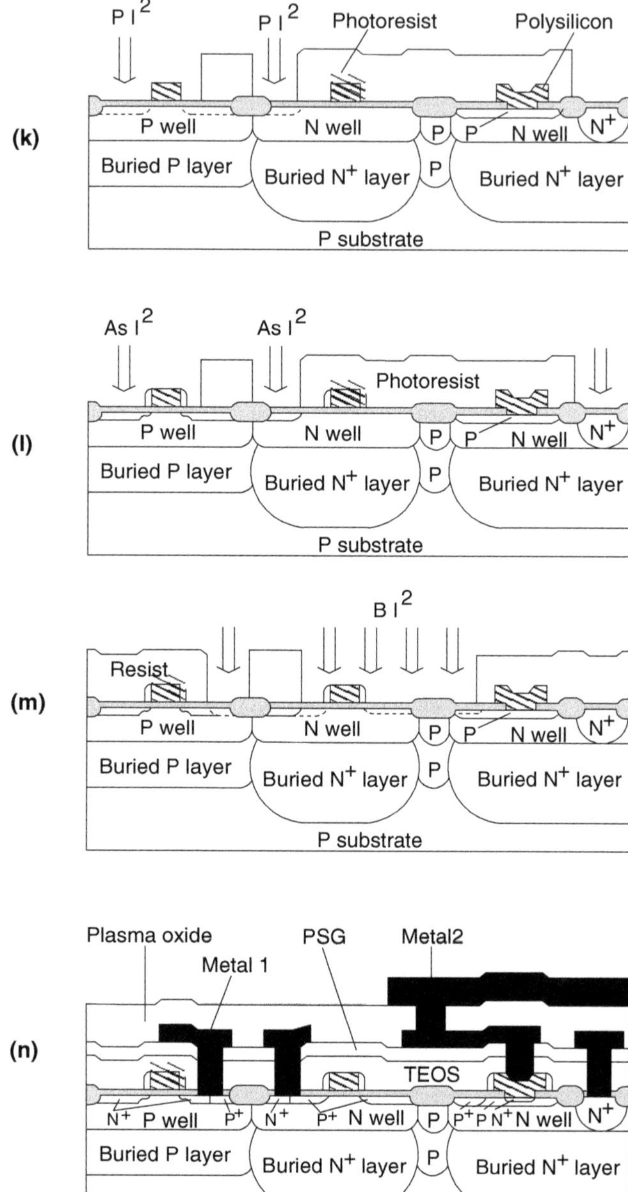

Fig. 2.43a–n. Process flow of an optimized high-performance BiCMOS process [159]

0.8 µm. The transit frequency of the NPN transistor was 11.5 GHz. A 0.5 µm BiCMOS process, for instance, is described in [170].

2.3.2 BiCMOS-Integrated Photodiodes

BiCMOS circuits enable the use of photodetectors described in Sects. 2.1 and 2.2. We, therefore, have to discuss only a few photodetectors here. BiCMOS-OEICs have already been introduced [171, 172]. The OEICs in both cases consisted of a PIN photodiode and an amplifier, which exploited only MOSFETs and no bipolar transistors. The BiCMOS technology, therefore, was chosen merely for the integration of the PIN photodiode. A standard BiCMOS technology with a minimum effective channel length of 0.45 µm without any modifications was used [171, 172]. This effective channel length corresponds to a drawn or nominal channel length of about 0.6 µm. The buried N^+-collector in Fig. 2.44 was used for the cathode of the PIN photodiode. The P^+–source/drain island served for the anode. The intrinsic zone of the PIN photodiode was formed by the N well and, therefore, had only a thickness of 0.7 µm. Consequently, the responsivity of the photodiode was only 0.07 A/W for a wavelength of 850 nm [171]. For a bias of 2.5 V across the $75 \times 75\,\mu m^2$ PIN photodiode with a capacitance of 1.8 pF, a −3dB bandwidth of 700 MHz was reported. The OEIC reached a bit rate of 531 Mb/s with a bit error rate of 10^{-9} and a sensitivity of −14.8 dBm. This bit rate was limited by the capacitance of the photodiode and the feedback resistor of 1.4 kΩ in the amplifier transimpedance input stage. The dark current of the photodiode was 10 nA for a reverse voltage of 2.5 V at room temperature.

In [172] a laser with a wavelength of 670 nm was used for the characterization of the same OEIC as in [171]. The data rate was increased to 622 Mb/s for this wavelength. No measured result for the responsivity was given in [172]. Instead, a three times larger responsivity value of approximately 0.2 A/W for $\lambda = 670$ nm was estimated due to the lower penetration depth than for

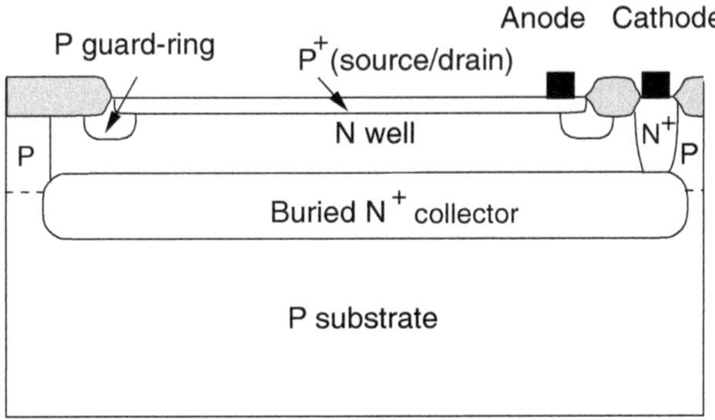

Fig. 2.44. PIN photodiode in a SBC-based BiCMOS technology [171]

$\lambda = 850$ nm. This estimation, however, may be doubtful, because possible destructive interference in the isolation and passivation stack is neglected. The main drawback for the photodiode used in [171,172] was its low responsivity due to the thin epitaxial layer in the SBC-based BiCMOS process.

In the following we will investigate an N^+/P-substrate photodiode in a CMOS-based BiCMOS technology. Figure 2.45 shows the cross section of this photodiode.

The process used for the fabrication of this photodiode is optimized with respect to a minimum number of masks and implements self-aligned wells. Therefore, there is a P well incorporated in the N^+/P-substrate photodiode. This leads to a thin N^+/P-substrate space-charge region and the effect of slow carrier diffusion is very pronounced (see Fig. 2.46). Rise and fall times of 26 and 28 ns, respectively, are determined due to the pronounced diffusion tail. The measured −3dB bandwidth is 6.7 MHz.

Another example for a photodetector in a standard BiCMOS technology will be added here. In a CMOS based (3-D) BiCMOS process, an N well can be used as the collector of a bipolar NPN transistor. This N well can also be

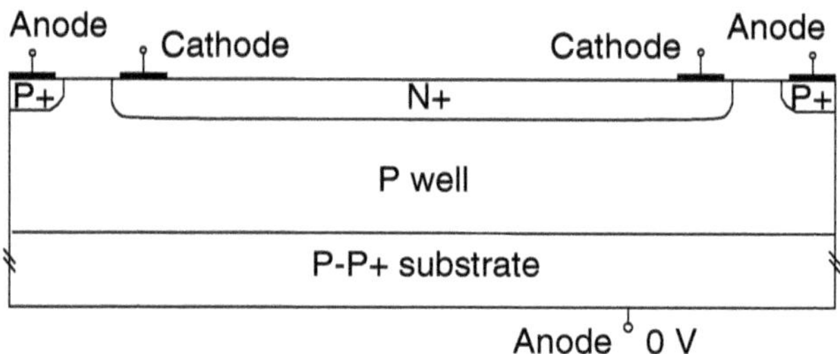

Fig. 2.45. Cross section of a N^+/P-substrate photodiode in self-adjusting well BiCMOS technology

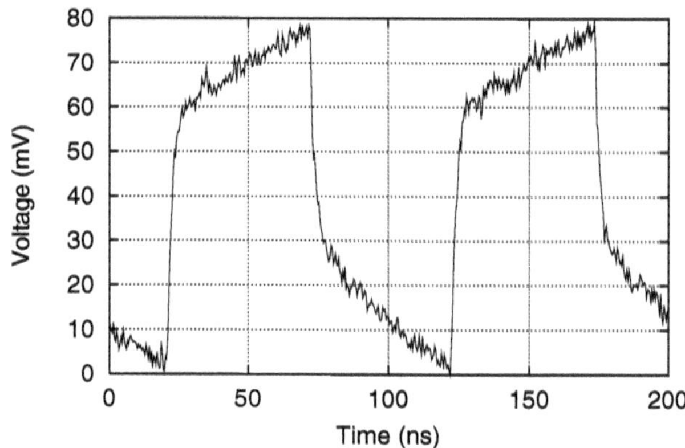

Fig. 2.46. Transient response of a N^+/P-substrate photodiode in a self-adjusting well BiCMOS technology

used in order to form the cathode of a so-called double photodiode (DPD). Figure 2.47 shows the cross section of such a DPD. The two anodes are connected to ground. The cathode is connected to the amplifier in the OEIC (see Fig. 6.69). Two PN junctions are vertically arranged. The N-well/P-substrate junction minimizes the effect of slow diffusion of photogenerated carriers from the substrate.

In addition to the two space-charge regions at the two vertically arranged PN junctions, an electric field is also present between the two space-charge regions (see Fig. 2.48) due to the doping gradient of the N well. Therefore, there is no contribution of slow carrier diffusion from the region between the two space-charge regions.

With an integrated polysilicon resistor of 1 kΩ, rise and fall times of 3.2 and 2.8 ns, respectively, were measured for the DPD with $\lambda = 638$ nm and $U_r = 2.5$ V [174]. There is no indication of a so-called diffusion tail in the photocurrent (Fig. 2.49).

The transient response of such a DPD is compared to that of a PIN photodiode with a P^+ anode at the surface from Sect. 2.2.5 in Fig. 2.50. The PIN photodiode possesses shorter rise and fall times. Furthermore, the behavior of a DPD with a CMOS N well is compared to that of a DPD with

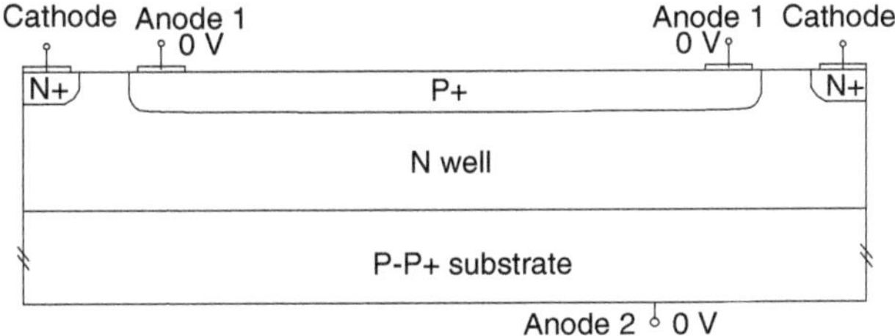

Fig. 2.47. Cross section of a double photodiode in BiCMOS technology [173]

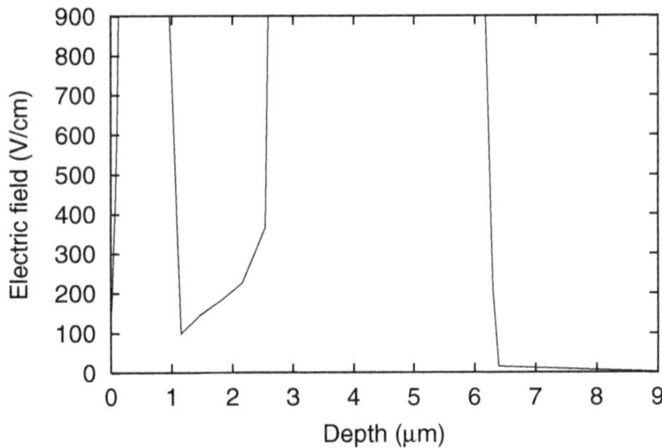

Fig. 2.48. Electric field in a double photodiode [67]

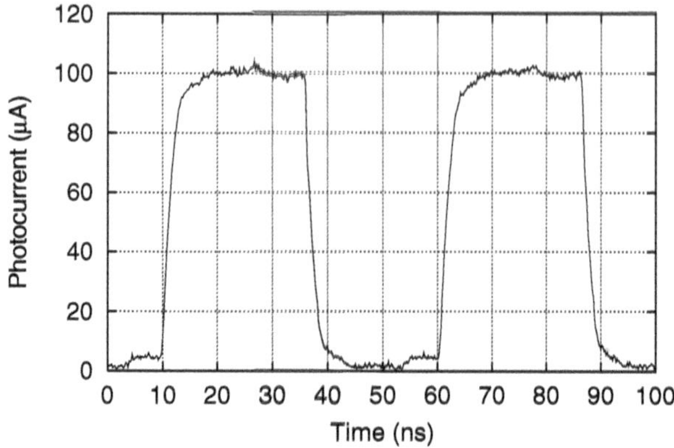

Fig. 2.49. Transient response of the double photodiode in BiCMOS technology shown in Fig. 2.47 [174]

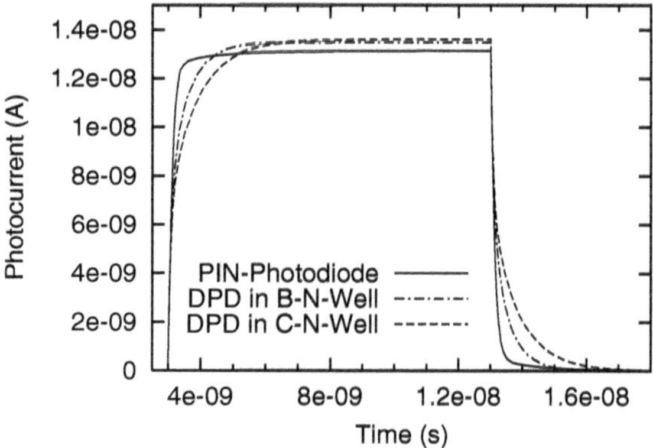

Fig. 2.50. Comparison of the transient response of double photodiodes with a bipolar N well and with a CMOS N well to that of a PIN-photodiode with $C_e = 5 \times 10^{13}\,\mathrm{cm}^{-3}$ ($\lambda = 638\,\mathrm{nm}$, $U_r = 2.5\,\mathrm{V}$)

a bipolar N well. The DPD with the bipolar N well shows shorter rise and fall times because the doping concentration in this well is lower than in the CMOS N well and the P^+/N-well space-charge region is thicker in the DPD with the bipolar N well.

With an integrated polysilicon resistor of $500\,\Omega$, rise and fall times of 1.8 and 1.9 ns, respectively, and a −3dB bandwidth of 156 MHz were measured for the DPD with $\lambda = 638\,\mathrm{nm}$ and $U_r = 2.5\,\mathrm{V}$ [173]. Later a bandwidth of 220 MHz was measured for an OEIC with a DPD of the size $50 \times 50\,\mu\mathrm{m}^2$ and with a simple transimpedance bipolar amplifier representing a low input impedance for the photocurrent of the DPD [175], leading to the conclusion that the rise and fall times of 1.8 and 1.9 ns as well as the bandwidth of 156 MHz were limited by the RC constant of load resistor and DPD capacitance.

For a DPD with the area of $23 \times 23\,\mu m^2$, finally rise and fall times of less than 0.37 ns and 0.70 ns, respectively, and a bandwidth of 367 MHz were determined in [67], where an OEIC with this smaller DPD and with a transimpedance bipolar amplifier has been investigated. The transient response of this OEIC with the small DPD is shown in Fig. 2.51. The faster speed of the smaller DPD can be explained with a lower series resistance in the N well of the DPD and with a lower space-charge capacitance due to its smaller size in addition to the low input impedance of the transimpedance amplifier. The values of 0.37 ns and 0.70 ns for the rise and fall times of the DPD determined in [67], therefore, better characterize the intrinsic speed of the DPD. With the conservative estimate a non-return-to-zero data rate $DR = 2/(3(t_r + t_f)) \geq 620$ Mb/s could be derived for the DPD [67].

Fast optoelectronic integrated circuits have been successfully realized by implementing the double photodiode [176, 177] in an industrial 0.8 μm BiCMOS process without any modifications. The main benefit of this photodiode is a very high bandwidth of about 230 MHz for a light sensitive area of $2700\,\mu m^2$. However, the capacitance of a DPD with an area of $2700\,\mu m^2$ is 876 fF at a reverse bias of 3.3 V, which is a drawback if large photodiodes are required.

Instead of the source/drain area of the PMOS transistor in a BiCMOS technology, the P-base of the NPN transistor can be used for the anode of a PIN photodiode with a thin I-layer (see Fig. 2.52). The N-epitaxial layer and the N^+-buried layer can be used instead of the N well. Connecting the P-base anode and the substrate contacts leads to a parallel circuit of the P-base/N-epi/N^+-buried layer PIN photodiode (UPD) and the N^+-buried-layer/P-substrate photodiode (LPD). The P-base layer simultaneously can be

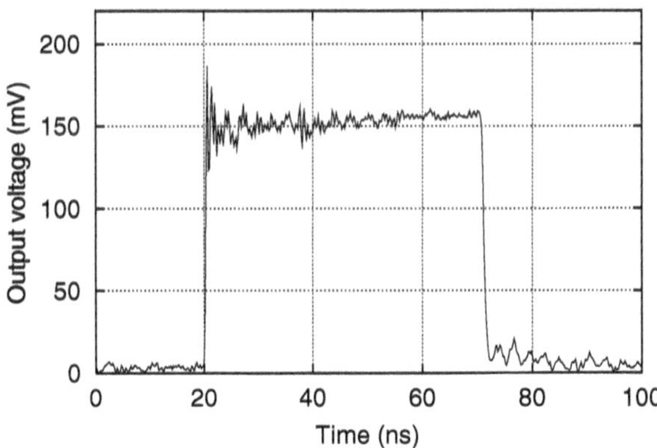

Fig. 2.51. Transient response of the double photodiode in BiCMOS technology with an area of $530\,\mu m^2$ [67]

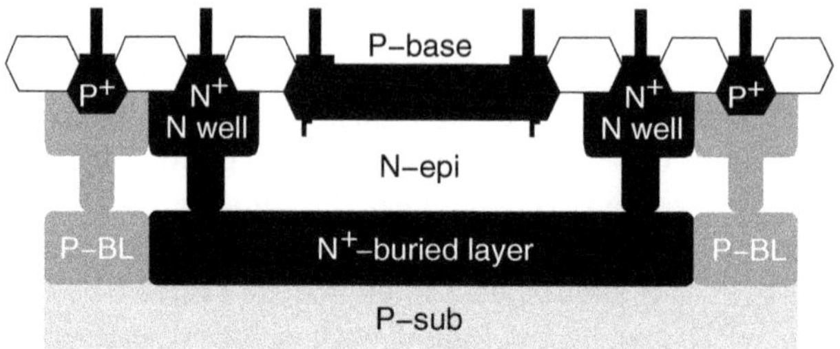

Fig. 2.52. Double photodiode in a SBC-based BiCMOS technology with a P-base anode [178]

used to shield the N^+-buried-layer/P-substrate photodiode against electric fields perpendicular to the photodiode surface. This photodiode is a double photodiode. Such a photodiode in a 0.6 µm BiCMOS technology was investigated in [178]. A responsivity of 0.4 A/W was observed for $\lambda = 780$ nm. This rather high value was due to the N^+/P-substrate junction in a depths of about 5 µm [178]. It should be noted that a responsivity measurement is a DC measurement. Therefore, a contribution of diffusing carriers from the substrate may have been present leading to such a high responsivity. Accordingly, the dynamic quantum efficiency may be lower than would correspond to this responsivity value.

Rise and fall times of 4–5 ns were reported for a similar double photodiode with a P^+-source/drain anode instead of the P-base anode for $\lambda = 780$ nm [179]. This is a quite fast response for a photodiode in standard BiCMOS technology with a responsivity of about 0.5 A/W at this rather long wavelength [179]. Rise and fall times of 0.13 ns were observed for the upper photodiode (UPD) for $\lambda = 660$ nm [179]. For $\lambda = 780$ nm the rise and fall times of the UPD were 1.4 ns and 1.8 ns, respectively [179]. The responsivities of the UPD for these two wavelengths were 0.23 A/W and 0.14 A/W, respectively. The lower photodiode (LPD) showed rise and fall times of 1.8 ns and 1.9 ns for $\lambda = 660$ nm [179]. For $\lambda = 780$ nm the rise and fall times of the LPD were 7 ns and 10 ns, respectively. The responsivities of the LPD for these two wavelengths were 0.17 A/W and 0.36 A/W, respectively.

Another photodiode which can be integrated in a BiCMOS process without any modifications is the N^+/N-well/P-substrate photodiode with a capacitance of only 112 fF for the same area (2700 µm^2) as that of the DPD with a capacitance of 876 fF [180].

In Fig. 2.53 the cross section of the N^+/N-well/P-substrate photodiode can be seen. The PN-junction is located between the N-well and the P-substrate. The depth of the pn-junction allows reliable operation at red laser light (λ = 638 nm). The bandwidth of the N^+/N-well/P-substrate photodiode for this wavelength is larger than 82 MHz [181]. In order to improve the quantum efficiency, the light sensitive area of the photodiode is covered by

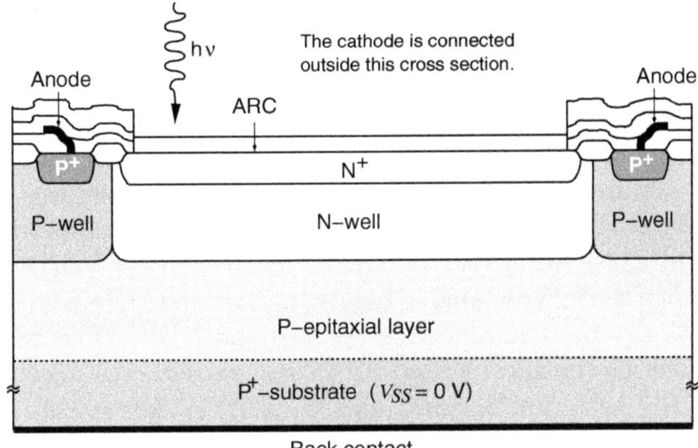

Fig. 2.53. Cross section of the N$^+$/N-well/P-substrate photodiode (ARC: antireflection coating, N$^+$-cathode contact not drawn)

an antireflection coating (ARC) and the same quantum efficiency as for the DPD is achieved.

The bandwidth of OEICs for application in optical storage systems rapidly increased recently. Also the same OEIC should be appropriate for application in CD-ROM, DVD, and future DVD as well as DVR with different wavelengths being used (780 nm, 650 nm, and 400 nm) in order to save development effort and lower the prices due to higher production volumes of the same OEIC. A photodiode for such a universally applicable OEIC was introduced in [182].

The photodiode implemented in a 0.6 μm standard BiCMOS process is shown in Fig. 2.54. In order to avoid slow diffusion of photogenerated carriers from the substrate for 780 and 650 nm, the photodiode formed by the N-epitaxial layer and the shallow P (SP) implant was used. The SP anode is interrupted, whereby the SP regions are connected electrically outside the light sensitive area. It is interrupted because of two reasons. First, the quantum efficiency for blue/UV light is higher than for a whole-area P-type region.

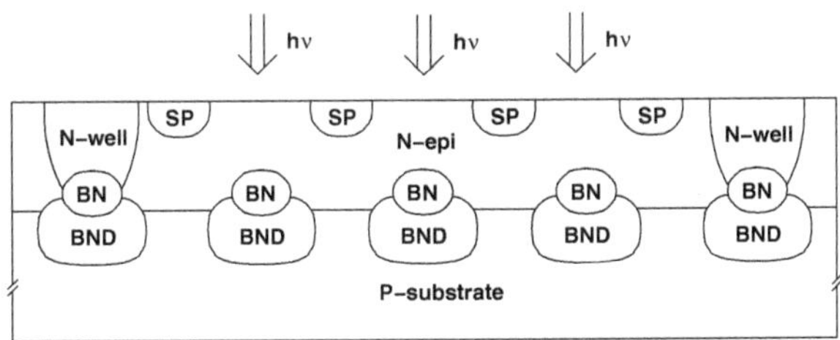

Fig. 2.54. Cross section of a shallow-P(SP)/N-epi/ photodiode in standard BiCMOS technology [182]

Second, the junction capacitance is lower than for a whole-area P-type region in the epitaxial layer yielding an improved signal-to-noise ratio (SNR) of the OEIC for all three wavelengths. The N-epi cathode has a low-ohmic connection to the supply voltage V_{cc} via the buried-N (BN) and buried-N-deep (BND) as well as via the N-well regions. The vertical doping profile of the BN/BND doping generates an electric field that accelerates the photogenerated carriers. The BN and BND regions are interrupted (and connected outside the light sensitive area) to increase the accelarating electric field.

A capacitance of 0.25 pF and a bandwidth of 450 MHz were mentioned for the photodiodes used in channels A–D of the OEIC. The measured responsivity values of the interrupted-SP/N-epi photodiode were 0.079 A/W for 780 nm, 0.169 A/W for 653 nm, and 0.120 A/W for 402 nm [182]. The capacitance and bandwidth values are impressive results for this sophisticated photodiode. It should be mentioned, however, that the relatively low responsivity values finally limit the sensitivity of OEICs implementing this photodiode.

The integration of photodiodes in BiCMOS technology, for which process modifications are necessary, is similar to the integration of photodiodes in bipolar or CMOS technology. The integration of PIN photodiodes in optimized high-performance BiCMOS technologies (see Fig. 2.42) requires a similar additional process complexity as the integration of PIN photodiodes in bipolar technology (see Fig. 2.7). The N well can be used for the collector of the NPN transistor in the BiCMOS process. The third mask for the doping of the collector in Fig. 2.7, therefore, is not necessary for the PIN-BiCMOS integration in a process with a structure as in Fig. 2.42. Consequently, two additional masks are necessary for PIN-BiCMOS integration in such a structure.

3 Detectors in Thin Crystalline Silicon Films

Photodiodes in thin Silicon-on-Insulator (SOI) films represent a possibility to avoid slow carrier diffusion for achieving high-speed operation. CMOS circuits in Silicon-on-Sapphire (SOS) thin films allow easy hybrid integration of III/V laser diodes, since the sapphire substrate is transparent and the light can be emitted downwards through the SOS film and the sapphire substrate. Upward light emission is therefore not necessary and the III/V substrate of the VCSELs does not have to be removed. Polyimid bonding of III/V laser diodes and photodiodes on silicon as an interesting wafer-scale process also shall be described in this chapter.

3.1 Photodiodes in Silicon-on-Insulator

Silicon-on-Insulator (SOI) has been investigated mainly as a material for CMOS technology as a substitute for bulk silicon [183]. The main advantages of thin-film SOI for CMOS are: (i) The die area consumed by device isolation can be less than for bulk material. (ii) The number of process steps can be reduced. (iii) The substrate current is suppressed. (iv) There is no latch-up effect. (v) Lower parasitic capacitances enable faster circuits and a reduction of the power consumption.

In addition to CMOS technology, SOI can be interesting for the integration of photodetectors. SOI was considered as a possibility to avoid the slow diffusion current, which is caused by carriers being generated by light with long wavelengths (in the region of 850 nm), from the substrate of integrated photodiodes. For photodetectors in a thin SOI layer, this light generates carriers in the silicon substrate wafer, which is isolated from the device SOI layer by the buried oxide. Therefore, SOI avoids the slow diffusion current known from photodiodes in bulk silicon.

A lateral PIN photodiode in a thin SOI layer was investigated first by Colinge [184]. The structure of this lateral PIN photodiode is shown in Fig. 3.1. The SOI layer was made by laser-recrystallization. The thickness of the silicon film was (440 ± 50) nm at the end of the process. The buried oxide layer had a thickness of 1 µm and the oxide on top of the silicon thin-film layer was 0.5 µm thick. The width of the intrinsic region was designed to be 7 µm. The N^+ and P^+ regions were each 8 µm wide. The carrier lifetime in the laser-recrystallized

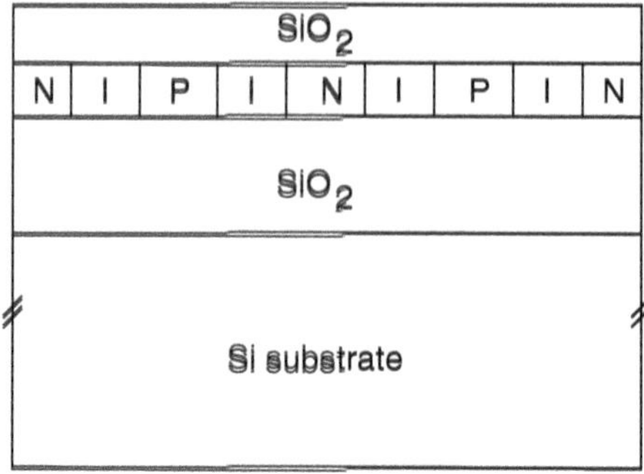

Fig. 3.1. Cross section of a lateral PIN photodetector in a thin-film SOI layer [184]

SOI film was found to be approximately 8 μs and the dark current of the lateral PIN photodiode was reported to be 0.1 pA per 1 μm junction length for a reverse voltage of 5 V. This value for the dark current is large and is due to the fabrication of the SOI layer by the laser-recrystallization technique, which results in a lower thin-film quality than bonded and etched-back SOI (BESOI). Due to the low thickness of the Si film and the buried and top oxide layers, interference effects were present in the spectral photoresponse. The maximum quantum efficiency of approximately 80 % was found at an interference maximum for green light. Due to the low thickness of the Si film, the quantum efficiency decreased strongly for longer wavelengths with smaller absorption coefficients. The quantum efficiency for $\lambda = 780$ nm for instance was approximately 9 % due to an interference maximum, whereas for $\lambda = 850$ nm the quantum efficiency was reduced to only approximately 2 %. No results for the transient response of the lateral PIN photodiode were reported.

Ghioni et al. [185] described a lateral PIN photodetector, which is compatible with trench isolation VLSI technology. A larger thickness of (2.8 ± 0.5) μm was chosen for the SOI layer in order to obtain a larger quantum efficiency than mentioned above. The resistivity of the N-type BESOI layer was in the range of 10 to 15 Ωcm, which corresponds to a doping concentration in the range of 4.5×10^{14} cm^{-3} to 3.0×10^{14} cm^{-3}. The thickness of the buried oxide was 1 μm. Due to the thickness of the SOI layer (2.8 μm) the trench isolation technique as shown in Fig. 3.2 was used. The interdigitated PIN detector biased at 3.5 V attained −3dB bandwidths in excess of 1 GHz with $\lambda = 840$ nm for a finger width of 2 μm and for finger distances of less than 4 μm. Photodetectors located in different positions on the wafer exhibited slightly different bandwidths (± 0.2 GHz), due to the thickness variations of the SOI layer. The dark current was less than 10 pA at a bias of 5 V for a photodetector area of 75×75 μm^2. The responsivity was 0.048 A/W ($\eta = 0.071$, quantum efficiency = 7.1 %) for a finger spacing of 1.5 μm and 0.091 A/W ($\eta = 0.135$) for a finger spacing of 7.0 μm. A ± 15 % fluctuation

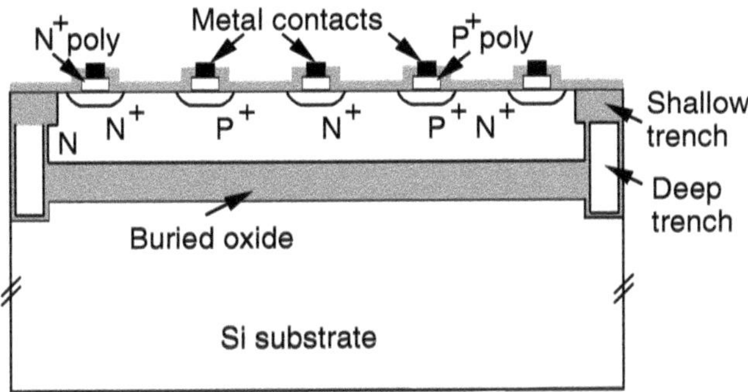

Fig. 3.2. Cross section of a trench isolated lateral PIN photodetector in a SOI layer [185]

was observed for these values due to the thickness variation of the SOI film over the wafer area. For the smaller finger spacing a larger amount of the incident light is reflected by the aluminum contacts and the quantum efficiency is lower than for the larger finger spacing.

Ghioni et al. [185] reported a calculated sensitivity of −22.5 dBm for a photoreceiver comprising a MOS transimpedance preamplifier and the lateral PIN photodetector in SOI with $R = 0.09$ A/W at a bit-error rate of 10^{-12}.

In order to increase the responsivity, interference effects were investigated. The stack consisting of the top passivating oxide, the SOI layer, and the buried oxide layer forms a Fabry–Perot cavity, with the passivating oxide as the top reflector and the buried oxide as the bottom reflector. In such a way the lateral PIN photodetector can be considered as a resonant photodetector that can be tuned to the desired wavelength in order to maximize the responsivity. The tuning can be done by carefully adjusting the thicknesses of the cavity layers. The conditions for a maximum responsivity are for normal incidence (with l, m and n being integer numbers) [185]:

(i) passivating oxide layer: $d_{\mathrm{PO}} = l{\cdot}\lambda/(2\bar{n})$;
(ii) SOI layer: $d_{\mathrm{SOI}} = m{\cdot}\lambda/(2\bar{n})$;
(iii) buried oxide layer: $d_{\mathrm{BO}} = n{\cdot}\lambda/(4\bar{n})$.

The responsivity of such a resonant structure depends strongly on the thickness of the active layer d_{SOI} (see Fig. 3.3). In the best case, the responsivity of such a resonant structure is more than doubled, but in the worst case, the responsivity of the SOI detector is reduced below that of a nonresonant detector (where no interference effects are taken into account and only reflection at the top oxide to SOI layer interface is considered in addition to the absorption of the SOI layer). The silicon layer incremental thickness between the best case and the worst case was reported to be only 50 nm. The responsivity for $d_{\mathrm{SOI}} = 2.74\,\mu$m, for instance, is 0.125 A/W ($\eta = 0.185$), whereas it is only 0.03 A/W ($\eta = 0.044$) for $d_{\mathrm{SOI}} = 2.79\,\mu$m in comparison to 0.055 A/W ($\eta = 0.081$) for a nonresonant detector with $d_{\mathrm{SOI}} = 2.79\,\mu$m. Therefore, an extremely good thickness control of less than $\pm 0.5\,\%$ is required for a typical active layer thickness of a few micrometers. An alternative design was suggested by Ghioni et al. [185], if such a strict control cannot be guaranteed:

(i) passivating oxide layer: $d_{PO} = (2l + 1)\cdot\lambda/(4\bar{n})$;
(ii) SOI layer: $d_{SOI} = m\cdot\lambda/(2\bar{n})$;
(iii) buried oxide layer: $d_{BO} = (2n + 1)\cdot\lambda/(4\bar{n})$.

Then a maximum responsivity of 0.105 A/W ($\eta = 0.155$) for $d_{SOI} = 2.74\,\mu m$ is obtained and the minimum responsivity of 0.057 A/W ($\eta = 0.084$) for $d_{SOI} = 2.79\,\mu m$ (see Fig. 3.4) is a little larger than the responsivity of 0.055 A/W for the nonresonant detector. A variation by a factor of approximately two, however, for the responsivity and therefore for the quantum efficiency remains for this design, also, because a thickness tolerance of less than 50 nm for a SOI layer thickness of several micrometers can hardly be obtained by BESOI.

Lateral PIN photodiodes in SOI films with a thickness of 3.75 µm were investigated in [186]. The quantum efficiency for 850 nm was 29 %. A bandwidth of 1.1 GHz was reported for a P-N spacing of 2 µm with a SOI film resistivity of 50–100 Ωcm at a photodiode reverse bias of 3 V [186]. For a reverse bias of 5 V and a SOI film thickness of 2.7 µm, the bandwidth was increased to 3.4 GHz. The quantum efficiency for 840 nm of the photodiodes in the 2.7 µmthick SOI film was 24 %. The reverse current of a $50 \times 50\,\mu m^2$ photodiode was 25 pA at a reverse bias of 5 V [186]. At a reverse bias of 9 V, photodiodes with a 250 nm finger spacing exhibited bandwidths of 15 and 8 GHz in SOI films with thicknesses of 0.2 and 2 µm, respectively [187]. The leakage current, however, was 500 µA for this 9 V bias. The quantum efficiencies for 850 nm were 2 and 12 %, respectively. Lateral PIN photodiodes in

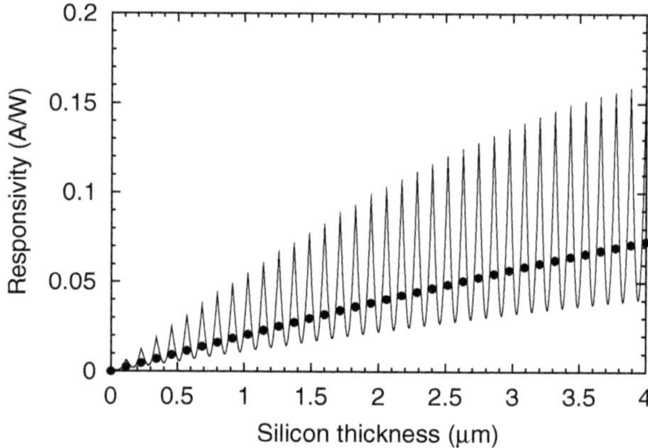

Fig. 3.3. Responsivity of a trench isolated lateral SOI PIN photodetector with a light-sensitive area of 50 % and with a half-wavelength top oxide layer versus SOI layer thickness. The *dotted curve* belongs to a nonresonant photodetector [185]

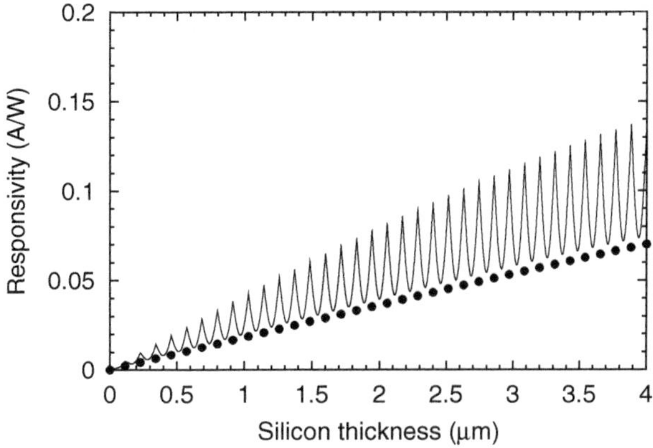

Fig. 3.4. Responsivity of a trench isolated lateral SOI PIN photodetector with a light-sensitive area of 50 % and with a quarter-wavelength top oxide layer versus SOI layer thickness. The *dotted curve* belongs to a nonresonant photodetector [185]

thin SOI films were implemented in OEICs, which operated up to data rates of 3.125 and even 5 Gb/s at a photodiode supply voltage of 24 V [188].

It can be concluded that despite several attempts to increase the quantum efficiency [185, 189–191], the main disadvantage of SOI photodetectors is the low quantum efficiency or responsivity. Therefore, internal amplification of the photocurrent is highly desirable.

The responsivity of a lateral PIN photodiode in a thin-film SIMOX layer was increased by the exploitation of the avalanche effect (in the opinion of the authors) for internal amplification of the photocurrent [192]. Figure 3.5 shows the structure of this photodetector for application in local area networks (LANs) such as the Gigabit Ethernet or Fibre Channel with a wavelength in the range of 780–850 nm. The N^+ and P^+ regions are 1 µm wide and the I-regions were 0.6 to 3.0 µm wide. The N^+-polysilicon gates with a thickness of 200 nm make the 50 nm thick silicon layer with an N-type doping concentration of $3 \times 10^{17}\,\mathrm{cm}^{-3}$ fully depleted at a supply voltage of 2.0 V. The polysilicon gates also can be used to adjust the dark current of the photodetector. On top of the gates, several micrometers of oxide were deposited. The capacitance of a 100-square µm photodiode was only 0.2 pF due to the thin SOI layer and the 110 nm thick buried oxide.

A photodetector with a size of $60 \times 60\,\mu\mathrm{m}^2$ for the use together with an optical fiber having a 50 µm core diameter was fabricated in a 0.25 µm CMOS technology. Without an antireflection coating, a responsivity of 0.4 A/W at $\lambda = 850$ nm was achieved at a reverse bias of only 2.0 V. A high electric field of $1\text{–}2 \times 10^5$ V/cm exists in the space-charge region at the P^+ side for this voltage bringing about an avalanche effect. The photocurrent is amplified due to the rise of the ionization rate in Si under a high electric field of more than 1×10^5 V/cm [193] achieving a responsivity of the 50 nm thick

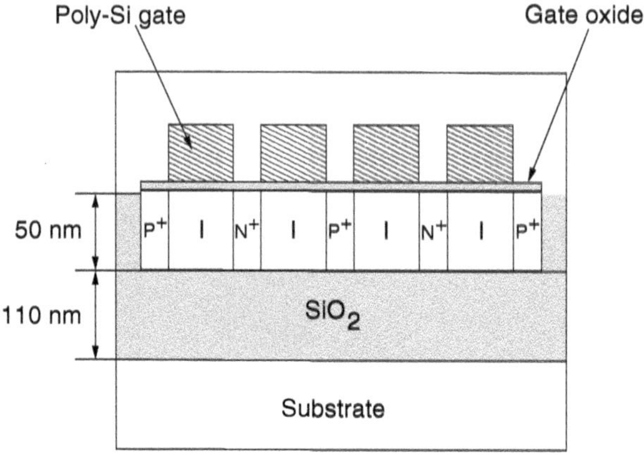

Fig. 3.5. Cross section of an avalanche photodiode in a thin-film SOI layer [192]

photodetector similar to that of a thick vertical PIN photodiode. A dark current of $0.2\,\mathrm{pA}/\mu\mathrm{m}^2$ at a polysilicon gate bias of $-0.8\,\mathrm{V}$ was reported for the photodetector. The amplifier used together with the lateral PIN avalanche photodiode in the OEIC for LAN applications will be discussed in Sect. 6.4.10 (Fig. 6.111).

3.2 Silicon on Sapphire

A hybrid integration approach representing a paradigm shift from traditional optoelectronic integration and packaging methods was described in [194,195]. A recent breakthrough in silicon-on-sapphire (SOS) CMOS VLSI technology offers simple and elegant solutions to many system integration and packaging challenges that have to be overcome when bulk silicon CMOS technologies are used. The substrate being present for bulk CMOS ICs absorbs at 850 nm and even at 980 nm wavelengths. Complex and expensive integration procedures like VCSEL substrate removal are therefore necessary to enable optical vias through the VCSEL substrate.

The optical transparency of sapphire substrates, its high thermal conductivity and the high-speed device characteristics of SOS CMOS circuits makes SOS technology an excellent choice for cost effective optoelectronic die-as-package (DASP) systems and for implementing high-performance optical interconnects [195]. SOS CMOS technology is a close relative to the more popular silicon-on-insulator (SOI) CMOS technology that was believed to become the standard technology for deep sub-micron VLSI [196,197]. SOS uses sapphire-aluminum oxide (Al_2O_3) instead of silicon as the substrate. Sapphire is an insulator and its thermal conductivity is higher than that of silicon dioxide. But what is more, sapphire is transparent from 200 nm to 5,500 nm. A 0.5 μm SOS CMOS process with an ultra-thin silicon layer hav-

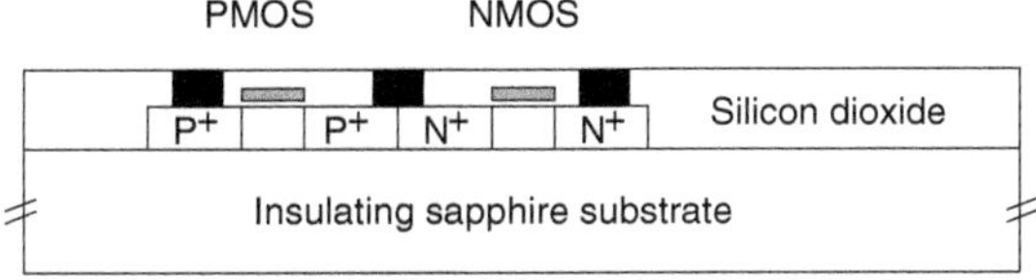

Fig. 3.6. Cross section of a CMOS inverter in ultra-thin SOS [195]

ing a thickness of only 100 nm is available from Peregrine via MOSIS foundry service [195]. The cross section of an SOS chip is shown in Fig. 3.6. The MOSFETs with low and zero threshold voltage have a lighter doping in the channel than the regular threshold voltage MOSFETs and are therefore fully depleted showing no kink effect in the I_D-V_{DS} characteristics in contrast to the the regular threshold voltage MOSFETs.

VCSELs were mounted on a SOS CMOS chip. The die of the SOS chips was not optically polished on the backside. In order to avoid severe light loss due to scattering from the back surface index matching was necessary. For that purpose the die was glued on a glass cover slide using index matching wax [195]. The glass cover slide was then mounted on a blank lithography mask and placed upside down in a mask aligner. Conductive epoxy was applied to each bondpad of the VCSEL die which was subsequently placed on the wafer in the mask aligner. To fix the VCSEL die on the chip, drops of UV curable epoxy were placed at the corners of the VCSEL die. After optical alignment the blank mask with the SOS CMOS die was lowered to make contact with the VCSEL die. Then the two dies were mechanically bonded together by exposing the assembly to UV light. Finally, the assembly was placed in an oven to cure the conductive epoxy for an hour at 60 °C. The VCSELs were operated at data rates above 1 Gb/s [195]. Metal-semiconductor-metal photodiodes also were bonded to SOS CMOS receivers. These optical receivers were tested at 1 Gb/s.

3.3 Polyimide Bonding

By using polyimide bonding GaAs PIN photodiodes and vertical cavity surface emitting laser diode (VCSEL) chips were hybridized to CMOS circuits [198]. Compared to other methods, this polyimide bonding has three advantages: (i) it is a wafer-scale process; (ii) it integrates both VCSELs and photodiodes simultaneously; and (iii) it needs no accurate alignment in the bonding step. The structure of a pixel with a GaAs PIN photodiode and a VCSEL is shown in Fig. 3.7.

To apply the polyimide bonding technique to an originally uneven CMOS wafer, the CMOS wafer surface has to be planarized before wafer bonding. Special VCSEL and photodetector (PD) attaching areas besides CMOS circuit areas were prepared. Only aluminum remained in these attaching areas

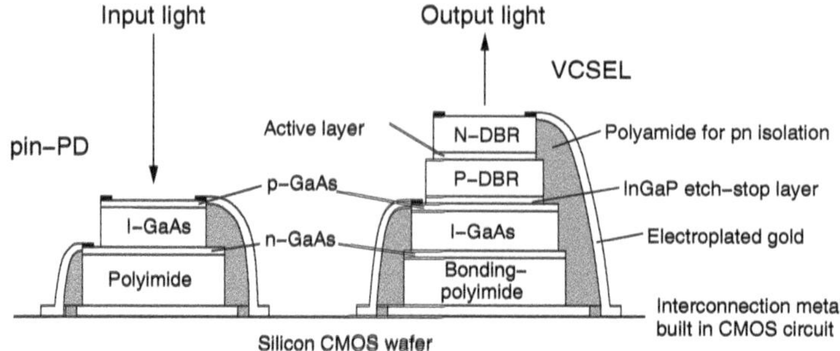

Fig. 3.7. Cross section of polyimide bonded CMOS/PD/VCSEL structure [198]

of the size $50 \times 50\,\mu\text{m}^2$. Gold was deposited onto the VCSEL attaching areas on the CMOS wafer to act as a heat pipe for the power dissipated in the VCSELs. Then the uneven CMOS surface was planarized with polyimide (of a different type from the bonding one). Planarization was completed by lapping and polishing the polyimide with ultra-fine aluminum oxide powder until the desired flatness was achieved. Reactive-ion etching with oxygen gas then was applied to expose the gold heat pipe areas. A surface flatness of $0.1\,\mu\text{m}$ was obtained [198].

After planarization, a GaAs wafer with epitaxial layers of VCSELs and photodiodes is bonded to the CMOS wafer using polyimide as an adhesive. After removing the GaAs substrate by mechanical lapping and selective wet etching, the GaAs epitaxial structure is divided into chip-size sections, so that the prefabricated marks in the CMOS wafer are exposed. Now, the photonic devices are contacted using the marks on the CMOS wafer for photolithography. Vertical electric interconnects between the photonic devices and the CMOS circuits are formed with electroplated gold. A thermal resistance of 400 K/W was reported for a polyimide bonded VCSEL with a $15\,\mu\text{m}$ upper distributed-Bragg-reflector (DBR) mesa diameter and a $40\,\mu\text{m}$ square lower-DBR-mesa resulting in a temperature rise of 6 K for a VCSEL power consumption of 15 mW [198].

4 SiGe Photodetectors

SiGe alloys allow the integration of infrared detectors on Si. The addition of Ge to Si also increases the absorption coefficient in the spectral range from 400–1000 nm, allows a reduction of the detector thickness, and, therefore, enables faster detectors than with pure Si. To exploit the advantages of SiGe alloys for Si-based photodetectors, however, the problem associated with the lattice constant mismatch has to be understood. Subsequently, this chapter describes an example of a SiGe OEIC.

4.1 Heteroepitaxial Growth

The growth and physical properties of $Si_{1-x}Ge_x$ heteroepitaxial layers on Si have been investigated for more than 20 years [199–203] finally producing a technology which is entering high-volume and large-scale manufacturing of heterojunction bipolar transistors (HBTs) [204] and SiGe-HBT-BiCMOS circuits [205, 206]. Optoelectronic applications of $Si_{1-x}Ge_x$/Si devices also have been developed. Medium-wavelength (1.3–1.55 μm) photodetectors for optical communication [207–209], 2–12 μm infrared photodetectors for two-dimensional focal plane arrays for thermal imaging and night vision [210–212], optical waveguides [213], and infrared light emitters for chip-to-chip optical interconnects [214, 215] have been suggested.

On the way to introducing optoelectronic devices into VLSI and ULSI to obtain a silicon-based *superchip* [216, 217], however, the following significant limitations of the SiGe/Si material system have been discovered [218]:

(i) Due to the large lattice constant of Ge, which is about 4.2 % larger than that of Si, a critical thickness of $Si_{1-x}Ge_x$ layers on Si for thermal stability against misfit dislocation generation due to lattice mismatch exists [219]. This critical thickness is only about 15 nm for a Ge fraction x of 0.2 with a resulting bandgap change of approximately 150 meV [220]. Due to this low critical thickness, the quantum efficiency of normal incidence photodetectors relying on photogeneration across the $Si_{1-x}Ge_x$ bandgap is severely limited.

(ii) When a $Si_{1-x}Ge_x$ layer with a thickness greatly exceeding the critical thickness is grown directly onto Si, the layer relaxes but forms a dislocation density at the surface of order $10^{12}\,cm^{-2}$ [221]. This dislocation density reduces the carrier mobility and increases leakage currents significantly and is far too large to allow economic yields of devices in production [206]. Although relaxed $Si_{1-x}Ge_x$ buffer layers with the bulk lattice constant of the $Si_{1-x}Ge_x$ alloy on top of the Si substrate can be introduced and pseudomorphic heterostructures, which have a lateral lattice constant larger than that of Si, can then be grown on top of the buffer [222, 223], these relaxed buffers still reduce the threading dislocation density only to below $10^4\,cm^{-2}$ for $Si_{0.8}Ge_{0.2}$ [221]. This density is already comparable to many III/V systems [206]. More recently, a number of annealing steps during growth have reduced the dislocation density to $10^2\,cm^{-2}$ for $Si_{0.8}Ge_{0.2}$ and $x < 0.2$ [224]. Dislocations, however, still pose a problem for higher Ge fractions and future large-scale optoelectronic applications, therefore, remain uncertain.

(iii) The dopant diffusion during thermal processing for device fabrication often degrades the nanometer precision in the placement of dopant atoms required for heterojunction devices, which can be achieved during the initial growth of heterostructures at temperatures below 700 °C [225, 226].

(iv) Under all combinations of strain, $Si_{1-x}Ge_x$ is an indirect-bandgap material, resulting in inefficient emission of light and inefficient detection due to a small absorption coefficient compared to pure Ge, GaAs, or InGaAs, at least for $x < 0.75$.

Let us continue with light absorption of SiGe alloys (point (iv)) in the next section.

4.2 Absorption Coefficient of SiGe Alloys

The exchange of Si atoms by Ge atoms increases the absorption coefficient. Furthermore the bandgap reduces with increasing Ge fraction and wavelengths longer than 1100 nm can be detected. Figure 4.1 shows the absorption coefficient for Ge fractions of 20, 50, and 75 %.

It is advantageous to use SiGe detectors instead of Si detectors, when longer wavelengths, a thinner absorbing layer, a higher quantum efficiency and/or a higher speed of the detector are needed. In the following, an example of a SiGe receiver trying to use these advantages will be described.

4.3 SiGe/Si PIN Hetero-Bipolar-Transistor Integration

SiGe photodiodes were described in [228,229], however, the amplifiers in these references employed pure Si transistor devices. SiGe/Si technology has developed rapidly in recent years and it has been demonstrated that it outperforms Si technology in terms of the speed of transistors [230]. Therefore, it is advantageous to combine SiGe/Si photodetectors with SiGe/Si transistors [231].

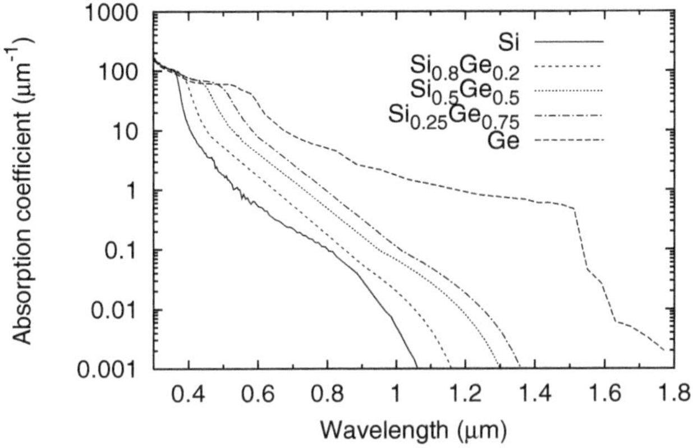

Fig. 4.1. Absorption coefficient for SiGe with different compositions [227]

Figure 4.2 shows the cross section of the first monolithically integrated SiGe/Si PIN photodiode and heterojunction bipolar transistor (HBT) front-end photoreceiver. The PIN-HBT structure was grown by one-step molecular beam epitaxy (MBE). Table 4.1 contains the layer compositions and doping concentrations of the SiGe-OEIC. The emitter and collector layers consist of Sb-doped Si. The base layer of the double heterojunction NPN HBT structure is a $Si_{1-x}Ge_x$ alloy with a smaller bandgap than that of Si. The Ge mole fraction in the base layer is graded from $x = 0.1$ at the emitter side to $x = 0.4$ at the collector side to accelerate the electrons traveling through the base towards the collector by a quasi-electric field. Spacer layers on both sides of the base minimize the effect of outdiffusion during epitaxy and processing.

The PIN photodiode is formed by the Si N^+-subcollector, by the Si N^- collector, and the SiGe P^+ base layers of the HBT (see Fig. 4.2). The I-absorption layer is formed by the Si N^- collector due to the one-step MBE

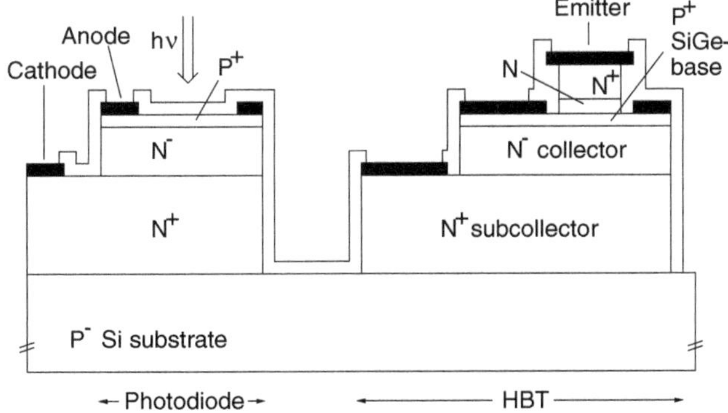

Fig. 4.2. Schematic cross section of a SiGe/Si PIN HBT photoreceiver [231]

Table 4.1. Layer compositions and doping concentrations of a SiGe-Si PIN-HBT OEIC

Layer	Material	Type	Doping (cm^{-3})	Thickness (nm)
Emitter Contact	Si	N^+	1×10^{19}	200
Emitter	Si	N	2×10^{18}	100
Spacer	$Si_{0.9}Ge_{0.1}$	I		1
Base	$Si_{1-x}Ge_x$ (x:0.1→0.4)	P^+	5×10^{19}	30
Spacer	$Si_{0.6}Ge_{0.4}$	I		10
Collector	Si	N^-	1×10^{16}	250
Subcollector	Si	N^+	1×10^{19}	1500
Substrate	Si	P^-	2×10^{12}	540 μm

growth, which provides advantages over regrowth such as better planarity, simpler processing, and higher yield [231]. The MBE growth temperatures for the collector and emitter layers were 415 °C. The base was grown at 550 °C. The growth rate was only 0.2 nm/s at these low temperatures.

The devices were isolated by mesa formation. The mesa size of the photodiode was $12 \times 13\,\mu m^2$. After MBE, the emitter contact was defined by evaporation, followed by emitter mesa formation with SF_6- and O_2-based dry and KOH-based wet etching. Minimal undercut and base over-etch were obtained by the two-step etch procedure, resulting in a low base access resistance. Then the base-collector mesa was formed by dry etching. The collector and PIN cathode contacts were formed by evaporation on the exposed highly doped subcollector layer. Another etch step of the subcollector layer was applied for the separation of the devices. After this mesa isolation a PECVD SiO_2 layer was deposited and the pad contacts were opened.

The dark current of the photodiode was about 0.1 μA at $U_{PIN} = 4$ V and 1 μA at $U_{PIN} = 9$ V. For the PIN photodiode with an I-layer thickness of 0.25 μm, a bandwidth of 450 MHz for $\lambda = 850$ nm and $U_{PIN} = 9$ V was measured. The PECVD SiO_2 layer with a thickness of 1.1 μm served as an antireflection coating leading to a measured responsivity of 0.3 A/W and an external quantum efficiency of 43 % for $U_{PIN} = 5$ V. Here, the SiGe anode did not increase the quantum efficiency by a significant amount compared to a Si anode. For a large enhancement of the quantum efficiency a high Ge fraction in the intrinsic zone of the PIN photodiode would have been necessary, which of course would introduce a high density of dislocations.

The bandwidth of 450 MHz was ascribed to the slow diffusion of carriers generated in the substrate and in the subcollector [231], because of the small I-layer thickness. We should not, however, accept this explanation, because the responsivity value of 0.3 A/W is too high for this explanation. It can be understood only when the cathode current, which is the sum of the subcollector/P^--substrate diode photocurrent and of the P^+-SiGe-anode/N-collector/N^+-subcollector diode photocurrent, was measured. The bandwidth then was not limited by carrier diffusion from the P^- substrate but by drift in the space-charge region of the subcollector/P^--substrate diode. The space-charge region extended far into the low doped P^--substrate with $N_A = 2 \times 10^{12}\,\mathrm{cm}^{-3}$ and the measured bandwidth of 450 MHz for $U_{PIN} = 9\,\mathrm{V}$ seems possible. The reported increase in the responsivity and in the bandwidth with increased reverse bias also supports the new explanation with drift instead of carrier diffusion.

5 Design of Integrated Optical Receiver Circuits

Most of the optoelectronic integrated circuits (OEICs) described in this book are analog circuits. We, therefore, will restrict ourselves to the design of analog integrated circuits. The exclusion of the design of digital integrated circuits, furthermore, seems to be justified because there are many books describing this topic (for instance [232–234]). The design of analog integrated circuits is, for instance, described in [235]. It will be, therefore, sufficient to give here an overview on the aspects being most relevant for the design of analog OEICs. Since photocurrents have to be converted to signal voltages in most applications, the transimpedance amplifier will be discussed in some detail, especially with respect to its implementation in OEICs. A new π model for such a transimpedance amplifier also will be introduced. Furtehrmore, the basics of electronic noise and sensitivity of optical receivers are applied to point out the possibilities of silicon OEICs.

5.1 Circuit Simulators and Transistor Models

Circuit simulation is useful for the development of integrated circuits in order to avoid long time periods for fabrication and high development costs. Actually, circuit simulation is absolutely necessary for the development of integrated circuits in advanced submicrometer silicon technologies, because analytical models and calculations do not cover short channel effects and, therefore, are not accurate enough. In addition, most of the circuits contain too many devices for analytical investigations.

Circuit simulators implement transistor models covering their temperature dependence as well as short-channel effects and guarantee the correct calculation of the operating point, which is the prerequisite for the simulation of frequency and transient responses, because the amplification of the circuit depends on the operating point of the circuit, which determines the voltages across all the transistor terminals and the currents flowing in all the devices. Therefore, the transconductance g_{m} of each transistor, which determines the amplification, and the junction capacitances in each transistor, which determine the frequency and transient responses, depend on the operating point of the circuit.

There are several widespread circuit simulation programs: for instance HSPICE originally developed by the University of California at Berkeley and ACCUSIM as part of the MENTOR design framework as well as SPECTRE-S as part of the CADENCE design framework. ACCUSIM and SPECTRE-S are part of the professional design frameworks being used by semiconductor chip manufacturers. Foundries and manufacturers of application specific integrated circuits (ASICs) also support these frameworks. The manufacturers of ASICs make the transistor models and parameters available to customers and design houses. One of the newest and widely used transistor models covering the small signal and large signal behavior for MOSFETs is BSIM3.3 [236]. This model was used here for the development of CMOS OEICs with ACCUSIM within the MENTOR framework and for the development of BiCMOS OEICs with SPECTRE-S within the CADENCE framework.

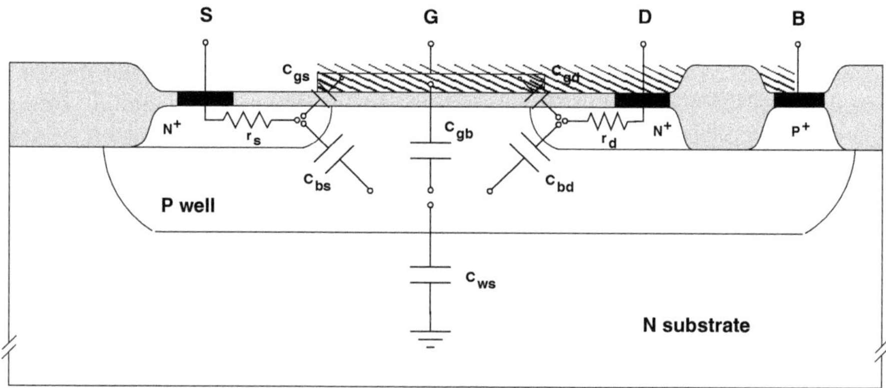

Fig. 5.1. Integrated-circuit MOS transistor structure including parasitic elements

The modeling of MOSFETs dates back to [237, 238]. The models consider parasitic capacitors and resistors within the MOS transistors shown in Fig. 5.1. The parasitic elements are included in the small-signal equivalent circuit (Fig. 5.2) underlying the transistor models implemented in modern circuit simulators. A small-signal model implicitly assumes a linear behavior. The parameters of the small-signal model are usually designated by lower case letters.

The conductances g_{bd} and g_{bs} are the conductances of the bulk-to-drain and bulk-to-source junctions. These conductances are normally very small, since these junctions are reverse-biased. The channel transconductances g_{m}, g_{mb}, and g_{ds} are defined as

$$g_{\mathrm{m}} = \frac{\partial i_{\mathrm{d}}}{\partial v_{\mathrm{gs}}}, \; g_{\mathrm{mb}} = \frac{\partial i_{\mathrm{d}}}{\partial v_{\mathrm{bs}}}, \; g_{\mathrm{ds}} = \frac{\partial i_{\mathrm{d}}}{\partial v_{\mathrm{ds}}} \tag{5.1}$$

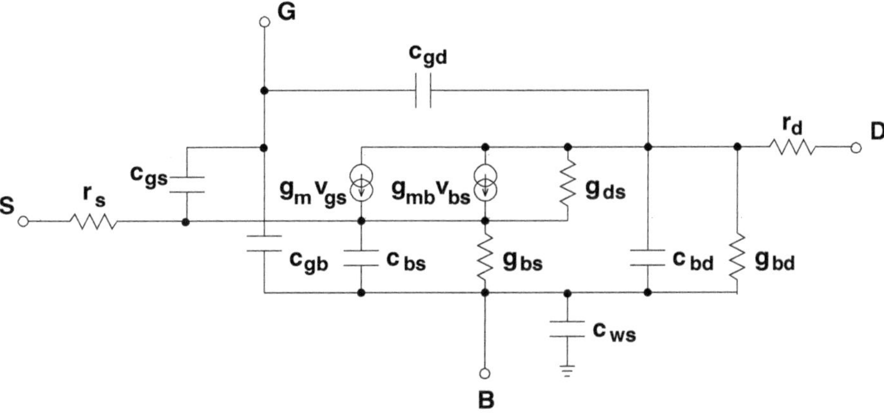

Fig. 5.2. Complete MOS transistor small-signal equivalent circuit [239]

at the quiescent point, i. e. at the operating point. The transconductance g_{m} gives rise to the desired amplification of a MOSFET. The so-called back-gate effect, arising from g_{mb}, however, reduces the amplification when source and bulk, i. e. well, cannot be connected as in the case when PMOS input transistors would be used in a difference amplifier on an N-type substrate, for instance. The conductance g_{ds} considers the channel length modulation effect for the transistor in saturation.

For bipolar transistors in HSPICE, level 1 can be used, for instance. SPECTRE-S, which was used in the development of BiCMOS OEICs, implements several high-current effects in addition to the large-signal model of Gummel and Poon. The simpler Ebers–Moll large-signal model is obtained, when certain parameters are left unspecified. The complete large-signal and small-signal models consider parasitic capacitors and resistors within the bipolar transistors shown in Fig. 5.3.

The complete small-signal equivalent circuit of a bipolar transistor considering the parasitic capacitors and resistors is shown in Fig. 5.4.

The parameter c_π not shown in Fig. 5.3 is the sum of the base-charging capacitance $c_{\mathrm{b}} = \tau_{\mathrm{F}} g_{\mathrm{m}}$ and the emitter-base depletion layer capacitance c_{je} [240]. For the diffusion transistor with a constant base doping concentration, the base transit time τ_{F} is equal to $Q_{\mathrm{E}}/I_{\mathrm{C}} = W_{\mathrm{B}}^2/(2D_{\mathrm{N}})$, where Q_{E} is the minority-carrier charge in the base, W_{B} is the base width, and D_{N} is the minority-carrier diffusion coefficient. The base-collector capacitance c_μ is the sum of $c_{\mu 1}$ and $c_{\mu 2}$ shown in Fig. 5.3. The collector-substrate capacitance c_{cs} is the sum of c_{cs1}, c_{cs2}, and c_{cs3}. The parameters r_π and r_μ are obtained easily:

$$r_\pi = \frac{\beta}{g_{\mathrm{m}}}, \tag{5.2}$$

and

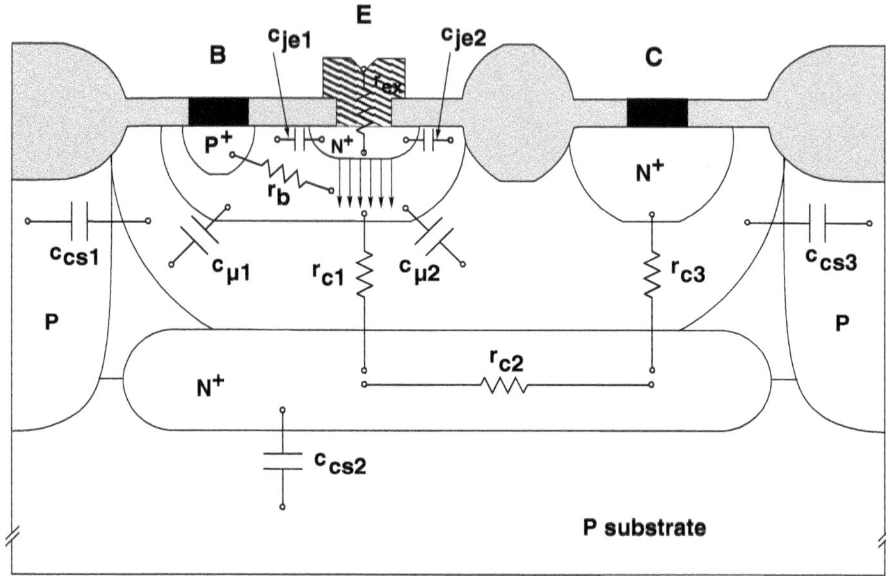

Fig. 5.3. Integrated-circuit NPN bipolar transistor structure including parasitic elements [240]

$$r_{\mu} = \frac{\Delta V_{\mathrm{CE}}}{\Delta I_{\mathrm{B}}} = \frac{\Delta V_{\mathrm{CE}}}{\Delta I_{\mathrm{C}}} \frac{\Delta I_{\mathrm{C}}}{\Delta I_{\mathrm{B}}} = r_{\mathrm{o}} \beta, \tag{5.3}$$

where r_{o} is the small-signal output resistance. The collector series resistance r_{c} is the sum of r_{c1}, r_{c2}, and r_{c3} (compare with Fig. 5.3). The emitter series resistance is represented by r_{ex}. The base series resistance r_{b} is one of the most important parameters.

The values for the parameters shown in Fig. 5.4 are available in the parameter file from the manufacturers of ASICs for several predesigned transistors.

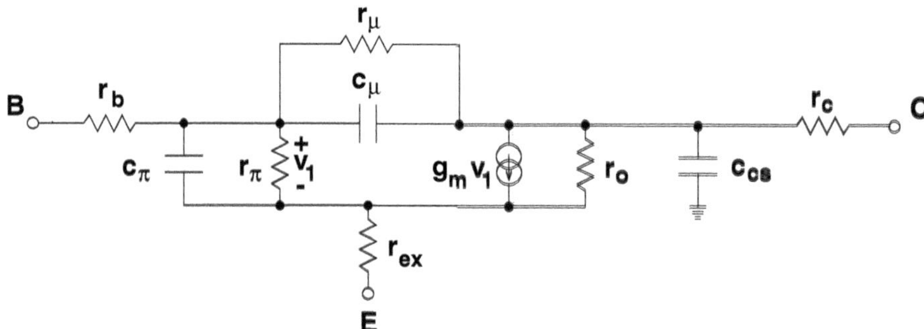

Fig. 5.4. Complete bipolar transistor small-signal equivalent circuit in a common-emitter circuit [240]

In addition to typical parameter values, parameter values for the so-called slow and fast cases are also available. These extreme parameter values have to be considered in order to guarantee the correct function of a circuit within the complete specified range for the fabrication process. Furthermore, the influence of the temperature on the performance of a circuit has to be investigated within the specified operating temperature range. For MOS circuits, the bandwidth usually reduces with increasing temperature and increases for temperatures lower than room temperature. Simulations have to be made in order to check whether the bandwidth is sufficient for the highest specified operating temperature and whether the circuit is immune against oscillations for the lowest specified operating temperature. The worst cases for MOS circuits are the combination of the highest specified operating temperature with the slow MOS transistor parameter values and the combination of the lowest specified operating temperature with the fast MOS transistor parameter values.

The current in bipolar transistors increases with temperature resulting in an increase of the circuit bandwidth with temperature. Therefore, the slow bipolar transistor parameter values have to be combined with the lowest specified operating temperature and the fast bipolar transistor parameter values have to be combined with the highest specified operating temperature in order to consider the worst cases for bipolar circuits.

In addition to the differences in the worst case behavior between MOSFETs and bipolar transistors, there are fundamental differences in the amplification and frequency behavior of MOS and bipolar transistors as well as analog MOS and bipolar circuits, which result from the fundamental difference of the output characteristics of the bipolar transistor $I_{\mathrm{C}}(U_{\mathrm{CE}}, U_{\mathrm{BE}})$ in the forward active region and of the MOS transistor $I_{\mathrm{D}}(U_{\mathrm{DS}}, U_{\mathrm{GS}})$ in the saturation region:

$$I_{\mathrm{C}} = I_{\mathrm{S}} \left(1 + \frac{U_{\mathrm{CE}}}{U_{\mathrm{Ea}}}\right) \exp[U_{\mathrm{BE}}/(k_{\mathrm{B}}T/q)], \tag{5.4}$$

$$I_{\mathrm{D}} = \frac{1}{2}\mu_{\mathrm{n}} C'_{\mathrm{ox}} \frac{W}{L}(1 + \lambda_{\mathrm{ch}} U_{\mathrm{DS}})(U_{\mathrm{GS}} - U_{\mathrm{Th}})^2. \tag{5.5}$$

These equations consider the Early effect, i. e. the increase of I_{C} with U_{CE} due to a decrease in the base width, by the incorporation of the Early voltage U_{Ea} and the channel length modulation effect, i. e. the increase in I_{D} because of the decrease in the effective channel length with an increase in the drain space-charge region width due to an increase in U_{DS} by the incorporation of λ_{ch}.

The differences between bipolar and MOS analog circuits are caused by the exponential dependence of the collector current I_{C} on the base voltage U_{BE} and by the quadratic dependence of the drain current I_{D} on the gate voltage U_{GS}. The transconductance of a bipolar transistor, therefore, is

$$g_{\mathrm{m}} = \frac{\partial I_{\mathrm{C}}}{\partial U_{\mathrm{BE}}} = \frac{I_{\mathrm{C}}}{k_{\mathrm{B}}T/q}, \tag{5.6}$$

whereas the transconductance of a MOSFET is

$$g_{\mathrm{m}} = \frac{\partial I_{\mathrm{D}}}{\partial U_{\mathrm{GS}}} = \frac{2I_{\mathrm{D}}}{U_{\mathrm{GS}} - U_{\mathrm{Th}}}. \tag{5.7}$$

When we assume that $I_{\mathrm{C}} = I_{\mathrm{D}}$ and that $U_{\mathrm{GS}} - U_{\mathrm{Th}} = 0.5\,\mathrm{V}$, then the transconductance of the bipolar transistor is 10 times larger than that of the MOSFET, because $k_{\mathrm{B}}T/q = 25\,\mathrm{mV}$. This means that bipolar amplifiers have a much larger amplification factor than MOS amplifiers, since the amplification factor is proportional to the transconductance for most types of amplifiers. Capacitive loads can be charged more rapidly by bipolar than by MOS transistors and the bipolar transistor is said to have the better driver capability. The bandwidth of bipolar amplifiers as a consequence is larger than the bandwidth of MOS amplifiers. For many operational amplifiers, as an example, the gain-bandwidth-product is proportional to the transconductance of the input transistor [235].

The design frameworks offer schematic entry for the definition of circuit diagrams, i. e. the circuit diagram can be drawn by placing device symbols and connecting them. Within the CADENCE framework, the schematic entry program is called COMPOSER and within the MENTOR framework there is the schematic entry program DESIGN ARCHITECT. It is, therefore, not necessary for a designer to write a netlist. The circuit simulator derives the netlist from the schematic. In such a way the number of possible errors is minimized.

5.2 Layout and Verification Tools

When the required specifications are met with the circuit diagram according to circuit simulations, the next step is the layout of the integrated circuit. The frameworks contain the layout tools necessary for this task. The chip manufacturers also make the process definition and design rule files available to the customer or design house. These files are necessary for the design rule check (DRC) of the layout in order to detect violations and to be able to correct them.

When the design rule check does not report any errors, there is still the possibility of deviations from the circuit diagram. Therefore, tools are included within the frameworks for the comparison of layout and schematic. These layout-versus-schematic (LVS) tools require the design rule file. LVS is one part of the design verification. A further verification tool is included within each framework, which enables the extraction of parasitic capacitances and resistances of interconnects from the layout. These parasitics were not included in the circuit diagram for prelayout circuit simulations. They can, however, deteriorate the behavior of the circuit. Therefore, the extraction tools allow us to add these parasitics to the circuit diagram or to include them in a netlist. In such a way the so-called postlayout simulations become possible. Usually, the bandwidth of the circuit calculated by postlayout simulations will be lower than initially determined by the prelayout simulations.

Since the OEICs described here are optimized (full custom) designs, no library cells could be used for the amplifiers, i. e. all mask levels had to be designed, and the layout was done mostly manually.

5.3 Design of OEICs

Figure 5.5 shows the flowchart for the development of OEICs. Starting from the technological flowchart of a bipolar, CMOS, or BiCMOS process, the PN junction of a diode or the two PN junctions of a transistor have to be selected for a photodetector. After this selection, the topology and the doping profiles of the photodetector can be calculated using a one- or two-dimensional process simulation program considering the technological flowchart. Such programs are, for instance, SUPREM3 [241] and TSUPREM4 [242]. This topology and these doping profiles can serve as an input for a device simulator, which offers a model for photogeneration and optical transmission of isolation and passivation stacks. Such a device simulator is, for instance, MEDICI [5]. For the device simulations, the operating conditions like photodetector bias, operating temperature, and optical wavelength have to be considered. With the help of the device simulations, rise and fall times of the detector photocurrent, bandwidth, quantum efficiency, and capacitance C_D (compare Fig. 1.17) can be extracted. For the example of the integration of PIN photodiodes in a CMOS process, the results of the device simulations were used to develop process modules for the implementation of shallow junctions in order to increase the internal quantum efficiency of the PIN photodiodes and for the implementation of an antireflection coating in order to increase the optical quantum efficiency. The device simulations also showed that the series resistance of the PIN photodiodes, which would result from the use of surface substrate contacts, can be eliminated by a back contact.

The most important aspect in the design flowchart of OEICs is the combination of photodetector device simulation and circuit simulation. Experience shows that the use of C_D and of a ramp with the rise time or fall time for the photocurrent leads to good results for transient circuit simulations (Fig. 5.6). For AC simulations of the frequency response, it is also sufficient to use the photodetector capacitance C_D. It should, however, be noted that a DC offset current has to be added to the AC current source, because negative photocurrents are not possible. C_D has to be added to the schematic of the amplifier and appropriate current sources have to be defined for the circuit simulations.

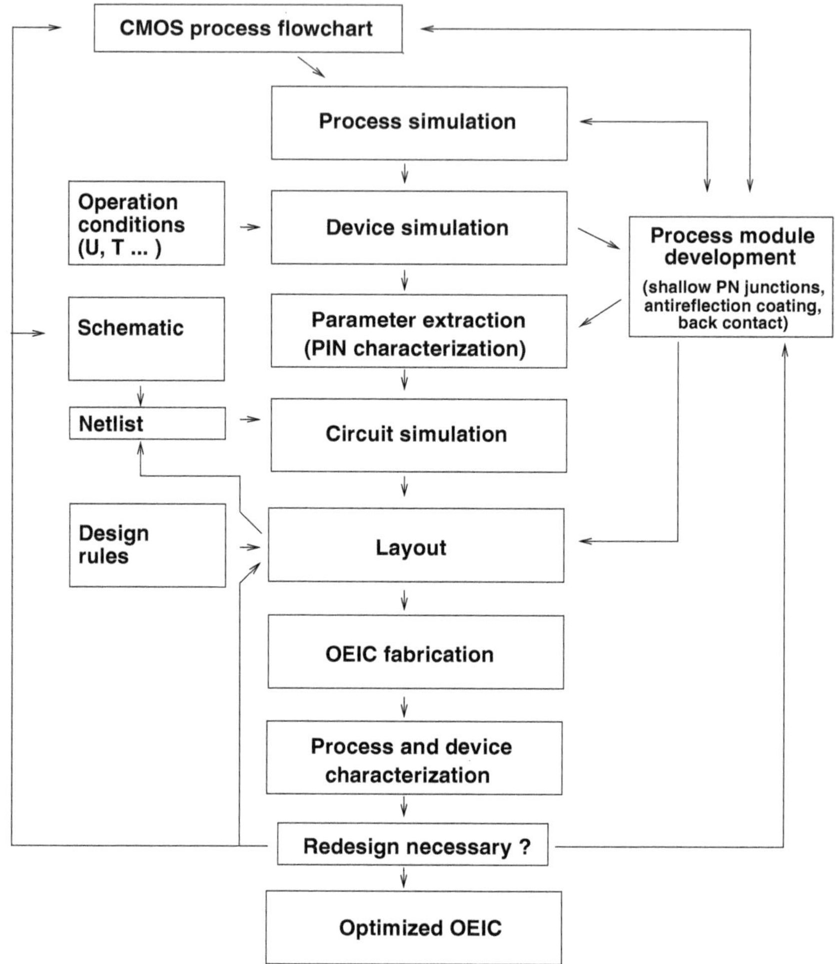

Fig. 5.5. Design flowchart for the development of OEICs. Here the integration of PIN photodiodes in CMOS OEICs is taken as an example

For some photodetectors, C_D alone may not be sufficient and it may be necessary to include the series resistor R_S (see Figs. 1.17 and 5.6)). It should be mentioned that this model may not be sufficient, when diffusion of photogenerated carriers is present in the photodetector. For this case, it may be suggested to store the photocurrent values calculated by the device simulator as a function of time in a file and to read this $I_{ph}(t)$ file by the circuit simulator in order to simulate the transient response of the OEIC. The method being most convenient concerning the combination of device and circuit simulation would be to simulate the OEIC, i. e. the photodetector together with the amplifier, using a circuit simulator on the device level. This method, how-

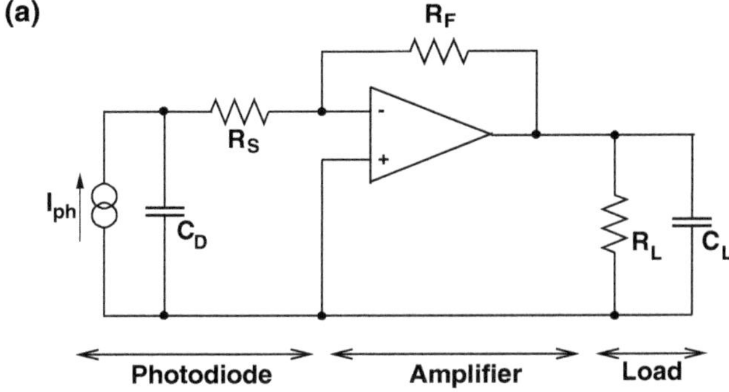

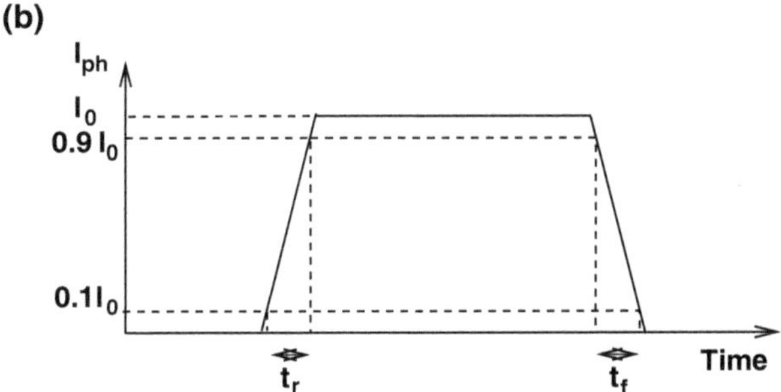

Fig. 5.6. (**a**) Equivalent circuit of a photodiode for the circuit simulation of an OEIC. (**b**) Photocurrent versus time for the consideration of rise and fall times obtained by device simulations or by measurements

ever, would be very CPU time intensive when all transistors in the amplifier have to be considered by device simulations like the photodetector.

Let us return to the design flowchart of OEICs (Fig. 5.5). Possibly the schematic of the amplifier has to be changed in order to meet the specifications of the OEICs. When these prelayout simulations predict that the specifications can be fulfilled, the layout can be drawn. For the integration of photodetectors, it is, of course, necessary to obey the design rules of the process and to add as few new design rules as possible for new process modules like the antireflection coating for instance. Due to the reduction of the doping concentration in the N^- epitaxial layer for the speed enhancement of the CMOS-integrated PIN photodiodes, the minimum distance of analog P-type wells had to be increased in order to avoid reach-through currents between P-type wells with different potentials. When the layout is finished, a design rule check is performed. After the correction of possible errors, the layout is compared with the netlist (layout versus schematic) for the detection of possible errors. Finally, when no errors are present or after they are corrected, parasitic capacitors of metal and polysilicon interconnects as well as series resistances of these interconnects are extracted and added to the netlist. For metal interconnects usually only the capacitances are extracted, because their resistances are negligible. Now, the postlayout simulation can be performed. When the specifications for the OEIC are no longer met, an iteration becomes necessary. Changes in the layout, in the netlist, or in the schematic may be necessary. Usually, it will be rather unlikely that the detector has to be modified. When all specifications are fulfilled, the so-called chip finishing is done, i. e. alignment masks and process control modules are added and arranged together with the chip layout on the reticles or on a wafer-size area. Then the masks are fabricated and the chips can be processed.

The OEICs produced have to be characterized. Information obtained by the process monitoring with the help of the process control modules will be used for the interpretation of the OEIC performance. The measured results have to be compared with the specifications for the OEIC. When all knowledge and know-how concerning the design of non-optoelectronic analog integrated circuits has been considered, the design of OEICs should be successful after the first OEIC fabrication as is most desirable and as is usually achieved in industrial chip design. When certain specifications are not met, a correction of the layout, of the schematic, of the process modules, or of the steps for photodetector integration in the technological flowchart may have to be made.

In the example of the PIN CMOS integration, the process modules, and the technological flowchart were developed perfectly at the first attempt. A redesign was only necessary to increase the sensitivity of the OEICs, i. e. of the active load resistors connected to the photodiodes, and to improve the group delay flatness in the specified frequency range [85].

5.4 Transimpedance Amplifier

A current/voltage (I/U) converter is also called a transimpedance circuit, because an output voltage being proportional to the input current is generated similarly as at an impedance. Each resistor is a current/voltage converter, with its output impedance being equal to its resistor value. In an active current/voltage converter, however, the output impedance is determined by an operational amplifier and, therefore, is independent of the proportionality factor between I and U.

A real current source can be split into an ideal current source with an impedance equal to infinity and into a parallel impedance $Z_{\mathrm{D}} = R_{\mathrm{D}} \| C_{\mathrm{D}}$. Without a series resistance, i. e. $R_{\mathrm{S}} = 0$, the equivalent circuit of a photodiode (Fig. 1.17) is equal to the equivalent circuit of the real current source in Fig. 5.7. In fact, the circuit in Fig. 5.7 is the most common amplifier used together with photodiodes. The compensation capacitor C_{F} is needed to avoid gain-peaking or to guarantee stability. The implementation of an operational transimpedance amplifier configuration is also very interesting for the optoelectronic integration especially when low offset voltages compared to a ref-

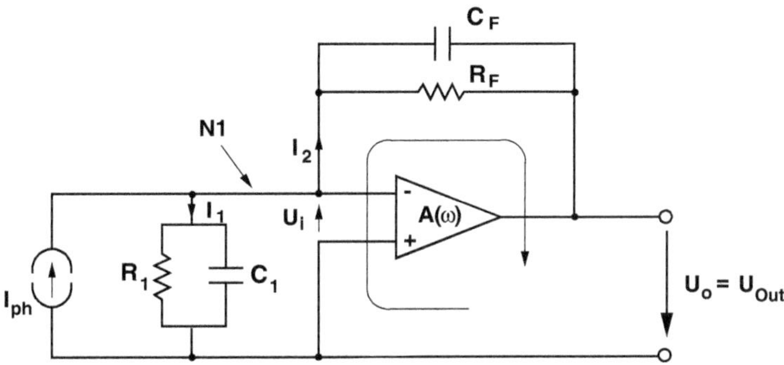

Fig. 5.7. Operational amplifier in transimpedance configuration for I/U conversion of a photocurrent

erence voltage are required. The properties of this configuration, therefore, will be discussed in detail here.

5.4.1 Frequency Response

The frequency response function $G(\omega)$ for the circuit shown in Fig. 5.7 can be derived thus:

$$U_\mathrm{i} + I_2 Z_\mathrm{F} + U_\mathrm{o} = 0, \tag{5.8}$$

$$I_\mathrm{ph} - I_1 - I_2 = 0 \tag{5.9}$$

with $Z_\mathrm{D} = R_\mathrm{D} \| C_\mathrm{D}$, $Z_\mathrm{F} = R_\mathrm{F} \| C_\mathrm{F}$, $U_\mathrm{i} = U_\mathrm{o}/A(\omega)$, and $I_1 = U_\mathrm{i}/Z_\mathrm{D} = U_\mathrm{o}/(A(\omega)Z_\mathrm{D})$ where $A(\omega)$ is the frequency dependent amplification factor of the operational amplifier. Inserting U_i into (5.8), results in

$$Z_\mathrm{F} I_2 + \left(1 + \frac{1}{A(\omega)}\right) U_\mathrm{o} = 0. \tag{5.10}$$

Inserting I_1 into (5.9), multiplying both sides of the equation by Z_F, and adding to (5.10) leads to the transfer function, which is the *transimpedance* and possesses the dimension Ω:

$$\frac{U_\mathrm{o}}{I_\mathrm{ph}} = G(\omega) = -\frac{Z_\mathrm{F}}{1 + (1/A(\omega))(1 + (Z_\mathrm{F}/Z_\mathrm{D}))}. \tag{5.11}$$

When we insert the frequency-dependent amplification of the operational amplifier $A(\omega)$ with the transit frequency ω_T

$$A(\omega) = \frac{A_0}{1 + \mathrm{i} A_0 \omega/\omega_\mathrm{T}} \tag{5.12}$$

into (5.11), we obtain with $A_0 >> 1$:

$$\frac{U_o}{I_{ph}} = G(\omega) = -\frac{Z_F}{1 + (Z_F/A_0 Z_D) + i(\omega/\omega_T)(1 + (Z_F/Z_D))}. \tag{5.13}$$

After the insertion of

$$Z_D = R_D \| C_D = \frac{R_D}{1 + i\omega R_D C_D} \tag{5.14}$$

and

$$Z_F = R_F \| C_F = \frac{R_F}{1 + i\omega R_F C_F}, \tag{5.15}$$

the transimpedance is:

$$G(\omega) = -\frac{R_F}{1 - (\omega^2/\omega_T) R_F (C_F + C_D) + i\omega[(1/\omega_T) + R_F(1/(R_D\omega_T) + C_F + C_D/A_0)]}. \tag{5.16}$$

For an ideal operational amplifier with $A_0 = \infty$ and $\omega_T = \infty$, (5.16) can be simplified considerably:

$$G(\omega) = -\frac{R_F}{1 + i\omega R_F C_F}. \tag{5.17}$$

This ideal transfer function obviously is that of a first order low pass filter. The −3dB frequency, therefore, is $f_{3dB} = 1/(2\pi R_F C_F)$. This is not surprising for the ideal operational amplifier, when we look at Fig. 5.7. This assumption of the ideal operational amplifier, however, is too simple an approximation of the real behavior of transimpedance amplifiers. The gain-peaking which occurs often in practice, for instance, is not included in the ideal transfer function. Especially for photodiodes with large capacitance C_D, the ω^2 term in the denominator of (5.16) must not be neglected. For a more thorough analysis, we therefore introduce the following abbreviations: $a = (1/\omega_T) + R_F[(1/R_D\omega_T) + C_F + (C_D/A_0)]$ and $b = (R_F/\omega_T)(C_F + C_D)$. The transfer function for the real operational amplifier then is:

$$\frac{U_o}{I_{ph}} = G(\omega) = -\frac{R_F}{1 - \omega^2 b + i\omega a}. \tag{5.18}$$

The magnitude of the transfer function can be determined from (5.18) to be:

$$\frac{|U_o|}{|I_{ph}|} = |G(\omega)| = \frac{R_F}{\sqrt{(1 - \omega^2 b)^2 + (\omega a)^2}}. \tag{5.19}$$

Equation (5.19) can have a maximum, i.e. a gain-peak (GP), when $(1 - \omega^2 b) = 0$, i. e. when $\omega_{GP} = \sqrt{1/b}$. The frequency for which the gain-peaking occurs then is

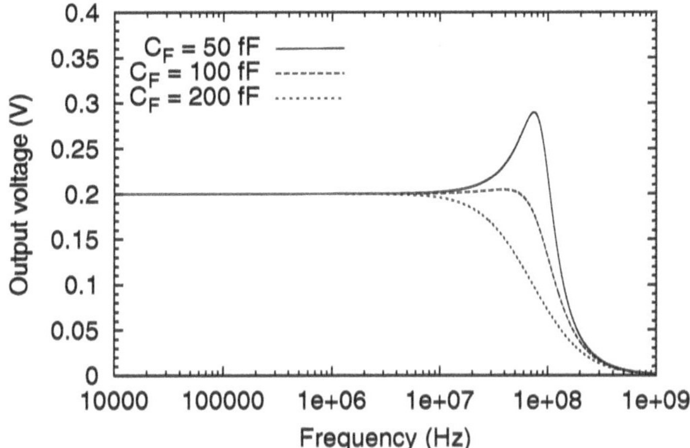

Fig. 5.8. Frequency response of an operational amplifier in transimpedance configuration for three different values of the feedback capacitance C_{F} ($I_{\mathrm{ph}} = 10\,\mu\mathrm{A}$, $R_{\mathrm{F}} = 20\,\mathrm{k}\Omega$, $R_{\mathrm{D}} = 10^9\,\Omega$, $\omega_{\mathrm{T}} = 6.28\,\mathrm{GHz}$ ($f_{\mathrm{T}} = 1\,\mathrm{GHz}$), $C_{\mathrm{D}} = 1\,\mathrm{pF}$, and $A_0 = 100$) calculated with (5.19)

$$f_{\mathrm{GP}} = \frac{\omega_{\mathrm{GP}}}{2\pi} = \frac{1}{2\pi\sqrt{b}} = \frac{1}{2\pi}\sqrt{\frac{\omega_{\mathrm{T}}}{R_{\mathrm{F}}(C_{\mathrm{F}} + C_{\mathrm{D}})}}. \tag{5.20}$$

If the gain-peaking is not desired, it can be eliminated for a photodiode as a current source by increasing C_{F}, because in this case C_{F} has a stronger effect on a than on b. This elimination of the gain-peaking, however, is at the expense of a reduced -3dB bandwidth (see Fig. 5.8).

For $C_{\mathrm{F}} = 50\,\mathrm{fF}$ (or less) a large gain-peak occurs at about 70 MHz. With $C_{\mathrm{F}} = 100\,\mathrm{fF}$, the gain-peak is already almost suppressed. Increasing C_{F} to 200 fF suppresses the gain-peak completely. The -3dB bandwidths for the three cases are about 120 MHz, 90 MHz, and 50 MHz, respectively. The case with $C_{\mathrm{F}} = 200\,\mathrm{fF}$ can practically be considered as over-compensated.

The influence of the depletion capacitance C_{D} of a photodiode on the frequency response is shown in Fig. 5.9. The gain-peak vanishes, when the capacitance of the photodiode is reduced from 1 pF to 0.1 pF. At the same time, the -3dB bandwidth increases from about 120 to 180 MHz. We can conclude that a PIN photodiode with a capacitance of the order of 100 fF is advantageous compared to a double photodiode with a capacitance of the order of 1 pF. In order to understand the advantage of a transimpedance amplifier compared to a voltage follower configuration (compare with Fig. 1.16) with the same sensitivity, we calculate the bandwidth of the load resistor and of the photodiode capacitance $f_{\mathrm{g}} = 1/(2\pi R_{\mathrm{F}} C_{\mathrm{D}})$. For $R_{\mathrm{F}} = 20\,\mathrm{k}\Omega$ and $C_{\mathrm{D}} = 100\,\mathrm{fF}$, we obtain the bandwidth $f_{\mathrm{g}} = 79.6\,\mathrm{MHz}$ for the voltage follower configuration. This bandwidth of the voltage follower is much lower than the bandwidth of about 180 MHz for the transimpedance configuration with

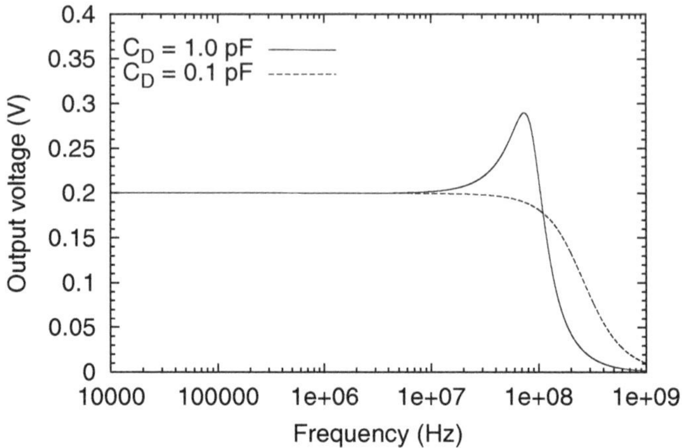

Fig. 5.9. Frequency response of an operational amplifier in transimpedance configuration for two different values of the photodiode capacitance C_D ($I_\mathrm{ph} = 10\,\mu\mathrm{A}$, $R_\mathrm{F} = 20\,\mathrm{k}\Omega$, $R_\mathrm{D} = 10^9\,\Omega$, $\omega_\mathrm{T} = 6.28\,\mathrm{GHz}$ ($f_\mathrm{T} = 1\,\mathrm{GHz}$), $C_\mathrm{F} = 50\,\mathrm{fF}$, and $A_0 = 100$) calculated with (5.19)

$R_\mathrm{F} = 20\,\mathrm{k}\Omega$ and $C_\mathrm{D} = 100\,\mathrm{fF}$. This example clearly demonstrates the advantage of the transimpedance amplifier. This advantage is due to the amplification factor $A(\omega)$, which reduces the effective photodiode capacitance in the transimpedance configuration with the help of the operational amplifier. Let us now return to the theory of the transimpedance amplifier.

The bandwidth can be determined from the condition

$$|G(\omega)| = G(\omega = 0)/\sqrt{2} = R_\mathrm{F}/\sqrt{2}, \tag{5.21}$$

which leads to

$$(1 - \omega_\mathrm{g}^2 b)^2 + (\omega_\mathrm{g} a)^2 = 2. \tag{5.22}$$

The single positive, real solution of this equation is [243]

$$f_\mathrm{g} = \frac{\omega_\mathrm{g}}{2\pi} = \frac{1}{2\pi}\sqrt{\frac{1}{2b^2}(2b - a^2) + \sqrt{\frac{1}{4b^4}(2b - a^2)^2 + \frac{1}{b^2}}}. \tag{5.23}$$

The bandwidth of the transimpedance configuration is not simply proportional to $1/R_\mathrm{F}$. Therefore, the frequency responses of the transimpedance configuration with $R_\mathrm{F} = 20\,\mathrm{k}\Omega$ and $R_\mathrm{F} = 40\,\mathrm{k}\Omega$ were calculated and compared in Fig. 5.10 in order to show the influence of R_F on the bandwidth. For $R_\mathrm{F} = 40\,\mathrm{k}\Omega$ the bandwidth reduces to about 100 MHz compared to 180 MHz for $R_\mathrm{F} = 20\,\mathrm{k}\Omega$.

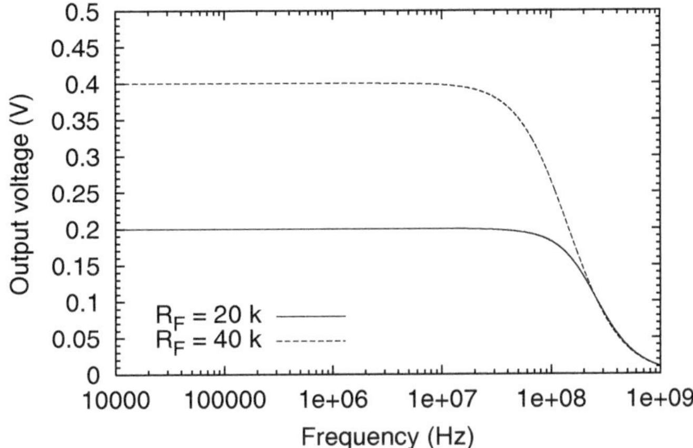

Fig. 5.10. Frequency response of an operational amplifier in transimpedance configuration for two different values of the feedback resistor R_F ($I_{ph} = 10\,\mu A$, $R_D = 10^9\,\Omega$, $\omega_T = 6.28\,GHz$ ($f_T = 1\,GHz$), $C_D = 100\,fF$, $C_F = 50\,fF$, and $A_0 = 100$) calculated with (5.19)

5.4.2 Phase and Group Delay

The phase, in degrees, can be obtained in general from

$$\Phi(\omega) = \frac{180^\circ}{\pi} \arctan\left(\frac{\mathrm{Im}(G(\omega))}{\mathrm{Re}(G(\omega))}\right). \tag{5.24}$$

For the operational amplifier in transimpedance configuration, we obtain

$$\Phi(\omega) = -\frac{180^\circ}{\pi} \arctan\left(\frac{\omega a}{1-\omega^2 b}\right) + 180^\circ. \tag{5.25}$$

In order to account for the inverted output signal, 180° are added for all frequencies. The pole in the argument of the arctan for $\omega = 1/\sqrt{b} = \omega_g$ causes a 180° step in the phase due to the periodicity of the arctan function. Fig. 5.11 shows the phase response of the operational amplifier in transimpedance configuration for the same parameters as in Fig. 5.8.

The influence of the photodiode capacitance on the phase response can be seen in Fig. 5.12. The phase does not change as much for a lower photodiode capacitance than for a larger photodiode capacitance when the frequency increases. A PIN photodiode again is advantageous compared to a double photodiode.

For certain applications of amplifiers, the group delay has to be constant in a wide frequency range. We, therefore, will discuss the group delay t_{GD}, which we define here as $t_{GD} = -\pi/180^\circ \cdot \partial\Phi(\omega)/\partial\omega$, for the transimpedance amplifier. First we introduce the substitution $z = \omega a/(1-\omega^2 b)$. Knowing that $\partial \arctan(z)/\partial z = 1/(1+z^2)$, we obtain from (5.25):

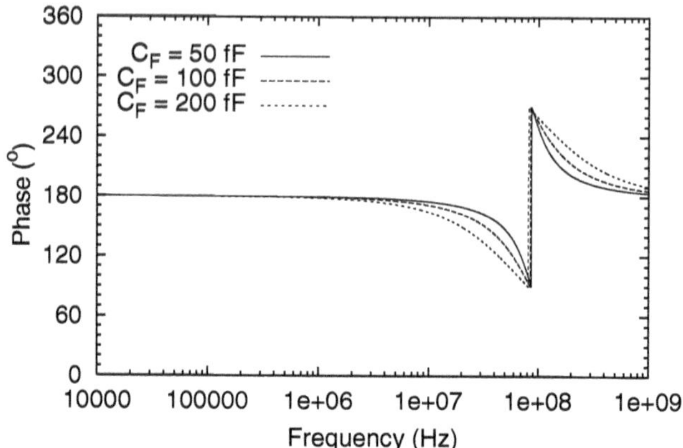

Fig. 5.11. Phase response of an operational amplifier in transimpedance configuration for three different values of the feedback capacitance C_F ($I_{ph} = 10\,\mu A$, $R_F = 20\,k\Omega$, $R_D = 10^9\,\Omega$, $\omega_T = 6.28\,GHz$ ($f_T = 1\,GHz$), $C_D = 1\,pF$, and $A_0 = 100$) calculated with (5.25)

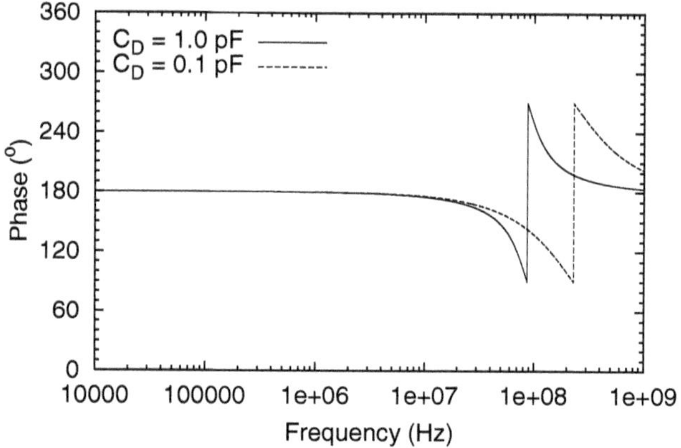

Fig. 5.12. Phase response of an operational amplifier in transimpedance configuration for two different values of the photodiode capacitance C_D ($I_{ph} = 10\,\mu A$, $R_F = 20\,k\Omega$, $R_D = 10^9\,\Omega$, $\omega_T = 6.28\,GHz$ ($f_T = 1\,GHz$), $C_F = 50\,fF$, and $A_0 = 100$) calculated with (5.25)

$$t_{GD} = \frac{a(1 + b\omega^2)}{(1 - b\omega^2)^2 + (a\omega)^2}. \qquad (5.26)$$

Figure 5.13 shows the group delay for the three cases depicted in Fig. 5.8. There is a peak in the group delay at f_{gp} for $C_F = 50\,fF$ and $C_F = 100\,fF$. This peak is suppressed with $C_F = 200\,fF$ leading to a decrease already at rather low frequencies. The optimum C_F value for a good group delay behavior lies between 100 and 200 fF.

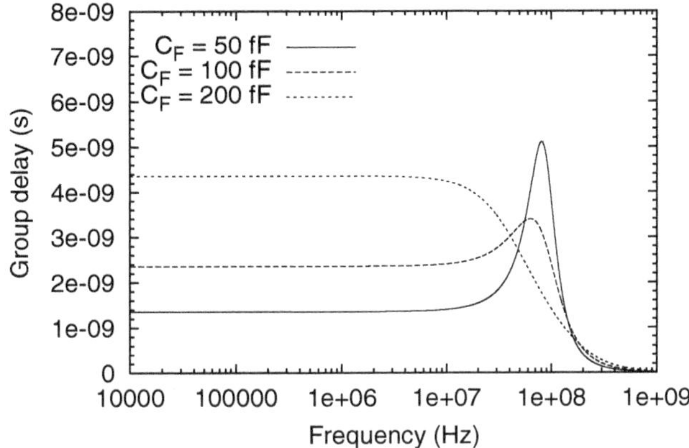

Fig. 5.13. Group delay of an operational amplifier in transimpedance configuration for three different values of the feedback capacitance C_{F} ($I_{\mathrm{ph}} = 10\,\mu\mathrm{A}$, $R_{\mathrm{F}} = 20\,\mathrm{k}\Omega$, $R_{\mathrm{D}} = 10^9\,\Omega$, $\omega_{\mathrm{T}} = 6.28\,\mathrm{GHz}$ ($f_{\mathrm{T}} = 1\,\mathrm{GHz}$), $C_{\mathrm{D}} = 1\,\mathrm{pF}$, and $A_0 = 100$) calculated with (5.26)

The influence of the photodiode capacitance on the group delay is illustrated in Fig. 5.14. For the large capacitance of a double photodiode of the order of 1 pF, a large peak results in the group delay. For the low capacitance of a PIN photodiode of the order of about 100 fF, a constant group delay is obtained for frequencies of up to 100 MHz. The PIN photodiode again is advantageous compared to the double photodiode.

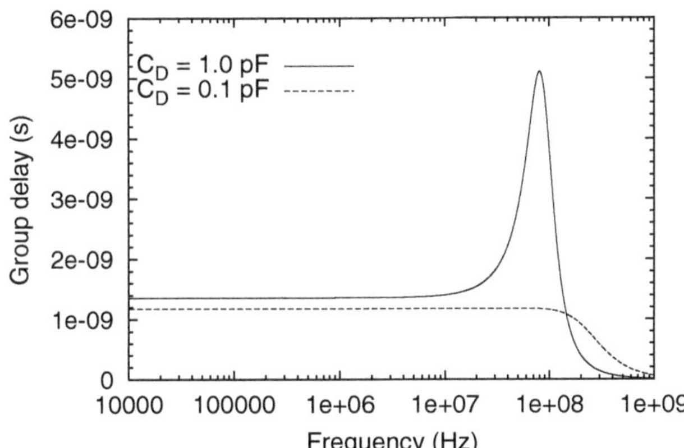

Fig. 5.14. Group delay of an operational amplifier in transimpedance configuration for two different values of the photodiode capacitance C_{D} ($I_{\mathrm{ph}} = 10\,\mu\mathrm{A}$, $R_{\mathrm{F}} = 20\,\mathrm{k}\Omega$, $R_{\mathrm{I}} = 10^9\,\Omega$, $\omega_{\mathrm{T}} = 6.28\,\mathrm{GHz}$ ($f_{\mathrm{T}} = 1\,\mathrm{GHz}$), $C_{\mathrm{F}} = 50\,\mathrm{fF}$, and $A_0 = 100$) calculated with (5.26)

5.4.3 Stability and Compensation

Above it was shown that a transfer function with two poles is obtained for the transimpedance amplifier. For such a function large values of C_D may lead to a so-called gain-peaking, an overshoot in the frequency response curve at high frequencies. Oscillations may result for a strong gain-peaking. A detailed analysis [244] leads to a value of

$$\begin{aligned} C_F &= \frac{\sqrt{2A_0\, R_F\, C_D\, \omega_c} - (1 + \omega_c\, R_F\, C_D)}{R_F\, \omega_c\, A_0} \\ &\approx \sqrt{\frac{2C_D}{R_F\, \omega_c\, A_0}} - \frac{C_D}{A_0} \end{aligned} \tag{5.27}$$

for the compensation capacitor, which is necessary in order to suppress gain-peaking. The symbol ω_c is the -3dB frequency of the amplifier; i. e. the gain of the amplifier can be expressed by $A(\omega) = A_0/(1 + i\omega/\omega_c)$. Usually equation (5.27) leads to unnecessarily high values of C_F [245]. If a phase margin of 45 ° is required, a thorough analysis of the transfer function [244] leads to

$$\begin{aligned} C_F &= \frac{\sqrt{\frac{1}{\sqrt{2}}\, A_0\, R_F\, C_D\, \omega_c} - (1 + \omega_c\, R_F\, C_D)}{R_F\, \omega_c\, A_0} \\ &\approx \sqrt{\frac{C_D}{\sqrt{2}\, R_F\, \omega_c\, A_0}} - \frac{C_D}{A_0}. \end{aligned} \tag{5.28}$$

In practical transimpedance amplifier implementations, for the feedback capacitance a value between those specified by equations (5.27) and (5.28) should be chosen.

5.4.4 Bandwidth of Transistor Transimpedance Amplifier Circuit

As we saw above, there is a too simple expression for the bandwidth of a transimpedance amplifier taking into account only the feedback resistor and the feedback capacitor: $f_{3dB} = 1/(2\pi R_F C_F)$. The bandwidth of a transimpedance amplifier can be influenced by other quantities. Let us assume a simple transistor transimpedance amplifier to investigate additional factors being important for its bandwidth. The circuit diagram of the simple transistor transimpedance amplifier is shown in Fig. 5.15. In the following we will derive an equation for its bandwidth, which considers the operating point of this amplifier.

In Fig. 5.16 a simplified small signal equivalent circuit of the transimpedance amplifier is depicted. Compared to R_1, the resistors r_{o1}, r_{c1} as well as the collector-substrate-capacitance C_{CS1} are negligible. Applying the Miller-theorem, $C_{\mu 1}$ can be converted to $C_M = C_{\mu 1}\,(1 + g_{m1}\, R_1) \approx 103\, C_{\mu 1}$

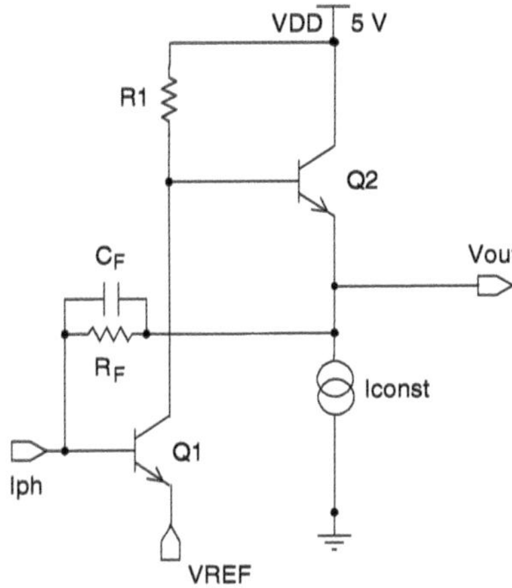

Fig. 5.15. Circuit diagram of a two-transistor transimpedance amplifier [244]

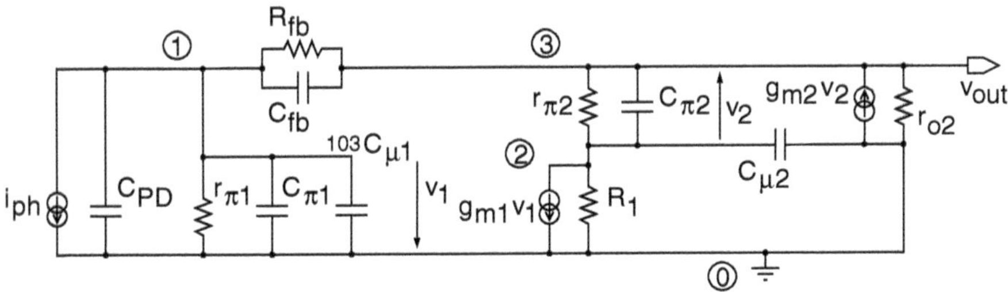

Fig. 5.16. Small-signal equivalent circuit of the first stage of the transimpedance amplifier depicted in Fig. 6.81 [244]. Node numbers are placed within circles.

($g_{\mathrm{m1}} = 1.92$ mS, $R_1 = 54.2\,\mathrm{k}\Omega$ for the amplifier shown in Fig. 6.81 [244]) parallel to $C_{\pi 1}$. Also r_{b2}, C_{CS2}, r_{b1} and r_{c2} need not be taken into consideration for the calculations.

For the derivation of the bandwidth, the matrix equation

$$\underbrace{\begin{pmatrix} -i_{\mathrm{ph}} \\ -g_{\mathrm{m1}}\, v_{\mathrm{n1}} \\ g_{\mathrm{m2}}\, v_{\mathrm{n2}} \end{pmatrix}}_{\mathcal{I}} = \mathcal{M} \cdot \underbrace{\begin{pmatrix} v_{\mathrm{n1}} \\ v_{\mathrm{n2}} \\ v_{\mathrm{out}} \end{pmatrix}}_{\mathcal{V}} \tag{5.29}$$

is set up according to the node potential method, where v_{out} is identical with v_{n3}. Here the resistors $r_{\pi 1}$, $r_{\pi 2}$ and r_{o2} are neglected due to their large values. $C_{\mu 2}$ is in the fF range and is also neglected.

In this circuit the matrix $\mathcal{M}$ can be written as

$$\mathcal{M} = \begin{pmatrix} \mathcal{M}_{11} & 0 & -\frac{1+\mathrm{i}\omega R_{\mathrm{F}} C_{\mathrm{F}}}{R_{\mathrm{F}}} \\ 0 & \mathrm{i}\omega\, C_{\pi 2} + G_1 & -\mathrm{i}\omega\, C_{\pi 2} \\ -\frac{1+\mathrm{i}\omega R_{\mathrm{F}} C_{\mathrm{F}}}{R_{\mathrm{F}}} & -\mathrm{i}\omega\, C_{\pi 2} & \mathcal{M}_{33} \end{pmatrix}, \tag{5.30}$$

with $\mathcal{M}_{11} = \mathrm{i}\omega\,(C_\mathrm{D} + C_{\pi 1} + (1 + g_{\mathrm{m}1}\,R_1)\,C_{\mu 1}) + (1 + \mathrm{i}\omega\,R_\mathrm{F}\,C_\mathrm{F})/R_\mathrm{F}$, $\mathcal{M}_{33} = \mathrm{i}\omega\,C_{\pi 2} + \frac{1+\mathrm{i}\omega\,R_F C_F}{R_F}$ and $G_1 = 1/R_1$. The voltages $v_{\mathrm{n}1}$, $v_{\mathrm{n}2}$, and v_out point from nodes 1 to 3 to the ground node 0 (see Fig. 5.16).

Using matrix

$$\mathcal{D} = \begin{pmatrix} 0 & 0 & 0 \\ -g_{\mathrm{m}1} & 0 & 0 \\ 0 & g_{\mathrm{m}2} & 0 \end{pmatrix}$$

the relationship

$$\mathcal{V} = (\mathcal{M} - \mathcal{D})^{-1} \cdot (-i_\mathrm{ph}\;0\;0\;0)^T \tag{5.31}$$

can be obtained.

Even now the complete transfer function $v_\mathrm{out}/i_\mathrm{ph}$ is too complicated for a complete analytical analysis. Therefore, the poles and zeros are calculated numerically. From the pole/zero diagram in Fig. 5.17 it can be seen that the zeros are located at $-56.6 \cdot 10^9\,\frac{1}{\mathrm{s}}$ and at $144.8 \cdot 10^9\,\frac{1}{\mathrm{s}}$. The poles can be found at $-0.47 \cdot 10^9\,\frac{1}{\mathrm{s}}$, $-(0.91 + i19.9) \cdot 10^9\,\frac{1}{\mathrm{s}}$ and at $-(0.91 - i19.9) \cdot 10^9\,\frac{1}{\mathrm{s}}$ [244]. Since the pole at $-0.47 \cdot 10^9\,\frac{1}{\mathrm{s}}$ is below the frequency of the other pole by a factor of about two and since the values for the zeros are comparatively high, it is safe to assume that this pole is dominant. Circuit simulations with a lowpass having the bandwidth f_1 and a second lowpass with the bandwidth $f_2 = 2 \times f_1$ in series resulted in a total bandwidth of $0.8 \times f_1$. The error caused by neglecting the second pole here, therefore, is about 20 %.

This means that in the denominator of the transfer function which can be obtained by solving equation (5.31), all terms containing higher order of the frequency ω can be eliminated. The remaining terms lead to the expression

$$f_{-3\,\mathrm{dB}} = \frac{1}{2\pi\left(R_\mathrm{F}\,C_\mathrm{F} + \frac{C_{\pi 2}}{g_{m2}} + \frac{1+g_{m2}R_F}{(1+g_{m1}R_1)g_{m2}}C_\mathrm{f}\right)} \tag{5.32}$$

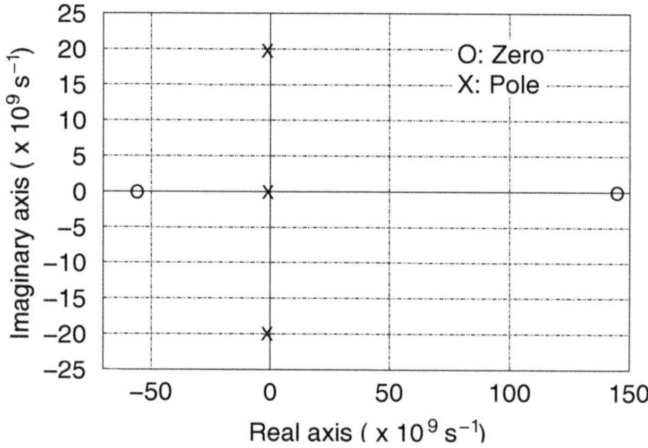

Fig. 5.17. Pole/zero diagram of the transimpedance amplifier

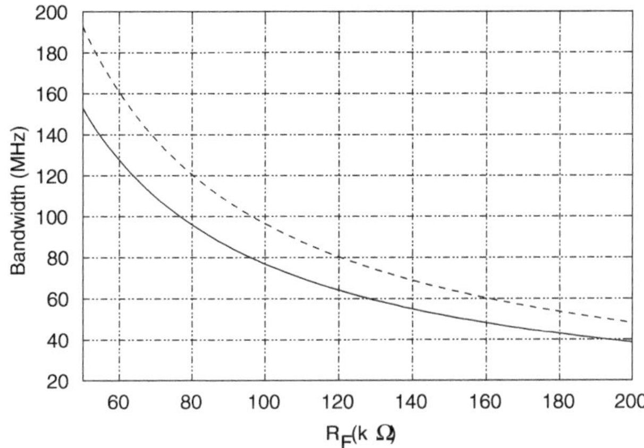

Fig. 5.18. Possible bandwidths of the transistor transimpedance amplifier as a function of R_F according to eq. (5.32) [solid line] and the simple expression $f_{3dB} = 1/(2\pi R_F C_F)$ [dashed line] for $C_F = 15\,fF$ and $C_D = 210\,fF$ [244]

with $C_f = (C_D + (1 + g_{m1}\,R_1)\,C_{\mu 1} + C_{\pi 1})$. It can be seen that the bandwidth depends on the values for the feedback capacitance C_F, on the resistor R_F as well as on the transistor parameters g_{m1}, g_{m2}, $C_{\mu 1}$, $C_{\pi 1}$, and $C_{\pi 2}$. In the same way the value of the photodiode capacitance C_D has to be taken into consideration in the equation. This means that when the photodiode area becomes lower, the bandwidth increases. The most important parameters are photodiode capacitance C_D, R_F, and C_F for a fixed operating point of the transistor amplifier. Fig. 5.18 shows the actual bandwidth according to eq. (5.32) in comparison with that calculated from the simple expression. Equation (5.32) considering the operating point of the TIA in Fig. 6.81 leads to about 25 % lower values than expected from the simple expression.

Inserting the transistor parameters for the nominal case taken from a circuit simulation at the operating point into Eq. (5.32), a bandwidth of 76 MHz can be calculated for the circuit in Fig. 6.81 in high sensitivity, i. e. switch S_1 open. A numerical postlayout simulation at a temperature of 25 °C and nominal model parameters revealed a 3 dB frequency of 68 MHz [244].

5.4.5 Bandwidth Limit of Integrated Transimpedance Amplifiers

In the following, we will derive an estimate for the maximum bandwidth of an integrated transimpedance amplifier resulting from R_F and its parasitic capacitance. An ideal operational amplifier is implicitly assumed. Furthermore it is presumed that C_F is not needed for compensation, i. e. that the parasitic capacitance of R_F is sufficient for compensation.

In analog integrated circuits, polysilicon resistors are used. These polysilicon resistors are fabricated on top of an oxide layer (field oxide, FOX)

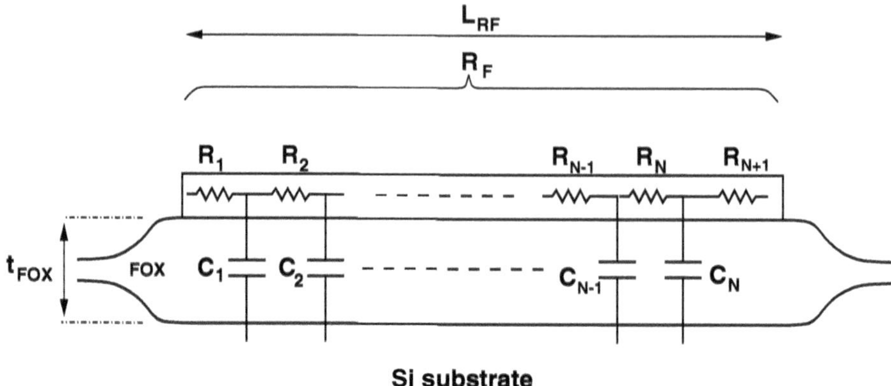

Fig. 5.19. R-C chain of an integrated polysilicon resistor

on the silicon substrate. It is therefore necessary to consider the polysilicon resistor like a chain of RC elements (see Fig. 5.19). The feedback resistor R_F, therefore, has to be split in $R_1 = 0.5R_F/N$, $R_i = R_F/N$ $(i=2\ldots N)$, $R_{N+1} = 0.5R_F/N$ and the total parasitic capacitance of R_F has to be split in $C_i = C_{RF}/N$, where $C_{RF} = A_{RF}/(\epsilon_{SiO2}\epsilon_0 t_{FOX})$ with the area $A_{RF} = L_{RF}\, W_{RF}$ of R_F. The value of R_F can be calculated from the sheet resistance $R_\square^{Poly}$ and the length L_{RF} and the width W_{RF} of R_F: $R_F = R_\square^{Poly} \cdot L_{RF}/W_{RF}$ assuming a linear polysilicon resistor. When R_F is meandered, for the corner squares, $R_\square^{Poly} \cdot 0.56$ has to be considered and A_{RF} has to be determined in a more complicated way. For simplicity, we, however, will assume a linear polysilicon resistor.

For the integrated resistor, the parasitic capacitance increases when the resistor value has to be increased (for $W_{RF}=$ constant, $L_{RF} \propto R_F$ and $A_{RF} \propto R_F$) in order to achieve a larger sensitivity S of the transimpedance amplifier. Let us, for simplicity, assume the time constant $\tau_{RF} = R_F \cdot C_{RF}$ for the polysilicon resistor. Then, when $R_F \propto S$ is increased, C_{RF} increases proportional to R_F and to S, resulting in $\tau_{RF} \propto S^2$. This leads to the conclusion that the bandwidth is proportional to $1/S^2$ for an integrated polysilicon resistor in a bipolar, CMOS, or BiCMOS transimpedance amplifier.

In order to illustrate this derived dependence of the bandwidth on S, circuit simulations have been performed with an R-C chain including $N = 10$ elements for $R_\square^{Poly} = 100\,\Omega_\square$ and $1000\,\Omega_\square$ with $R_F = 10\,k\Omega$, $20\,k\Omega$, and $40\,k\Omega$ for a polysilicon width $W_{RF} = 2\,\mu m$ and $C_{RF}/A_{RF} = 0.1\,fF/\mu m^2$. The fringe capacitance of the polysilicon resistor has been neglected in this example. The -3dB bandwidths obtained are listed in Table 5.1.

These results show that for $R_\square^{Poly} = 100\,\Omega_\square$, polysilicon resistors with values of $10\,k\Omega$ and $20\,k\Omega$ will not limit the bandwidths of integrated transimpedance amplifiers. For a value of $40\,k\Omega$, however, the polysilicon resistor reduces the bandwidth to 61 MHz i. e. below the value of 90 MHz, which

Table 5.1. Simulated bandwidths of R-C chain for an integrated polysilicon resistor

$R_\square$ ($\Omega_\square$)	$f_{3dB}(R_F = 10\,k\Omega)$ (MHz)	$f_{3dB}(R_F = 20\,k\Omega)$ (MHz)	$f_{3dB}(R_F = 40\,k\Omega)$ (MHz)
100	980	245	61
1000	9800	2450	610

was obtained for a BiCMOS OEIC for optical storage systems [176]. For $R_\square^{Poly} = 1000\,\Omega_\square$, finally, polysilicon resistors with values of up to $40\,k\Omega$ possess larger -3dB bandwidths than common integrated high-speed operational amplifiers and will, therefore, not limit the bandwidths of integrated transimpedance amplifiers.

5.4.6 New π-Model for Integrated Transimpedance Amplifier

A transmission-line approach was presented to derive a new π-model for integrated transimpedance amplifiers [246]. Due to their strong capacitive coupling to substrate ground, polysilicon resistors used in the feedback circuit contribute to the frequency response of the overall amplifier. Limitations in bandwidth, signal distortion and even instability may result. A novel, exact model of the resistor based on the transmission-line theory was suggested to describe the behavior of the overall transimpedance amplifier. A simple and descriptive approximation model, derived from the exact one, was used to find design formulas to optimize the overall amplifier performance. Though possible in general, this was demonstrated in case of an idealized amplifier with high gain, both using the exact and the simplified model to show its usefulness. The model was also verified for a real amplifier by comparison of simulated results for the approximation model and for the exact grounded uniformly distributed RC-line (URC) model.

OEICs are increasingly needed for optical storage (OS) systems like CD-ROM, Digital-Versatile-Disk (DVD) and Digital-Video-Recording (DVR) as well as for opto-electrical front-ends in fiber receivers. In both cases, the gain-bandwidth product, i. e. the transimpedance-bandwidth product is an important design parameter. In practice, the feedback resistors used in TransImpedance Amplifiers (TIAs) have values in the order of $100\,k\Omega$ for OS-OEICs down to the order of $1\,k\Omega$ for fiber receivers, while the bandwidth reaches from below 100 MHz to several GHz. Made of polysilicon, these integrated resistors have considerable stray capacitances to ground, i. e. they depend strongly on frequency. The parasitic stray capacitance of $100\,k\Omega$ feedback resistors in OS-OEICs has to be considered, if signal frequencies reach 100 MHz or more even when the sheet resistance of the (high-resistivity) polysilicon is on the order of 1 kΩ (see last section). Fiber receivers often have to be realized for system-on-chip applications in digital deep-sub-μm CMOS or in low-cost digital CMOS technologies where only gate polysilicon with a sheet resistance of much less than $100\,\Omega_\square$ is available. Therefore, the parasitic capacitance of feedback resistors in the order of several kΩ also has to be considered in fiber receivers for Gb/s-application. Capacitive coupling has to be taken into account, and its negative effects on stability are generally reduced using a feedback capacitor in parallel with the feedback resistor.

Transmission-line approaches up to now only were performed for metallic interconnects in chips. In [247], a so-called single grounded uniformly distributed RC line was described for integrator applications using a two-port formulation. An improved T-model and an improved Π-model consisting of one resistor and two capacitors was suggested for the modeling of on-chip interconnects [248]. A π model also consisting of one resistor and two capacitors was suggested to model the effective capacitance for the RC interconnect of CMOS gates [249]. An admittance parameter formulation was used to describe a capacitive double layers uniformly distributed RC line [250].

The goal of this section is to analyze the overall TIA taking the frequency-dependent feedback resistor into account. Using the transmission-line model, a simple approximation for the integrated feedback resistor, a novel π equivalent circuit will be derived. A very simple approximation from this model implying only a resistor and an inductor makes it possible to understand the basic behavior of the feedback circuit. A comparison with the exact model is also made. The new π equivalent circuit model as well as its simple approximation allow the optimization of an integrated transimpedance amplifier independent of and before using a circuit simulator.

Modeling Amplifier and Feedback Network

To study the amplifier with feedback network in general, it is necessary to introduce at first the major terms and parameters describing the nature of amplifier and feedback network. Fig. 5.20a shows a simple transimpedance amplifier consisting of the photodiode's current source $I(s)$ in parallel with the photodiode's capacitance C_D, the amplifier (with finite complex gain $A(s)$, with the input impedance Zin(s), and with the output impedance Zout(s)), and the feedback resistor R_F described by its y-parameters Y(s) in parallel with the feedback capacitor C_F. $Vin(s)$ and $Vout(s)$ are the input and output voltages, respectively, both depending on the complex frequency s. The behavior of the whole amplifier can now be described using an ideal amplifier shown in Fig. 5.20b and the transimpedance $Z(s)$

$$Z(s) = \frac{1}{\frac{y_{11m}(s)}{A(s)} - y_{12m}(s)} \,, \tag{5.33}$$

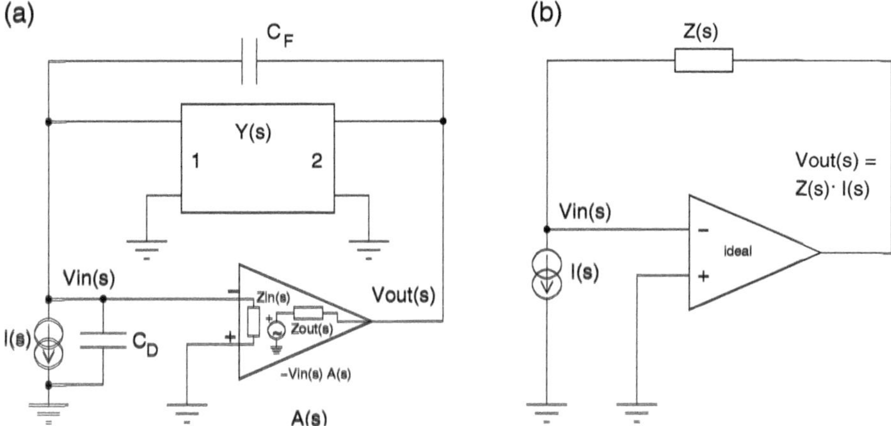

Fig. 5.20. Transimpedance amplifier consisting of real amplifier and modelled, non-ideal resistor **(a)**. Equivalent representation with ideal amplifier and the transimpedance Z(s) **(b)**

with the modified y-parameters Ym(s), consisting of the feedback resistor's y-parameters y(s), the feedback capacitor C_F, the photodiode capacitance C_D and the amplifier input impedance Zin(s).

The amplifier's output impedance Zout(s) is assumed to be small compared to the feedback network input impedance (which acts as a load). This means that

$$|Z_{\mathrm{out}}(s)| \ll \frac{1}{|y_{22\mathrm{m}}(s) - \frac{y_{21\mathrm{m}}(s)}{A(s)}|} \,. \tag{5.34}$$

Equivalent Circuit Representation for Polysilicon Resistor

The polysilicon resistor has typically a width of a few micrometers and a thickness of a few hundred nanometers. The field oxide is about 0.4 μm to 0.5 μm thick. The resistor is covered by several micrometer thick oxide and passivation layers. The length l of the polysilicon resistor is of the order 10 to 100 times longer than its widths w. Therefore, the polysilicon resistor behaves like a lossy micro-strip line in a nearly homogenous medium. With the energy concentrated in the field oxide and its distribution being independent from frequency up to several GHz (due to the small dimensions compared to wavelength), the capacitance of the field oxide is constant. The large (intended) series resistance is the major reason for strong dispersion and frequency dependency. Because $l >> w$, the resistor can be considered as a one-dimensional lossy transmission line lying between its connections with an accuracy similar to that of micro-strip lines of a few percent [251]. From transmission-line theory, the general form for line impedance Z_{w} and the propagation constant γ are known as [251, 252]

$$Z_{\mathrm{w}}(s) = \sqrt{\frac{R' + sL'}{G' + sC'}} \quad \text{and} \tag{5.35}$$

$$\gamma(s) = \sqrt{(R' + sL') \cdot (G' + sC')} \ , \tag{5.36}$$

with R' and L' representing the resistance and inductance per length unit, respectively, and G' and C' denoting the conductance and capacitance to ground per length unit, respectively. Taking into account that for a polysilicon resistor realized in a technology with smallest structure sizes of about 1 µm, $R' \approx 1\,\mathrm{k\Omega/\mu m}$, $L' \approx 300\,\mathrm{fH/\mu m}$, $C' \approx 0.2\,\mathrm{fF/\mu m}$, and $G' < 1\,\mathrm{nS/\mu m}$, it follows that $|sL'| << |R'|$ and $|G'| << |sC'|$ up to several GHz. A series resistance between C' and ground caused by the finite substrate conductivity can be neglected for frequencies below 10 GHz, typically [246]. Therefore, it is possible to consider the polysilicon resistor as a grounded uniformly distributed RC-line (URC-line). Introducing the time constant τ

$$\tau = R' \cdot C' \cdot l^2 \ , \tag{5.37}$$

the normalized frequency Ω

$$\Omega = \omega \cdot \tau \ , \tag{5.38}$$

as well as the normalized complex frequency S

$$S = s \cdot \tau \ , \tag{5.39}$$

the Y-parameters of the polysilicon resistor are [252]

$$Y(S) = \frac{1}{R' \cdot l} \cdot \begin{pmatrix} \frac{\sqrt{S}}{\tanh(\sqrt{S})} & -\frac{\sqrt{S}}{\sinh(\sqrt{S})} \\ -\frac{\sqrt{S}}{\sinh(\sqrt{S})} & \frac{\sqrt{S}}{\tanh(\sqrt{S})} \end{pmatrix} \ . \tag{5.40}$$

This is the *exact* frequency representation of a polysilicon resistor with Y-parameters. The matrix is symmetric. It is easily shown that the absolute value of all four elements in brackets y11, y12, y21 and y22 converge to 1/(R'·l) if S comes close to zero. The element's poles lie on the negative real axis, their zeros in infinity.

π Equivalent Circuit — Exact Form and Approximation

The new π equivalent circuit (see Fig. 5.21b) was chosen for two reasons: (i) it is easily created from the Y-matrix, and (ii) further components can simply be added in parallel (e. g. the feedback capacitor, see below). Fig. 5.21a shows the exact π equivalent circuit using frequency dependent impedances. The frequency-dependence of these impedances $Z1(S)$ and $Z2(S)$ is rather complicated, because both are irrational functions of S. Therefore, an exact, but simple network representation with only a few components being valid for all frequencies is not existing [246].

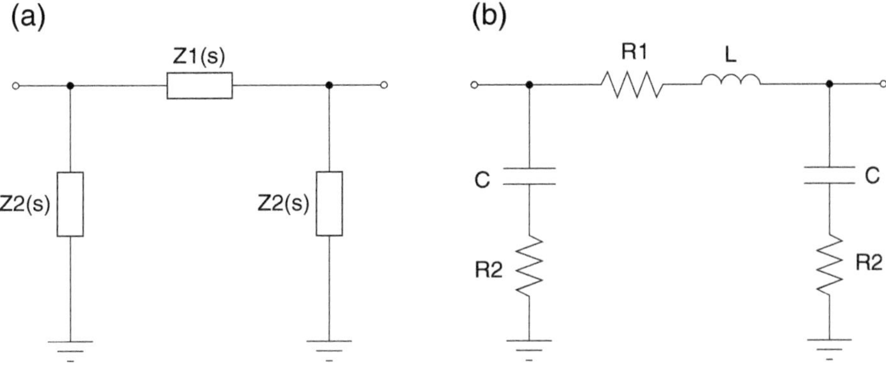

Fig. 5.21. Polysilicon resistor in exact π equivalent circuit representation **(a)** and in approximated π equivalent circuit representation **(b)**

In Fig. 5.21b a simple network with lumped elements and constant (i. e. frequency independent) component values is used. The simple network is equivalent to the exact one for DC and at low frequencies. The component values are determined in the following:

Using (5.40), it follows that

$$Z1(S) = R' \cdot l \cdot \frac{\sinh(\sqrt{S})}{\sqrt{S}} \quad \text{and} \tag{5.41}$$

$$Z2(S) = R' \cdot l \cdot \frac{\sinh(\sqrt{S})}{\sqrt{S} \cdot (\cosh(\sqrt{S}) - 1)} \quad . \tag{5.42}$$

Now the expressions for Z1(S) and Z2(S) have to be simplified using the truncated taylor series expansion for hyperbolic sine and hyperbolic cosine. From (5.41), Z1(S) can be written as

$$Z1(S) = R' \cdot l \cdot \frac{\sum_{k=0}^{\infty} \frac{(\sqrt{S})^{(2k+1)}}{(2k+1)!}}{\sqrt{S}} \quad \text{or} \tag{5.43}$$

$$Z1(S) = R' \cdot l \cdot \left(1 + \frac{S}{3!} + \frac{S^2}{5!} + ...\right) \quad . \tag{5.44}$$

Truncating (5.44) after the linear term of S and using (5.37) and (5.39), one finds $Z1(S)$ consisting of a resistor $R1$ in series with an inductor L with their values

$$R1 = R' \cdot l \quad \text{and} \tag{5.45}$$

$$L = \frac{R' \cdot l}{6} \cdot \frac{S}{s} = \frac{R'^2 \cdot C' \cdot l^3}{6} \quad . \tag{5.46}$$

Similarly, $Z2(S)$ can be expressed from (5.42) as

$$Z2(S) = R' \cdot l \cdot \frac{\sum_{k=0}^{\infty} \frac{(\sqrt{S})^{2k+1}}{(2k+1)!}}{\sqrt{S} \cdot \left(\sum_{m=0}^{\infty} \frac{(\sqrt{S})^{2k}}{(2k)!} - 1\right)} \tag{5.47}$$

or

$$Z2(S) = R' \cdot l \cdot \frac{1 + \frac{S}{3!} + \frac{S^2}{5!} + \ldots}{\frac{S}{2!} + \frac{S^2}{4!} + \ldots} \tag{5.48}$$

Truncating (5.48) after the linear terms and dividing by S finally leads to

$$Z2(S) \cong 2 \cdot R' \cdot l \cdot \left(\frac{1}{S} + \frac{1}{6}\right) \quad . \tag{5.49}$$

$Z2(S)$ consists of a capacitor C in series with a resistor $R2$ according to

$$C = \frac{1}{2 \cdot R' \cdot l} \cdot \frac{S}{s} = \frac{C' \cdot l}{2} \quad \text{and} \tag{5.50}$$

$$R2 = \frac{R' \cdot l}{3} \quad . \tag{5.51}$$

Figure 5.21b shows the final circuit. A slightly better agreement between approximation and exact solution, however, was obtained with the following modified expression, found with the help of MATHCAD simulations [246]:

$$R2 = \frac{R' \cdot l}{6} \quad . \tag{5.52}$$

The approximation for both impedances is better than 10 %, if the normalized frequency Ω does not exceed a value of 4. Keeping in mind that Ω increases with the time constant τ according to (5.38), the latter turns out to be a central design parameter, which has to be kept small. Consequences for the physical design parameters of the polysilicon resistor will be discussed below.

Stability

In general, returning to (5.33), the whole amplifier is stable, if the trans-impedance $Z(s)$ has no poles in the right half-plane of s. This is equivalent to the denominator of (5.33) having its zeros only in the left half-plane of s.

To give an imagination, a very simple case is studied. It is assumed that there are no photodiode capacitance C_D, no feedback capacitance C_F, and that the amplifier has finite, but frequency independent gain A0 (i. e. no phase change). Amplifier input impedance Zin(s) should be infinite, while output impedance Zout(s) is zero. Then, using the normalized frequency S from (5.39), (5.33) may be written as

$$Z(S) = \frac{1}{\frac{y11(S)}{A0} - y_{12}(S)} \quad . \tag{5.53}$$

Setting the denominator of (5.53) equal to zero to find the poles of (5.53) and using (5.40) leads to

$$\frac{\sqrt{S}}{tanh(\sqrt{S}) \cdot A0} + \frac{\sqrt{S}}{sinh(\sqrt{S})} = 0 \tag{5.54}$$

or

$$cosh(\sqrt{S}) + A0 = 0 \ . \tag{5.55}$$

Solving (5.55) for A_0 leads to the following: The stability limit with two zeros of (5.55) lying on the imaginary axis is reached if the gain is

$$A_{0,\mathrm{max}} = cosh(\pi) = 11.6 \tag{5.56}$$

at a frequency of

$$S_{\mathrm{max}} = \pm j \cdot 2\pi^2 \ . \tag{5.57}$$

Critical damping leading to a step response without ringing (see also below) is accomplished if (index "c" for critical)

$$A_{0,\mathrm{c}} = 1 \tag{5.58}$$

at a frequency of

$$S_{\mathrm{c}} = -\pi^2 \ . \tag{5.59}$$

Further poles of (5.53) lie on the negative real axis with the next one $4{\cdot}\pi^2$ apart from the dominant one and thus being neglected. Therefore the gain has to lie between the limits given by (5.56) and (5.58) corresponding to a LRC resonant circuit reaching from critical damping to undamped oscillations. Because of the small gain necessary for damped oscillations, the result is rather unsatisfying. If the amplifier's gain $A(s)$ is frequency dependent, higher gains are also possible. Better results are found using a feedback capacitor described in the next section.

Optimum Feedback Capacitor

From Fig. 5.21b, it can be seen that the transfer impedance $Z1(\Omega)$ shows an inductive behavior. This leads to phase differences between photodiode current $I(s)$ and output voltage Vout(s) of an OEIC amplifier. To compensate this and to maintain stability, the effect of a feedback capacitor will be investigated. To obtain an upper performance limit, the optimum feedback capacitor in terms of output voltage rise time and overshoot will be determined for an idealized large-gain large-bandwidth amplifier. Because of the large gain $A(s)$, the transimpedance $Z(s)$ only consists of $Z1(s)$ representing the polysilicon resistor in parallel with C_{F}, and C_{D} has no effect.

Using the approximation for $Z1(s)$, the feedback circuit becomes a simple LRC-resonant circuit having no ringing at critical damping (index "c") for

$$C_c = \frac{4 \cdot L}{R^2} \tag{5.60}$$

Applied to our case and taking the actual expressions for the components from (5.45) and (5.46), the result from (5.60) is

$$C_{F,c} = \frac{2}{3} \cdot C' \cdot l \ . \tag{5.61}$$

C_F is necessary even in case of infinite gain (see Fig. 5.22).

PSPICE simulations show that the rise time t_r (measured from 10 % to 90 % of the final value) of a series resonant circuit depending on the capacitance C around critical damping is approximately

$$t_r = \tau \cdot \left(\frac{C_F}{C_{F,c}}\right)^{1.5} \ . \tag{5.62}$$

The 3dB-bandwidth B can be estimated from the well-known equation

$$B \cong \frac{0.35}{t_r} = \frac{0.35}{\tau} \cdot \left(\frac{C_{F,c}}{C_F}\right)^{1.5} \ . \tag{5.63}$$

The accuracy of (5.62) and (5.63) is better than 10 % for $C_F = 0.5 \cdot C_{F,c}$ to $C_F = 1.5 \cdot C_{F,c}$ and better than 5 % for $C_F = C_{F,c}$. Improvements in bandwidth are only possible if (little) overshoot is allowed. Therefore, the simplified model is useful within the normal range of applications with no or a small amount of ringing.

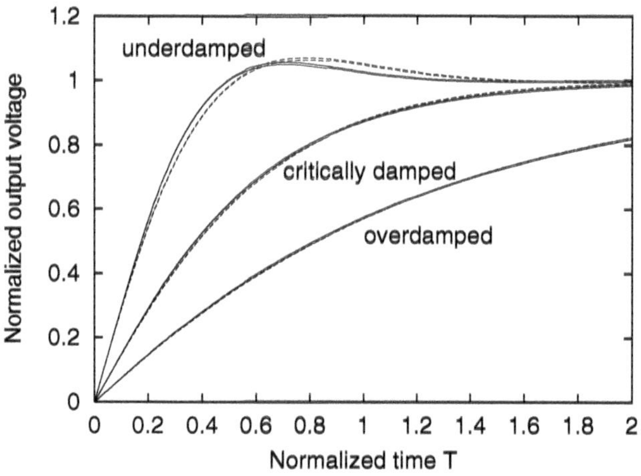

Fig. 5.22. Step response of an ideal (large gain) transimpedance amplifier with three different dampings depending on the feedback capacitor C_F: underdamped ($C_F = 0.5 \cdot C_{F,c}$), critically damped ($C_F = C_{F,c}$) and overdamped ($C_F = 2 \cdot C_{F,c}$). Both the exact solution (solid line) and the approximation (dashed line) are shown [246]

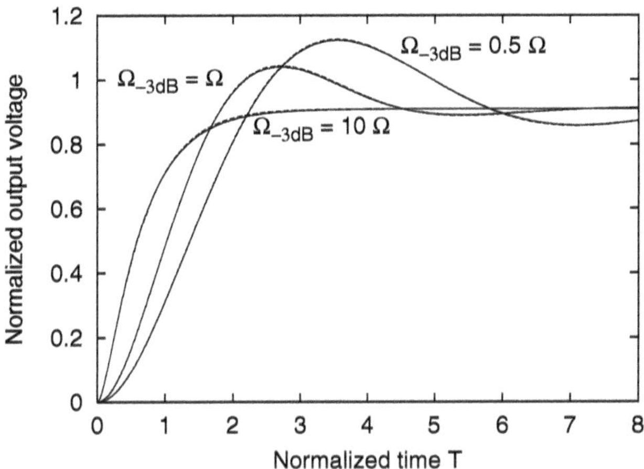

Fig. 5.23. Step response of a transimpedance amplifier with finite gain $A_0 = 10$, a photodiode capacitance $C_D = 5 \cdot C'l$ and a feedback capacitor $C_F = 0.5 \cdot C_{F,c}$ shown for three different amplifier bandwidths. Differences between exact solution (solid line) and the approximation (dashed line) are not visible [246]

In practice, using finite gain amplifiers with finite bandwidth, a lower or higher value for C_F may be necessary and C_D plays a role. Figure 5.23 shows an example in such a more real context with finite amplifier gain and bandwidth (Ω_{-3dB} is the normalized -3dB-bandwidth) in presence of the photodiode capacitance C_D.

Because of finite gain, now the complete model of Fig. 5.21b is used. The step responses shown in this figure were created using PSPICE, whereby the "exact" solution was calculated using an RC-chain of 20 elements. The example is for demonstration of the accuracy of the new π-model and is not optimized in terms of speed or ringing.

Design Requirements for Polysilicon Resistor

The resistor made of polysilicon with the DC-value $R = R'l$, the length l and the width w is characterized by the so-called "resistance per square unit", sheet resistance, $R_\square$ or R_{SQ}, which has the dimension of Ohms per square. The resistor is insulated from the conducting substrate ground via a field oxide of thickness d having the relative dielectric constant ϵ_r.

Remembering the key design parameter, the time constant τ (5.37),

$$\tau = R' \cdot C' \cdot l^2 = R' \cdot l \cdot C' \cdot l \ , \tag{5.64}$$

the product $C' \cdot l$ has to be minimized. The DC-resistance is

$$R = R' \cdot l = R_{SQ} \cdot \frac{l}{w} \ , \tag{5.65}$$

while the capacitance C is

$$C = C' \cdot l = \epsilon_0 \cdot \epsilon_r \cdot \frac{(w + \Delta w) \cdot l}{d} \quad , \tag{5.66}$$

with Δw taking the edge capacitance of the polysilicon resistor into account and ϵ_0 denoting the dielectric constant of vacuum. To obtain a required resistor R with a minimum width w, given in the design rules, and a R_{SQ}, specified for the chosen process, the resistor's length l is to be made equal to $R \cdot w / R_{SQ}$. Then, (5.37) can be written as

$$\tau = \epsilon_0 \cdot \epsilon_r \cdot \frac{(w + \Delta w) \cdot w}{d} \cdot \frac{R^2}{R_{SQ}} \quad . \tag{5.67}$$

As we know already from the last section, equation (5.67) shows that the width w should be kept as small as possible and that the sheet resistance of the resistor polysilicon, a parameter in semiconductor manufacturing, should be chosen as high as possible, i. e. analog process with high-resistive polysilicon are clearly preferable to digital processes with low-resistive polysilicon only. The dielectric constant ϵ_r as well as the thickness d often cannot be changed in the process. Of course, a low overall DC-resistance R also keeps the time constant τ small.

Finally, using (5.63) and (5.67), the expression

$$B \cdot R \propto \frac{1}{R} \quad , \tag{5.68}$$

follows for the resistor-bandwidth product, which means that low resistor values are preferable. In fact, assuming the same resistor-bandwith product, a DVD-amplifier with its large feedback resistor suffers more from the negative effects described herein than most fiber receivers.

To give some impression, Table 5.2 shows a numerical example of a 60 kΩ polysilicon resistor with a sheet resistance of 1 kΩ□, and a width of 3 µm for

Table 5.2. Practical example

Parameters of polysilicon resistor	
Resistor R (DC value)	60 kΩ
Length l	180 µm
Capacitive loading C‘	0.26 fF/µm
Calculated important parameters	
Feedback capacitor (critical damping)	31 fF
Time constant τ (= rise time at critical damping)	2.8 ns
Bandwidth B (critical damping)	125 MHz

instance. Such a resistor is useable in OEIC-applications for optical storage systems like DVD.

This example belongs to the curves for critical damping of a high-gain amplifier shown in Fig. 5.22. The accuracy of these results is better than 5 %. A bandwidth of the OEIC with a 60 kΩ feedback resistor, having the mentioned width and length, of 125 MHz is possible. It should be mentioned that this is the maximum achieveable bandwidth. Real amplifiers will reduce it at least slightly.

5.4.7 Shunt-Shunt Feedback

When the feedback in a transimpedance amplifier only consists of an ohmic resistor, the name transresistance amplifier can be used instead of transimpedance amplifier. Such a transresistance amplifier is formed by a shunt-shunt feedback configuration (Fig. 5.24). The parameters i_{S} and R_{D} are the photocurrent and the parallel resistance of a photodiode. The input port voltages are the same and the output port voltages are the same for the amplifier and feedback two-ports. The y-parameters, therefore, are appropriate for analyzing this shunt-shunt feedback configuration [253].

The amplifier and the feedback network are represented by their individual y-parameters:

$$\begin{aligned} i_1^{\mathrm{A}} &= y_{11}^{\mathrm{A}} v_1 + y_{12}^{\mathrm{A}} v_2 \\ i_2^{\mathrm{A}} &= y_{21}^{\mathrm{A}} v_1 + y_{22}^{\mathrm{A}} v_2 \end{aligned} \tag{5.69}$$

and

$$\begin{aligned} i_1^{\mathrm{F}} &= y_{11}^{\mathrm{F}} v_1 + y_{12}^{\mathrm{F}} v_2 \\ i_2^{\mathrm{F}} &= y_{21}^{\mathrm{F}} v_1 + y_{22}^{\mathrm{F}} v_2 \end{aligned} \tag{5.70}$$

where the superscript A indicates the amplifier and the superscipt F indicates the feedback network.

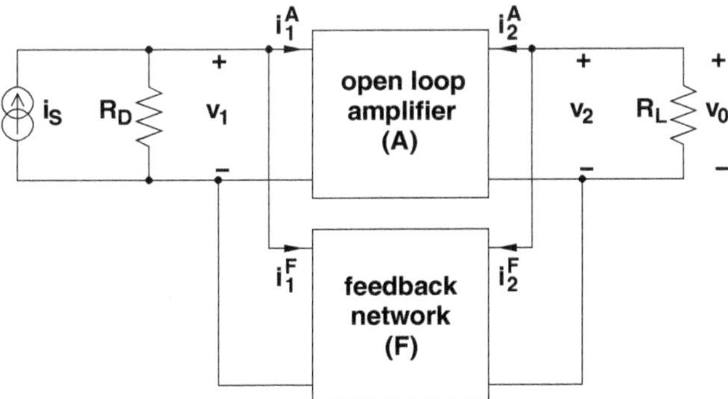

Fig. 5.24. Shunt-shunt feedback amplifier [253]

The total input current i_1 and the total output current i_2 can be written as

$$i_1 = i_1^{\mathrm{A}} + i_1^{\mathrm{F}} \quad \text{and}$$
$$i_2 = i_2^{\mathrm{A}} + i_2^{\mathrm{F}} , \tag{5.71}$$

Substituting equations (5.69) and (5.70) into (5.71) yields the two-port desciption for the shunt-shunt feedback amplifier:

$$i_1 = (y_{11}^{\mathrm{A}} + y_{11}^{\mathrm{F}})v_1 + (y_{12}^{\mathrm{A}} + y_{12}^{\mathrm{F}})v_2$$
$$i_2 = (y_{21}^{\mathrm{A}} + y_{21}^{\mathrm{F}})v_1 + (y_{22}^{\mathrm{A}} + y_{22}^{\mathrm{F}})v_2 \tag{5.72}$$

A more compact notation can be achieved by defining

$$y_{ij}^{\mathrm{T}} = y_{ij}^{\mathrm{A}} + y_{ij}^{\mathrm{F}}$$

and

$$i_1 = y_{11}^{\mathrm{T}} v_1 + y_{12}^{\mathrm{T}} v_2$$
$$i_2 = y_{21}^{\mathrm{T}} v_1 + y_{22}^{\mathrm{T}} v_2 \tag{5.73}$$

because the corresponding parameters of both networks again appear together in Eq. (5.72). Since the input current is determined stronger by the feedback network than by the amplifier output and since the amplifier (and not the feedback network) drives the load, it is allowed to assume

$$y_{12}^{\mathrm{F}} \gg y_{12}^{\mathrm{A}} \quad \text{and}$$
$$y_{21}^{\mathrm{A}} \gg y_{21}^{\mathrm{F}} \tag{5.74}$$

and Eq. (5.73) can be simplified to

$$i_1 = y_{11}^{\mathrm{T}} v_1 + y_{12}^{\mathrm{F}} v_2$$
$$i_2 = y_{21}^{\mathrm{A}} v_1 + y_{22}^{\mathrm{T}} v_2 . \tag{5.75}$$

The closed-loop gain of the shunt-shunt feedback amplifier considering the effects of R_{D} and R_{L} can now be found with the help of Eq. (5.75). At the input port and output port in Fig. 5.24, v_1 and i_1 plus v_2 and i_2 are related by

$$i_1 = i_{\mathrm{S}} - v_1 G_{\mathrm{D}} \quad \text{and}$$
$$i_2 = -G_{\mathrm{L}} v_2 . \tag{5.76}$$

Substituting Eq. (5.76) into (5.75) yields

$$i_{\mathrm{S}} = (G_{\mathrm{D}} + y_{11}^{\mathrm{T}})v_1 + y_{12}^{\mathrm{F}} v_2$$
$$0 = y_{21}^{\mathrm{A}} v_1 + (y_{22}^{\mathrm{T}} + G_{\mathrm{L}})v_2 . \tag{5.77}$$

The closed-loop transresistance can be obtained from (5.77) by solving for v_2 in terms of i_S:

$$A_{TR} = \frac{v_2}{i_S} = \frac{y_{21}^A}{y_{21}^A y_{12}^F - (G_D + y_{11}^T)(y_{22}^T + G_L)} \,. \tag{5.78}$$

Rearranging (5.78) into the standard form for a feedback amplifier gives

$$A_{TR} = \frac{v_2}{i_S} = \frac{\dfrac{-y_{21}^A}{(G_D + y_{11}^T)(y_{22}^T + G_L)}}{1 + \dfrac{-y_{21}^A}{(G_D + y_{11}^T)(y_{22}^T + G_L)} y_{12}^F} = \frac{A}{1 + A\beta_F} \tag{5.79}$$

where

$$A = \frac{v_0}{i_S} = -\frac{y_{21}^A}{(G_D + y_{11}^T)(y_{22}^T + G_L)} \qquad \text{and} \qquad \beta_F = y_{12}^F \,, \tag{5.80}$$

where the feedback parameter β_F has the dimension Ω^{-1} [253]. These two equations are interpreted in Figs. 5.25 and 5.26 for the case of shunt-shunt feedback. Figure 5.25 shows the feedback amplifier with an explicit representation of the two-port parameters of the feedback network with $y_{21}^F = 0$. According to eqs. (5.79) and (5.80) the gain of the amplifier A has to be calculated including the effects of y_{11}^F, y_{22}^F, R_D ($R_D = 1 / G_D$), and R_L ($R_L = 1 / G_L$).

A schematic representation of these equations is given by redrawing the amplifier as in the circuit in Fig. 5.26. The position of the feedback circuit elements y_{11}^F and y_{22}^F has been changed, but the overall circuit is once again the same. The amplifier A-circuit now includes y_{11}^F, y_{22}^F, R_D, and R_L, whereas the feedback network consists only of y_{12}^F.

Figures 5.27, 5.28, 5.29, and 5.30 illustrate the analysis technique in more detail. The three required y-parameters of the feedback network are found based on their individual definitions [253].

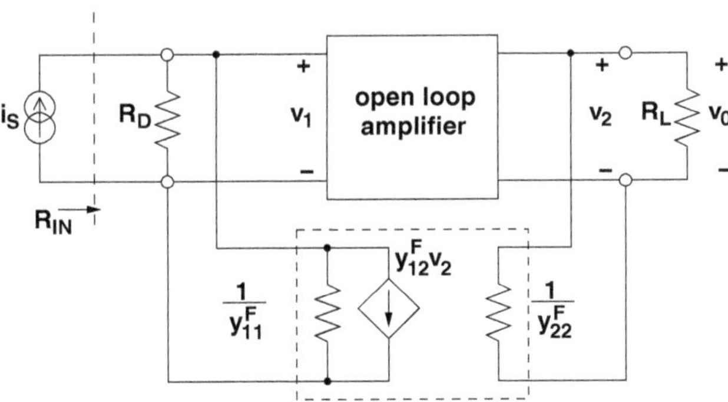

Fig. 5.25. Illustration of the amplifier described by equations (5.79) and (5.80)

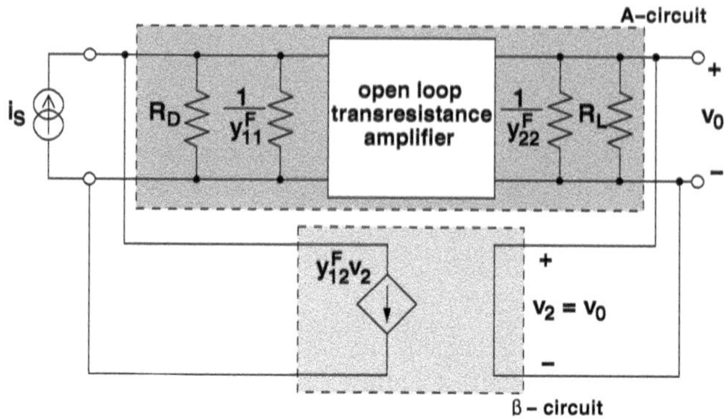

Fig. 5.26. Illustration of the feedback circuit described by equations (5.79) and (5.80) [253]

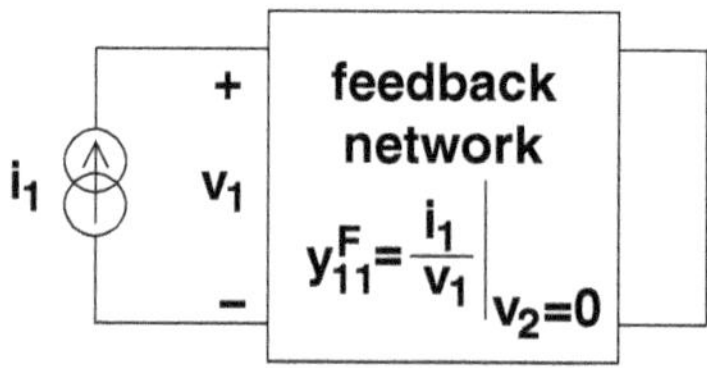

Fig. 5.27. First feedback circuit

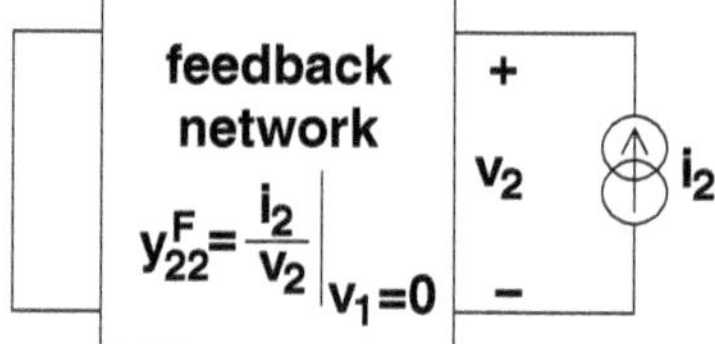

Fig. 5.28. Second feedback circuit

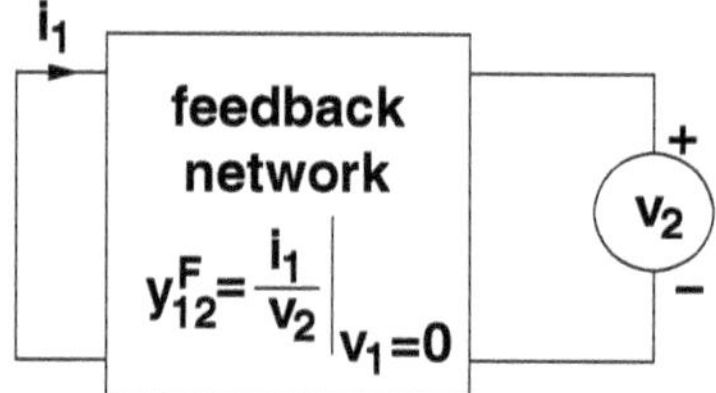

Fig. 5.29. Third feedback circuit

The determination of the y-parameter y^F_{11} with a shorted output is illustrated with Fig. 5.27. Figure 5.28 depicts the definition of the y-parameter y^F_{22}, where the input is shorted. Figure 5.29 shows the definition of the y-parameter y^F_{12}.

Then the transresistance of the open-loop amplifier A is calculated from the circuit shown in Fig. 5.30, which includes the loading effects of y^F_{11}, y^F_{22}, R_D, and R_L.

The gain finally is calculated directly from the A-circuit (Fig. 5.30).

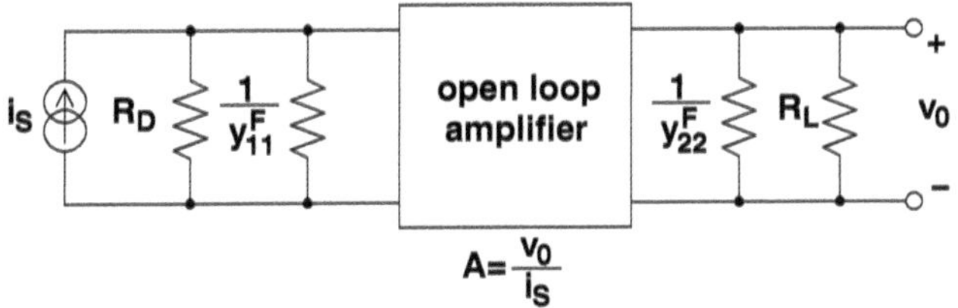

Fig. 5.30. A-circuit for the shunt-shunt feedback amplifier

5.4.8 Input and Output Resistance

Input Resistance

The input resistance R_{IN} of the closed loop shunt-shunt feedback amplifier (Fig. 5.25) can be calculated using the two-port description in (5.77) [253]. The input resistance is

$$R_{\mathrm{IN}} = \frac{v_1}{i_{\mathrm{S}}} . \tag{5.81}$$

Solution of (5.77) for i_{S} in terms of v_1 gives

$$i_{\mathrm{S}} = (G_{\mathrm{D}} + y_{11}^{\mathrm{T}})v_1 + y_{12}^{\mathrm{F}} \frac{-y_{21}^{\mathrm{A}}}{(y_{22}^{\mathrm{T}} + G_{\mathrm{L}})} v_1 \tag{5.82}$$

which can be rearranged as

$$R_{\mathrm{IN}} = \frac{1}{(G_{\mathrm{D}} + y_{11}^{\mathrm{T}}) \left[1 + \frac{-y_{21}^{\mathrm{A}}}{(G_{\mathrm{D}} + y_{11}^{\mathrm{T}})(y_{22}^{\mathrm{T}} + G_{\mathrm{L}})} y_{12}^{\mathrm{F}} \right]} = \frac{\left(\frac{1}{G_{\mathrm{D}} + y_{11}^{\mathrm{T}}} \right)}{1 + A\beta_{\mathrm{F}}} = \frac{R_{\mathrm{IN}}^{\mathrm{A}}}{1 + A\beta_{\mathrm{F}}} . \tag{5.83}$$

Shunt feedback reduces the resistance at the port by the factor $(1 + A\beta_{\mathrm{F}})$. As the loop gain approaches infinity — for an ideal operational amplifier, for example — the input resistance of the closed-loop transresistance amplifier approaches zero. For gigabit optical receivers, however, infinity is by far not reached.

Output Resistance

The output resistance of the closed-loop amplifier can be obtained in a similar manner using the circuit shown in Fig. 5.31 [253].

Here we start with (5.75) and apply a test source i_{x} to the output of the amplifier:

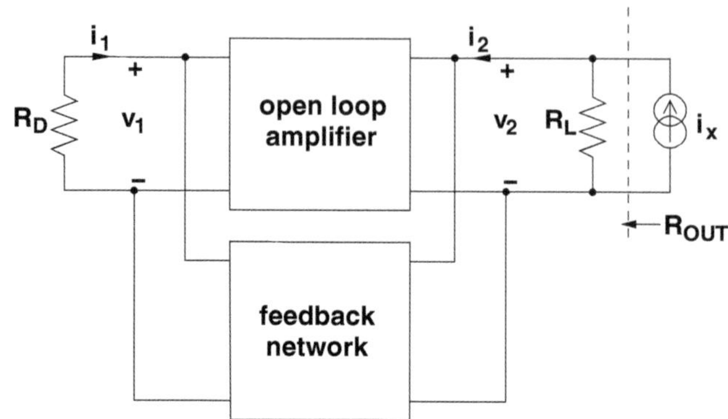

Fig. 5.31. Output resistance of the shunt-shunt feedback amplifier [253]

$$\begin{aligned} i_1 &= y_{11}^{\mathrm{T}} v_1 + y_{12}^{\mathrm{F}} v_2 \\ i_2 &= y_{21}^{\mathrm{A}} v_1 + y_{22}^{\mathrm{T}} v_2 \,. \end{aligned} \tag{5.84}$$

The voltage and current at the input port and at the output port are related by

$$\begin{aligned} i_1 &= -G_{\mathrm{D}} v_1 \\ i_2 &= i_{\mathrm{x}} - G_{\mathrm{L}} v_2 \,. \end{aligned} \tag{5.85}$$

Substituting (5.85) into (5.84) results in

$$\begin{aligned} 0 &= (G_{\mathrm{D}} + y_{11}^{\mathrm{T}}) v_1 + y_{12}^{\mathrm{F}} v_{\mathrm{x}} \\ i_{\mathrm{x}} &= y_{21}^{\mathrm{A}} v_1 + (y_{22}^{\mathrm{T}} + G_{\mathrm{L}}) v_{\mathrm{x}} \,. \end{aligned} \tag{5.86}$$

When we solve for i_{x} in terms of v_{x}, we obtain

$$i_{\mathrm{x}} = y_{21}^{\mathrm{A}} \frac{-y_{12}^{\mathrm{F}}}{(G_{\mathrm{D}} + y_{11}^{\mathrm{T}})} v_{\mathrm{x}} + (y_{22}^{\mathrm{T}} + G_{\mathrm{L}}) v_{\mathrm{x}} \,. \tag{5.87}$$

Rearranging (5.87) yields the result for the output resistance of the overall amplifier:

$$\begin{aligned} R_{\mathrm{OUT}} &= \frac{v_{\mathrm{x}}}{i_{\mathrm{x}}} = \frac{1}{(y_{22}^{\mathrm{T}} + G_{\mathrm{L}}) \left[1 + \dfrac{-y_{21}^{\mathrm{A}}}{(G_{\mathrm{D}} + y_{11}^{\mathrm{T}})(y_{22}^{\mathrm{T}} + G_{\mathrm{L}})} y_{12}^{\mathrm{F}} \right]} \\ &= \frac{\left(\dfrac{1}{y_{22}^{\mathrm{T}} + G_{\mathrm{L}}} \right)}{1 + A\beta_{\mathrm{F}}} = \frac{R_{\mathrm{OUT}}^{\mathrm{A}}}{1 + A\beta_{\mathrm{F}}} \,. \end{aligned} \tag{5.88}$$

The output resistance of the closed-loop amplifier is equal to the output resistance of the A-circuit decreased by the amount of feedback $(1 + A\beta_{\mathrm{F}})$. In the ideal case — approximately reached by operational amplifiers — the output resistance of the transresistance amplifier approaches zero when the loop gain approaches infinity.

5.4.9 Noise and Sensitivity

Electronic noise has to be considered in the design of optical receivers. The received signal has to be stronger than noise in order to avoid transmission errors. In the following, the basics of electronic noise will be described. The noise of a FET photoreceiver preamplifier will be explained in some detail because it will also be shown that its noise can be used also for noise analysis of receivers with transimpedance amplifiers. Finally the sensitivity of optical receivers will be introduced.

Noise of a photodiode with FET preamplifier

Figure 5.32 shows the circuit of a FET photoreceiver preamplifier [1]. The resistor R_L biases the depletion-type junction field-effect transistor and sinks the leakage current of the photodiode. And most impartant, the photocurrent is converted to a voltage signal at R_L.

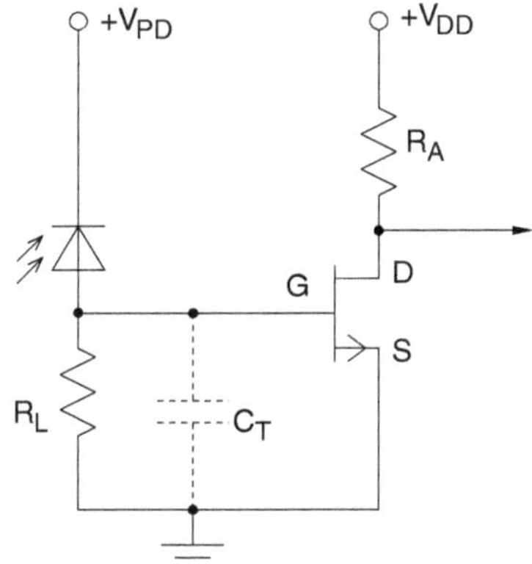

Fig. 5.32. Photodiode with FET amplifier

The capacitance C_T

$$C_\mathrm{T} = C_\mathrm{D} + C_\mathrm{GS} + C_\mathrm{S} \tag{5.89}$$

is the sum of the photodiode capacitance C_D, the gate capacitance C_GS, and the capacitance C_S of the metal line between photodiode and gate of the FET. The gate-drain and therefore the Miller capacitance is neglected in this simple consideration.

The main sources of electronic noise in the circuit of Fig. 5.32 are the thermal noise of resistor R_L and the thermal noise of the FET's channel resistance. We neglect the shot noise due to the gate leakage current. The thermal resistor noise is described by

$$< \delta i_\mathrm{th}^2(f) > \Delta f \;=\; \frac{4k_\mathrm{B}T}{R_\mathrm{L}}\Delta f \tag{5.90}$$

and the thermal noise of the FET's channel resistance is expressed by using the FET's transconductance g_m

$$< \delta i_{\mathrm{Ko}}^2(f) > \Delta f = 4k_{\mathrm{B}}T\Gamma_{\mathrm{F}}g_{\mathrm{m}}\Delta f \ . \tag{5.91}$$

The factor Γ_{F} is about 2/3 for silicon MOSFETs operating in the saturation region. We have to relate the channel noise of the transistor to the input of the amplifier, where we obtain an equivalent voltage fluctuation of

$$< \delta V_{\mathrm{Ki}}^2(f) > \Delta f = (4k_{\mathrm{B}}T\Gamma_{\mathrm{F}}/g_{\mathrm{m}})\Delta f \ , \tag{5.92}$$

which is present across the resistor R_{L} and the capacitance C_{T}. The corresponding equivalent current fluctuation is obtained by multiplication with the squared magnitude of the complex conductance

$$< \delta i_{\mathrm{Ki}}^2(f) > \Delta f = < \delta V_{\mathrm{Ki}}^2(f) > \left[\frac{1}{R_{\mathrm{L}}^2} + (2\pi f C_{\mathrm{T}})^2\right] \Delta f \ . \tag{5.93}$$

The total squared spectral fluctuation of the equivalent input noise current is the sum of the contributions

$$\begin{aligned} < \delta i^2(f) > \Delta f &= < \delta i_{\mathrm{th}}^2 + \delta i_{\mathrm{Ki}}^2 > \Delta f \\ &= \left[\frac{4k_{\mathrm{B}}T}{R_{\mathrm{L}}}\left(1 + \frac{\Gamma_{\mathrm{F}}}{g_{\mathrm{m}}}R_{\mathrm{L}}\right) + 16\pi^2 k_{\mathrm{B}}T\Gamma_{\mathrm{F}}f^2\frac{C_{\mathrm{T}}^2}{g_{\mathrm{m}}}\Delta f\right] \ . \end{aligned} \tag{5.94}$$

The circuit usually will be designed that $g_{\mathrm{m}}R_{\mathrm{L}} \gg 1$ is valid and it is obtained:

$$< \delta i^2(f) > \Delta f = \left[4k_{\mathrm{B}}T\left(\frac{1}{R_{\mathrm{L}}} + 4\pi^2 f^2 \Gamma_{\mathrm{F}}\frac{C_{\mathrm{T}}^2}{g_{\mathrm{m}}}\right)\right] \Delta f \ . \tag{5.95}$$

This relation shows that the input noise is determined by the factor $C_{\mathrm{T}}^2/g_{\mathrm{m}}$ especially at high frequencies. Low capacitances in the input circuit are necessary for low-noise operation. Here OEICs have clear advantages since they avoid the bondpad capacitances.

Noise of a Photodiode with Transimpedance Amplifier

The advantage of a transimpedance amplifier over the PIN FET preamplifier shown in Fig. 5.32, is that the input resistance is lower and to obtain the same bandwidth the feedback resistor R_{F} respective the input-node capacitance can be larger. We will see in the following, that the transimpedance amplifier has also advantages concerning noise. The first advantage is of course that the noise current of a larger resistor ($R_{\mathrm{F}} > R_{\mathrm{L}}$) is lower. To see the second advantage, we investigate the transimpedance amplifier in Fig. 5.33.

According to [254] the equivalent input noise current generator i_{i} can be calculated using

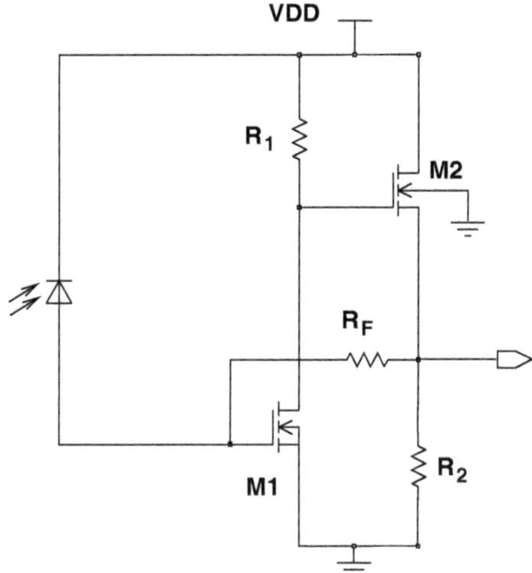

Fig. 5.33. Photodiode with MOS transimpedance amplifier

$$\bar{i_\mathrm{i}^2} = \bar{i_\mathrm{ia}^2} + 4k_\mathrm{B}T\frac{1}{R_\mathrm{F}+R_2}\Delta f \quad , \tag{5.96}$$

where i_ia is the input noise current of the amplifier, i. e. of the input transistor. That the sum of R_F and R_2 determines the noise current can be understood easily from the series circuit of the two resistors (see Fig. 5.33). Equation 5.96 illustrates the second advantage of the transimpedance amplifier concerning noise: the total equivalent input noise current is even lower than that of R_F due to R_2. This difference, however, may not be very large in practice since R_2 often is much lower than R_F. Therefore, it is a good approximation to use the noise analysis of the PIN FET preamplifier for the transimpedance amplifier with R_F instead of R_L [255].

Sensitivity for Digital Signals

Usually binary optical signals are used in optical data transmission leading to two discrete photocurrents $< i_0 >$ and $< i_1 >$ (Fig. 5.34). The expected value $< i_0 >$ represents the logical zero and $< i_1 >$ belongs to the logical one. It is allowed to assume that $< i_1 >$ is larger than $< i_0 >$. The photocurrent has a certain distribution around the expected or mean values due to the current noise introduced above. The time slot T defines the bit rate B. For simplicity, an amplifier with ideal low-pass characterisitics shall be assumed. The bandwidth of this amplifier will be assumed as $B/2$ [1]. The following derivation, however, can easily be done also for a bandwidth of $2B/3$ or $3B/4$ often used in practice.

The variance of the input noise current δi_ges is obtained by integration of the spectral noise density (5.94) and considering also shot noise due to the leakage current i_L of the photodiode [1]:

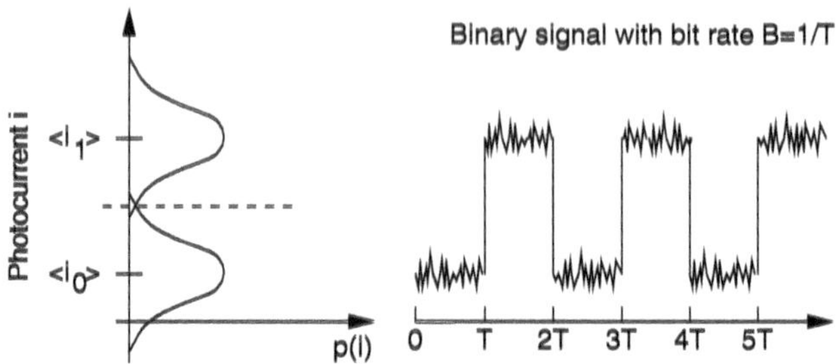

Fig. 5.34. Noisy photocurrent signal with the bit rate B and the distribution $p(i)$ of the instant i values around the logic levels $< i_0 >$ and $< i_1 >$ [1]

$$< \delta i^2_{\text{ges}} > = \int_0^{B/2} < \delta i^2(f) > df = \frac{2k_{\text{B}}TB}{R_{\text{L}}} + qi_{\text{L}}B + \frac{2k_{\text{B}}T\Gamma_{\text{F}}\pi^2 C^2_{\text{T}} B^3}{3g_{\text{m}}} , \tag{5.97}$$

where $g_{\text{m}}R_{\text{L}} \gg 1$ has been used. We assume that the instant current values have a gaussian distribution around the expected values $< i_0 >$ and $< i_1 >$. The probability density for a logical zero, therefore, is

$$p_0(i) = \frac{1}{\sqrt{2\pi < \delta i^2_{\text{ges}} >}} \, exp\left(-\frac{(i- < i_0 >)^2}{2 < \delta i^2_{\text{ges}} >}\right) . \tag{5.98}$$

For a logical one

$$p_1(i) = \frac{1}{\sqrt{2\pi < \delta i^2_{\text{ges}} >}} \, exp\left(-\frac{(i- < i_1 >)^2}{2 < \delta i^2_{\text{ges}} >}\right) \tag{5.99}$$

is obtained correspondingly. For equally distributed logical zeros and ones, the mean value of the current in respect to time is:

$$D_{\text{t}} = (< i_1 > + < i_0 >)/2 . \tag{5.100}$$

This value is defined as the decision threshold. For a current $i \leq D_{\text{t}}$, the detected signal is considered as logical zero. The probability of a wrong decision is given by the bit error rate BER for the logical zero

$$BER_0 = \int_{Dt}^{\infty} p_0(i)di = \frac{1}{\sqrt{2\pi}} \int_Q^{\infty} exp\left(-u^2/2\right) du , \tag{5.101}$$

with $u = (i- < i_0 >)/\sqrt{< \delta i^2_{\text{ges}} >}$ and where the noise distance

$$Q = \frac{< i_1 > - < i_0 >}{2\sqrt{< \delta i^2_{\text{ges}} >}} \tag{5.102}$$

also has been introduced. For the applied assumptions it can be shown that the bit error rate for the logical one is equal to that for the logical zero

$$BER_1 = \int_{-\infty}^{Drmt} p_1(i)di = BER_0 = BER \ . \qquad (5.103)$$

Therefore the bit error rate in general obeys to

$$BER = \frac{1}{\sqrt{2\pi}} \int_Q^{\infty} exp\left(-u^2/2\right) \approx \frac{1}{\sqrt{2\pi}} exp\left(-\frac{Q^2}{2}\right)\left(1-\frac{1}{Q^2}\right) \ . \qquad (5.104)$$

The error made by the approximation is less than 1 % for $Q \geq 2$. The relation between bit error rate (BER) and the noise distance Q is shown in Fig. 5.35. For instance, BER $= 10^{-9}$ holds at $Q = 6$ and BER $= 1.3 \times 10^{12}$ for $Q = 7$.

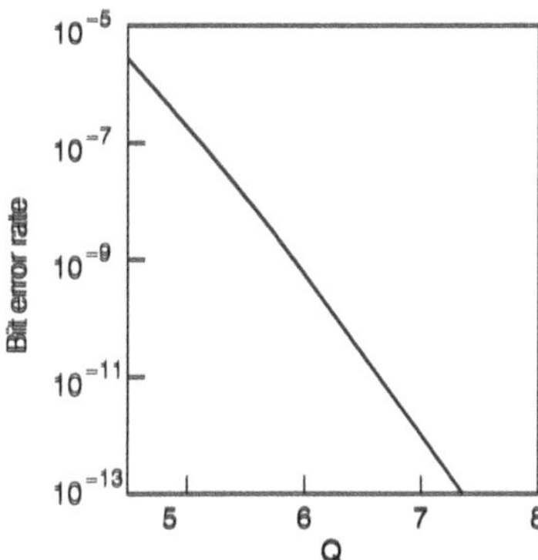

Fig. 5.35. Bit error rate in dependence of the noise distance Q [1]

Modern so-called communication analyzers or digital sampling oscilloscopes can be used to determine the noise distance Q. For this purpose, eye diagrams are taken with a set-up shown in Fig. 5.36.

The receiver under test in Fig. 5.36 has an analog output capable of driving the 50 Ω input of the digital sampling oscilloscope. The digital sampling oscilloscope or communication analyzer determines the histograms shown in Fig. 5.37. The mean values $< i_0 >$ and $< i_1 >$ as well as their variances can be read from the display of the communication analyzer. From the variances, $< \delta i^2_{ges} >$ and finally the noise distance Q is calculated by the communication analyzer.

When the receiver under test has a digital output the set-up shown in Fig. 5.38 can be used for a bit error analysis. A bit error analyzer also can be used when the analog output of an optical receiver has a larger output voltage than the sensitivity of the bit error analyzer. The bit error analyzer digitally compares the received bits with the sent bits and counts errors. A bit error

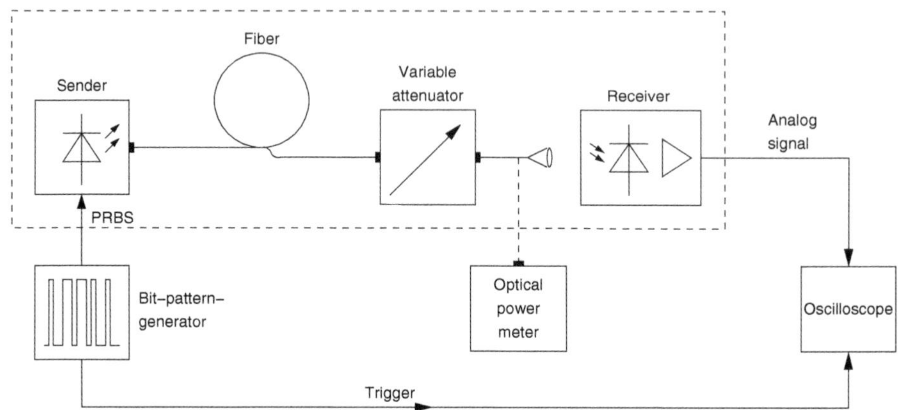

Fig. 5.36. Measurement set-up for eye diagram measurements

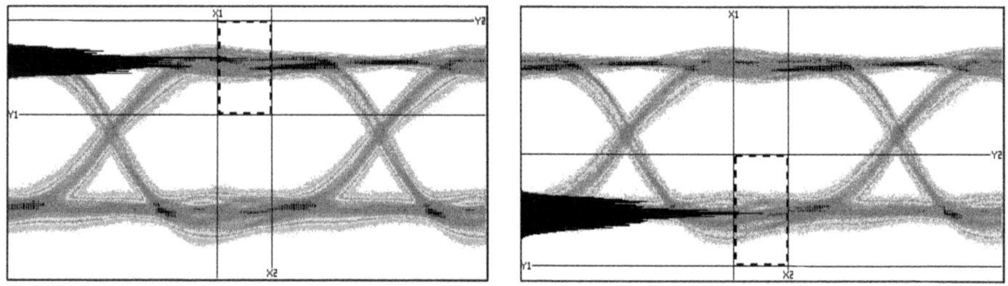

Fig. 5.37. Eye diagrams with histograms for the "1" (left) and "0" (right)

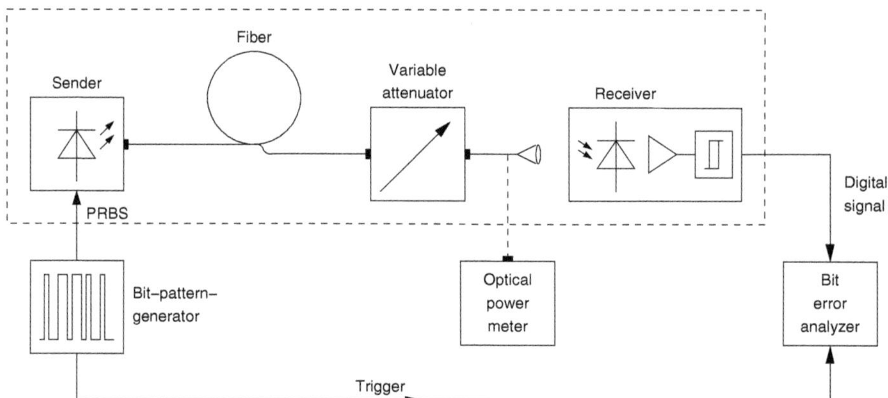

Fig. 5.38. Measurement set-up for digital bit error analysis

analyzer results in a more accurate characterization of optical receivers than the determination of Q with the set-up of Fig. 5.36.

The decision threshold D_t is in the middle between the "1"- and "0"-photocurrents. For bit sequences with equally distributed "1"s and "0"s it, therefore, can be expressed by the mean optical power $\eta < P >$, which is converted into photocurrent:

$$D_\mathrm{t} = \frac{< i_1 > + < i_0 >}{2} = \eta \frac{q}{\hbar\omega} < P > \quad . \tag{5.105}$$

The ratio r of the two photocurrent levels can be defined:

$$r = < i_0 > / < i_1 > \quad . \tag{5.106}$$

With this definition, the optical power of a binary signal necessary to achieve a certain required bit error rate (i. e. the corresponding Q-factor) results:

$$< P > = \frac{1+r}{1-r}\frac{\hbar\omega}{\eta q} Q \sqrt{< \delta i_\mathrm{ges}^2 >} \quad . \tag{5.107}$$

This quantity is called sensitivity of the optical receiver and it is usually expressed in dBm ($10 \times log(< P > / 1\,\mathrm{mW})$). It becomes lowest for $r=0$, i. e. $< i_0 > = 0$ or a vanishing "0" optical power. In this case it is

$$< P > = \frac{\hbar\omega}{\eta q} Q \sqrt{< \delta i_\mathrm{ges}^2 >} \quad . \tag{5.108}$$

When we insert (5.97) into (5.108), we obtain the lowest possible average optical power for a PIN-FET input stage

$$< P > = \frac{\hbar\omega}{\eta q} Q \sqrt{\frac{2k_\mathrm{B}TB}{R_\mathrm{L}} + qi_\mathrm{L}B + \frac{2k_\mathrm{B}T\Gamma_\mathrm{F}\pi^2 C_\mathrm{T}^2 B^3}{3g_\mathrm{m}}} \quad . \tag{5.109}$$

In reality, laser diodes have to be biased above the threshold to allow fast data rates. The extinction ratio ($=1/r$) is therefore less than infinity in practice and a somewhat larger $< P >$ is necessary than the minimum value for $r=0$.

When we consider that $\omega = 2\pi c_0/\lambda$, the sensitivity depends on the wavelength. For instance

$$10\,log(\frac{P^{650}}{P^{1300}}) = 10\,log(\frac{1300}{650}) = 10\,log(2) = 3\,\mathrm{dB} \tag{5.110}$$

shows that at an optical wavelength of 650 nm, a twice as large average optical power is needed than at 1300 nm for the same BER. For a silicon OEIC detecting a 650 nm signal this means, that the sensitivity value (e. g. −20 dBm) looks 3 dB worse than for a III/V-photodiode in combination with an amplifier possessing the same equivalent input noise current and operating with

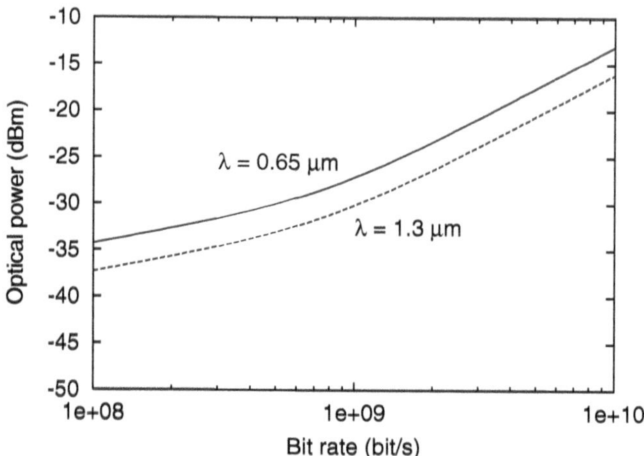

Fig. 5.39. Sensitivity of an optical receiver for two different wavelengths ($\eta = 0.5$, $Q = 6$ i.e. $BER = 10^{-9}$, $R_L = 3.2\,\text{k}\Omega$, $i_L = 10\,\text{pA}$, $\Gamma_F = 2/3$, $C_T = 0.5\,\text{pF}$, and $g_m = 1\,\text{mS}$) calculated with (5.109)

1300 nm (e.g. −23 dBm). In comparison to $\lambda = 1540\,\text{nm}$, which is the most frequently used wavelength for long-haul optical fiber communication, the sensitivity for $\lambda = 650\,\text{nm}$ is 3.75 dB worse.

Figure 5.39 shows this 3 dB penalty for a wavelength $\lambda = 650\,\text{nm}$ compared to $\lambda = 1300\,\text{nm}$. This figure also illustrates that the noise contribution of the resistor R_L (or R_F) dominates up to about 500 Mb/s for the parameters given in the figure caption. At higher bit rates, the slope of the curves increases indicating that the FET noise dominates there for the parameters listed in the caption of Fig. 5.39. According to Fig. 5.39, a silicon OEIC can achieve a sensitivity of about −27 dBm for $\lambda = 650\,\text{nm}$ at a bit rate of 1.0 Gb/s in the NRZ transmission mode.

6 Examples of Optoelectronic Integrated Circuits

In this chapter the full variety of silicon receiver OEICs in digital and analog techniques will be introduced. Examples of optical receivers range from low-power synchronous digital circuits for massively parallel optical interconnects and three-dimensional optical storage to asynchronous Gb/s fiber receivers. Hybrid integrated laser drivers are included as examples of optical emitters. Low-offset analog OEICs for two-dimensional optical storage systems like digital-versatile-disk (DVD) and digital-video-recording (DVR) will be described as well as image sensors. Among various optical sensors for industrial and medical applications, smart pixel sensors as well as distance measurement circuits leading to 3D cameras and paving the way to innovative cameras-on-a-chip (CoC) will be presented. Techniques like integrated voltage-up-converters and the four-quarter POF receiver approach for speed enhancement of SOEICs will also be introduced. Newest developments of burst-mode and deep-sub-μm receivers are additionally addressed. A comparison of fiber and optical interconnect receivers plus an innovative optoelectronical PLL circuit as well as a summary are finally included.

6.1 Digital CMOS Circuits

In this section, the properties of digital CMOS OEICs are described because of their potential application in massively parallel optical interconnects and volume holographic optical memories. Synchronous circuits and among them one implementing a spatially-modulated-light detector are appropriate for such purposes. Furthermore, asynchronous photoreceivers for application in the testing of digital CMOS circuits on the wafer level deserve to be described here.

6.1.1 Synchronous Circuits

For the application in massively parallel optical interconnects between VLSI chips a novel monolithic optoelectronic receiver system was presented in a standard 0.7 μm N-well CMOS technology [256]. The circuit, which requires two clock signals, *reset* (RST) and *store* (STORE), and therefore is a synchronous circuit, is shown in Fig. 6.1.

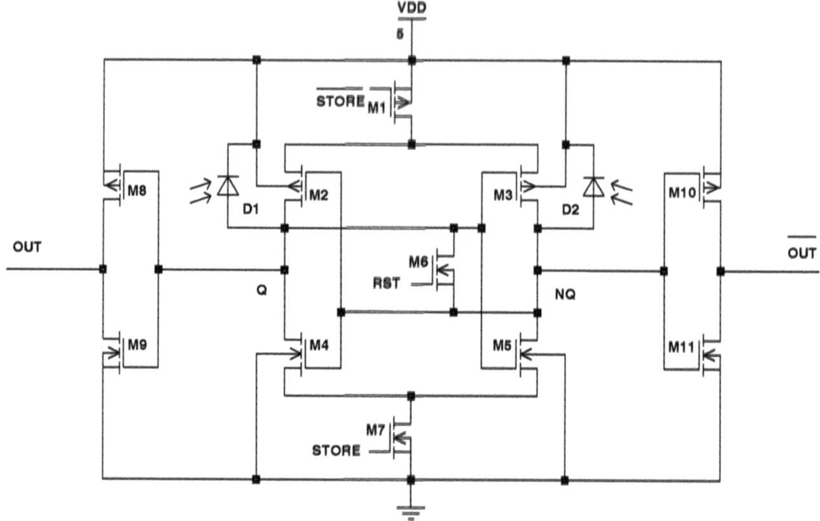

Fig. 6.1. CMOS synchronous photoreceiver circuit [256]

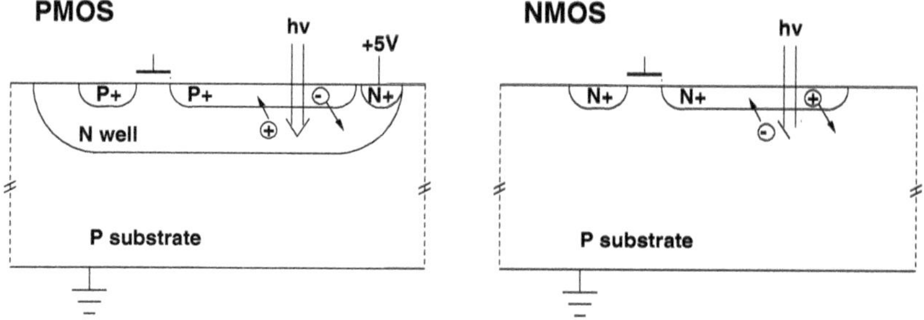

Fig. 6.2. Cross section of MOS transistors with enlarged drain areas as photodiodes [256]

The heart of the synchronous receiver is a CMOS bistable flip-flop, which acts as a sense-amplifier. The flip-flop is formed by two CMOS inverters, M2/M4 and M3/M5. The input of each inverter is connected to the output of the other inverter. In such a way a dual state can be stored. In order to use the flip-flop as a light receiver, two photodiodes were added to the drains of the P-channel transistors M2 and M3. In fact the diodes were not separated from the transistors. The diodes were obtained by extending the drains to an area of $15 \times 15\,\mu m^2$. The P^+ drain island and the N well, therefore, form the photodiode (Fig. 6.2).

The timing of the synchronous receiver is shown in Fig. 6.3. Before the flip-flop can receive, i. e. store, a new bit, a reset signal has to be applied to M6. During this reset signal, the flip-flop is deactivated by a *low* voltage level at the gate of M7 and by a *high* voltage level at the gate of M1. M6 is conducting during *reset* and forces the nodes Q and NQ to the same potential. After the reset phase, M6 is switched off and light can fall into one of the diodes, let us say into D1. Electron–hole pairs are generated at the P^+N junction between the drain of M2 and the N well. The N well is biased at VDD = 5 V. The potential of the drain can be assumed to be at a floating level of approximately VDD/2. The P^+N photodiode is biased in the reverse direction and the photogenerated carriers are separated in the electric field region of the PN junction effectively. The photogenerated electrons drift into the N well and the photogenerated holes drift into the P^+ region. There is also a slower contribution of diffusing minority carriers to the photocurrent, because the light penetrates also in deeper field-free regions of the N well. As a consequence of the photocurrents, the potential of node Q becomes more positive than that of node NQ. When the supply voltages are applied to the flip-flop via M1 and M7, the sense-amplifier flip-flop begins to work. A more positive input signal of the inverter M3/M5 results in a more negative output signal at node NQ, which causes a higher output signal of the inverter M2/M4 at node Q. This amplifying ring process continues until digital levels, i. e. VDD and ground levels are obtained at the two output nodes Q and NQ of the flip-flop, respectively. It was reported that this final stable state was reached after approximately 3 ns from the beginning of the light incidence [256].

A minimum light input energy of 176 fJ, which corresponds to an optical power of 59 μW within a time interval of 3 ns, was needed for an optical wavelength of 830 nm in order to obtain a correct decision of the sense-amplifier flip-flop. This minimum light energy causes a voltage change of 264 mV at

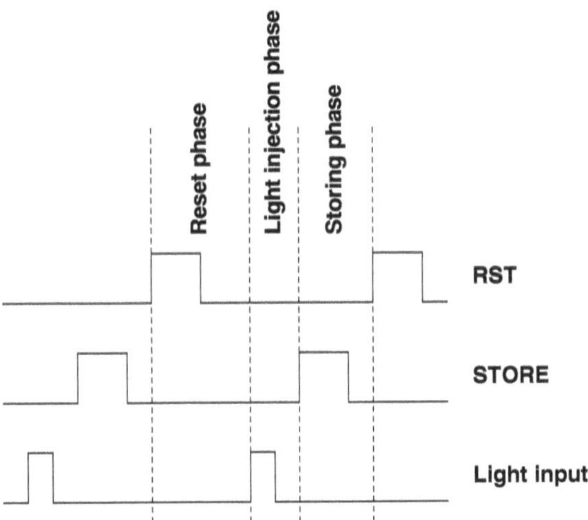

Fig. 6.3. Timing diagram of a synchronous CMOS photoreceiver [256]

node Q [256]. This voltage change is amplified by the flip-flop to a digital level. With the PMOS version described above a bit rate of 120 Mb/s was obtained. In the NMOS version, i. e. using the N^+-drain to P-substrate diodes of the transistors M4 and M5 as photodiodes (see Figs. 6.1 and 6.2), a maximum bit rate of 180 Mb/s was reported. This higher speed of the NMOS version of the synchronous receiver can be explained by the faster diffusion of electrons in the P substrate. The PMOS version is slower because holes are the minority carriers in the N well and holes are diffusing slower than electrons due to the lower hole mobility.

For massively parallel optical interconnects it is necessary to minimize the die area of the receivers. The receiver circuit shown in Fig. 6.1 occupied an area of $55 \times 24\,\mu m^2$ without the output buffers. The synchronous receiver with the sense-amplifier flip-flop is especially interesting for application in massively parallel optical interconnects due to its very small area consumption.

The function of this receiver is called synchronous because the clock signals *reset* and *store* are needed. These signals have to be transmitted in additional optical fibers, for which asynchronous receivers are needed, or have to be supplied electrically.

In the following, another application of sense-amplifier flip-flops, where the reset and store signals are readily available, will be described. Sense-amplifier flip-flops can readily be used for the read process of page-oriented optical memories (POMs), like volume holographic storage systems [257]. Such optical memories need highly parallel read circuits. A small size of each detector and each amplifier is therefore essential. The clock signals for reset and store (or latch) of the sense-amplifier flip-flop are readily available in optical memories and these signals do not have to be extracted for this application. The POMs offer the potential for high capacity, random access times from 10 to 100 μs, and page sizes up to 1 Mb, yielding data transfer rates of 10–100 Gb/s. Maximum output data rates of 250 Mb/s for CCD arrays [116] are insufficient for POM systems. Pixel circuits, like that in Fig. 6.4 with sense-amplifier flip-flops, combine a large speed, a high gain, and a small size. They can be embedded in each pixel yielding active receivers in a highly parallel arrangement. The circuit of a pixel in a photodetector array for a page-oriented optical memory (Fig. 6.4) combines two flip-flops (latches) in order to achieve a high gain.

The operation of the circuit is controlled by reset $\overline{\mathrm{RST}}$ and latch $\overline{\mathrm{LAT}}$. The P latch with transistors P1 and P2 is isolated from the N latch with N1 and N2 via P3 and P4 for $\overline{\mathrm{LAT}} = 1$. The pull-down devices N3 and N4 are conducting and the N latch is reset for $\overline{\mathrm{LAT}} = 1$. Next $\overline{\mathrm{RST}}$ is set to 0. During this reset, the photosensitive inputs A and B are shorted through P5 and $V_{\mathrm{A}} = V_{\mathrm{B}} = V_{\mathrm{RST}} \approx VDD + V_{\mathrm{Th,PMOS}}$. When $\overline{\mathrm{RST}}$ is taken high again, parasitic capacitances hold nodes A and B at V_{RST} putting the P latch in a metastable state. A photocurrent I_{ph} at one of the differential inputs causes

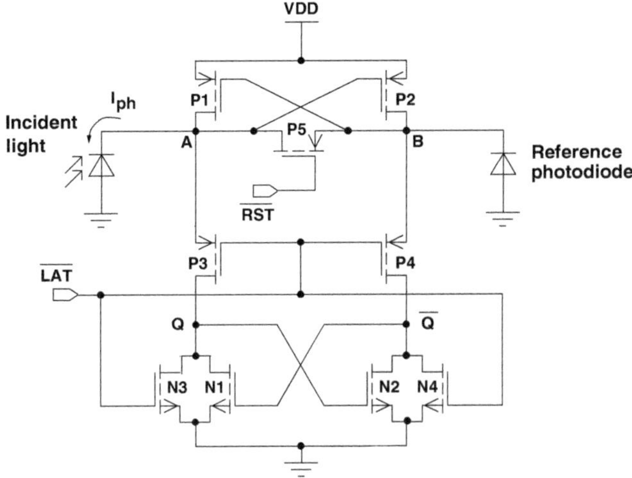

Fig. 6.4. Photoreceiver circuit for smart photodetector array [258]

the corresponding input voltage to drop. Positive feedback in the P latch increases the differential voltage. When $\overline{\text{LAT}}$ is changed to 0, this differential voltage is amplified by the N latch and the signal is acquired at Q and $\overline{\text{Q}}$. P1 and P2 should be small in size for a high sensitivity to small photocurrents. The static power consumption of the circuit in Fig. 6.4 is determined by leakage currents. The dynamic power consumption mainly depends on the capacitance of the photodiode. In [258], the N-well to P-substrate photodiode in a 0.35 μm CMOS process was used with a photosensitive N-well area of $50 \times 50\,\mu\text{m}^2$. The capacitance of this photodiode was approximately 1 pF. The size of the photoreceiver circuit in Fig. 6.4 without the photodiodes was $20 \times 22\,\mu\text{m}^2$. An optical power of 2.5 μW for $\lambda = 839$ nm was necessary to toggle the receiver. With an assumed photodiode responsivity of 0.3 A/W, a switching energy of 150 fJ was estimated [258]. A single-pixel data rate of 245 Mb/s was determined for the circuit in Fig. 6.4. With the circuitry necessary for error correction, a maximum pixel number of approximately 26 700 per 0.35 μm CMOS chip was projected in [258]. The bit rate per chip for corrected data was estimated to be 102 Gb/s stemming from the high parallelism.

A sense-amplifier receiver for optical data link applications and parallel optical interconnects also was used together with the SML detector (see Sect. 2.2.4). The circuit diagram of this sense-amplifier receiver [71] is shown in Fig. 6.5.

The receiver is reset during the negative phase of the clock when M8 and M9 are on and connect the immediate as well as the reference detector (both represented by diodes in Fig. 6.5) to V_{Bias}. When the clock turns high, both detector parts are left floating. Now the receiver is sensitive to light incidence.

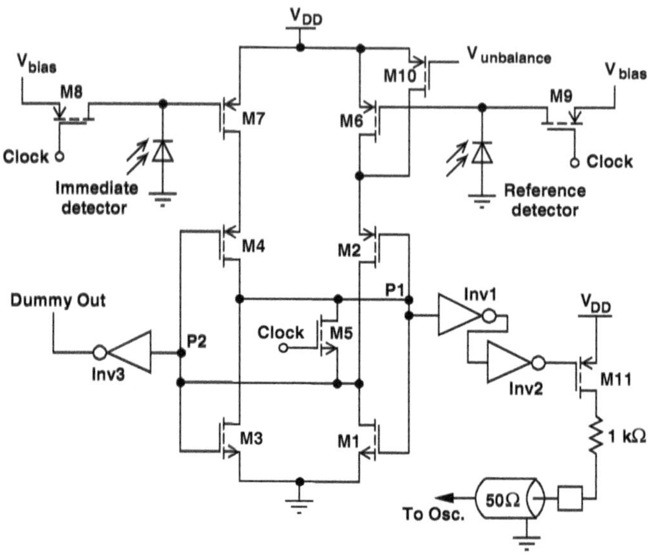

Fig. 6.5. Circuit of a CMOS synchronous receiver with SML detector and sense amplifier [71]

Both detector parts integrate carriers that reach the detector junctions. The potentials at both cathodes drop proportional to the integrated charges. The charge in the immediate detector, however, is larger (drifting and a part of diffusing carriers are integrated) than that in the reference detector (only a part of diffusing carriers is collected). Consequently, M7 has a larger $|V_{GS}|$ than M6. The drain current of M7 is larger than that of M6, which brings the sense-amplifier out of equilibrium. This equilibrium was established during the positive phase of the clock, when M5 was on. When light was received, the cross-coupled inverters M3, M4 and M1, M2 deliver a digital HIGH and a digital LOW at their outputs. When, there was no light incidence, both potentials at the gates of M6 and M7 are the same and noise determines the direction into which the inverter pair flips. To avoid this problem, an unbalance is built in by M10. An appropriate gate potential ($V_{\text{unbalance}}$) of M10 determines the direction into which the inverter pair flips, when there was no light incidence. With light incidence, it flips into the other direction. This unbalance measure implements a threshold. The incident light has to overcome this threshold to bring the receiver into the state "received light".

The inverters Inv1 and Inv3 are buffers to extract the received state. Actually for extraction only one inverter would be sufficient, however, for a better performance of the receiver a symmetric load on nodes P1 and P2 is required. Via Inv2, PMOS transistor M11 drives a voltage divider 21:1 to use an oscilloscope with a 50 Ω input impedance for characterization purposes.

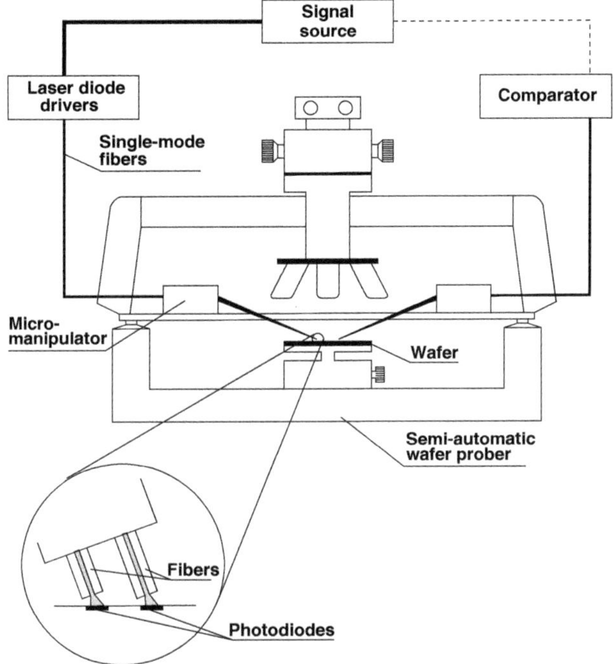

Fig. 6.6. Test equipment using optical inputs on a wafer level [260]

With a 0.8 µm CMOS technology, a bit rate of 155 Mb/s with an optical pulse power of 5.6 µW for $\lambda = 860$ nm was possible. The size of the SML detector was $70 \times 70\,\mu\text{m}^2$ and the sense amplifier (without output driver) fitted into the same area. The power consumption of the sense-amplifier receiver for a bit rate of 180 Mb/s was 0.5 mW [71].

6.1.2 Asynchronous Circuits

Compact and fast photoreceivers with on-chip photodiodes in standard CMOS technology have been developed as optical inputs for testing of digital circuits [259]. Novel CMOS circuits work at increased speed and they cannot be tested on the wafer level at their working speed with conventional electric needle contacts. In order to overcome this limitation, optical inputs can be used. The light is coupled into photodiodes on the chips via optical fibers being adjusted on a wafer prober (Fig. 6.6). The outputs of the chip having lower frequency were contacted with electric needles or capacitively with a special probe sensor. The output signals are checked for correctness by the comparator in a test equipment (Fig. 6.6).

In a 1.5 µm CMOS technology an input frequency of 250 MHz was obtained for an optical wavelength of 635 nm with an optical receiver requiring a die area of only $70 \times 70\,\mu\text{m}^2$. Maximum input frequencies of 243 and 187 MHz were reported for the wavelengths of 685 and 787 nm, respectively. The N^+-source/drain to P-substrate diode in the N-well CMOS process was used for the photodiodes with an area of $20 \times 20\,\mu m^2$. The speed of the circuit test on the wafer was increased several times compared to the conventional electrical input technique using this optical input technique. This performance could be obtained with a current comparator circuit with a statically supplied reference photodiode. The circuit diagram of this photoreceiver is shown in Fig. 6.7.

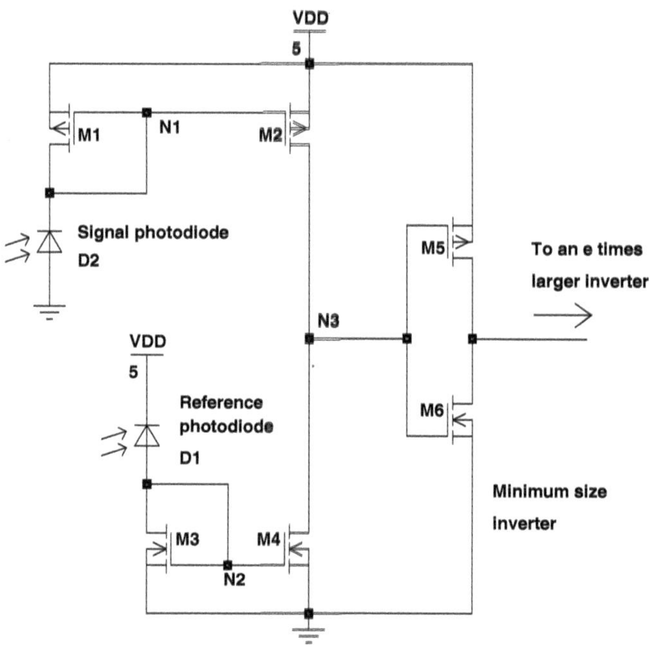

Fig. 6.7. CMOS current comparator photoreceiver circuit [259]

The transistors M1, M2 and M3, M4, respectively, form current mirrors, which amplify the photocurrents of the signal and reference photodiodes. The transient response has its optimum for a width ratio W2/W1 of 2 to 2.5 [259]. The amplified photocurrents are compared to each other by the transistors M2 and M4. The advantage of this configuration is its good sensitivity to small differences in the photocurrents due to the high impedance node N3. Due to the high drain resistances a small difference in the currents results in a large voltage change. The advantage of this photoreceiver clearly is that the photocurrent of the N^+P signal photodiode does not have to drop much to obtain a logical *zero* at the output node N3 of the current comparator, when the light is switched off for a short period, i. e. a large contribution of the slow diffusion current to the photocurrent is possible. It has to be mentioned, however, that large photocurrents and large optical powers are necessary to obtain fast current mirror amplifiers. The optical power used for chip testing was of the order of 1 mW [259]. Another advantage of this current comparator circuit is its robustness for the application in automatic chip testers. Compared to the synchronous receiver shown in Fig. 6.1, the current comparator circuit shown in Fig. 6.7 works asynchronously and, therefore, does not need overhead circuitry for timing. The speed of the current comparator circuit in a 1.5 μm CMOS technology is comparable to the speed of the synchronous receiver in a 0.7 μm CMOS technology for the optical wavelength of 787 nm. The speed of the current comparator circuit, therefore, can be improved considerably by using a technology with a smaller gate length. The signal photodiode feeding an N-channel current mirror and the reference photodiode feeding the P-channel current mirror probably would allow a further increase in the input frequency.

6.2 Digital BiCMOS Circuits

The principle of a current comparator circuit shown in Fig. 6.7 for a CMOS photoreceiver was also applied to a BiCMOS photoreceiver for the high speed testing of frequency dividers on the wafer level [260]. Figure 6.8 shows the BiCMOS version of a current comparator. In the 1.2 μm BiCMOS process fast NPN transistors were available, which preferably were used in the signal current mirror. The slower P-channel current mirrors are kept for the reference path.

The circuit shown in Fig. 6.8 was used for a high speed test of a BiCMOS frequency divider on the wafer level. N^+ to P-substrate and P^+ to N-well photodiodes with areas of $20 \times 20\,\mu m^2$ were used. A frequency of 800 MHz was successfully fed into the BiCMOS frequency divider optically with a wavelength of 635 nm via a single-mode fiber. The optical output power of the commercially available semiconductor laser used in [260] was of the order of 1 mW. The area consumption of an optical input was reported to be less than $70 \times 70\,\mu m^2$.

6.3 Laser Driver Circuits

A simple two-transistor CMOS driver (Fig. 6.9) based on a current-shunting principle, which provides a low-area, tunable power circuit with a measured small-signal bandwidth of 2 GHz in 0.5 μm CMOS technology, has been implemented in a flip-chip bonded CMOS VCSEL chip [261].

The PMOS transistor is used to supply an adjustable current through the laser, and the NMOS transistor is used to quickly shunt the current into and out of the VCSEL for digital operation. The multimode VCSELs could be operated at 1.25 Gb/s from below the laser threshold in this digital

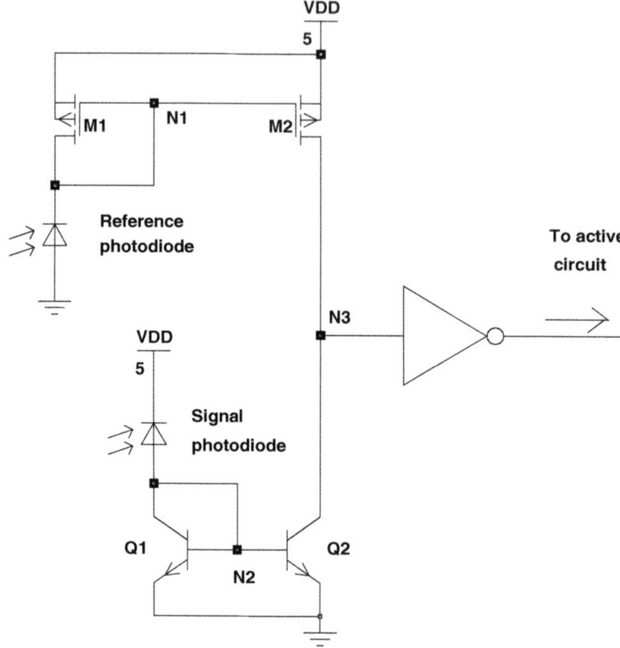

Fig. 6.8. BiCMOS current comparator photoreceiver circuit [260]

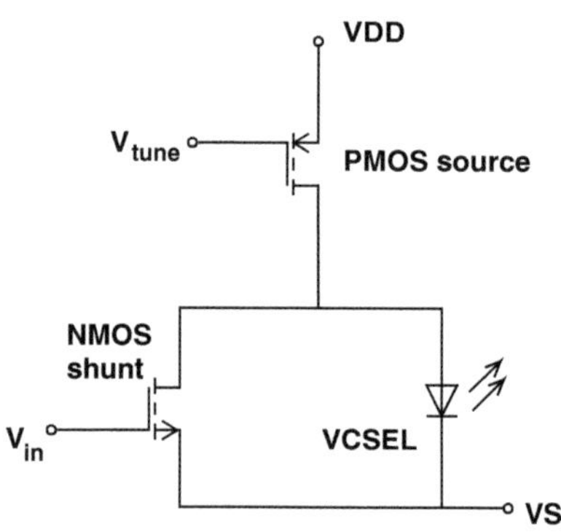

Fig. 6.9. CMOS shunt laser driver circuit [261]

operation. In this case, the eye pattern indicated a turn-on delay of the lasers of about 180 ps. For a bit error rate of less than 10^{-11} at 1.25 Gb/s, the total power consumption of one driver and laser was about 17.5 mW at an optical power of −6.9 dBm. The NMOS transistor used as a small signal modulator, reducing the laser current only by a small amount and especially not below the threshold current, allowed to verify a bandwidth of the order of 2 GHz. This low-power high-speed operation demonstrates the utility and potential of the flip-chip bonding technique for optical interconnect technology.

Another hybrid CMOS-VCSEL transmitter has been introduced [262]. An array of 8 GaAs-AlAs lasers with an InGaAs triple-quantum-well active region was wire-bonded to driver circuits in a 1.0 μm CMOS technology. The threshold current I_{th} of the lasers was 2.7 mA and the peak optical output power exceeded 1 mW without heat sinking. Typical turn-on voltages $V_{ld}(I_{th})$ were (3 ± 0.5) V. The driver circuit is shown in Fig. 6.10.

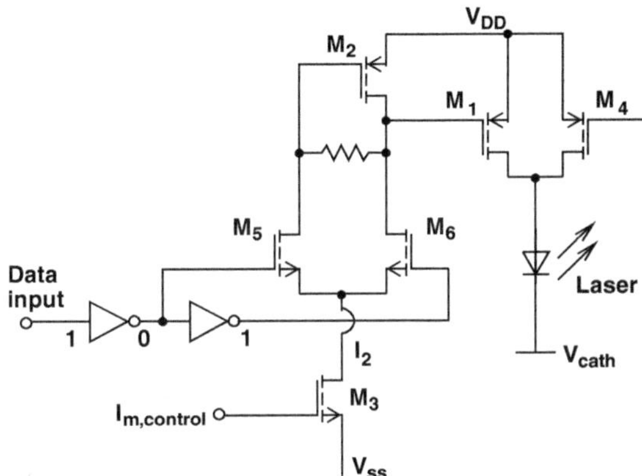

Fig. 6.10. CMOS laser driver circuit [262]

Since the laser array had one common cathode, only the lasers' anode terminals are independent and the modulation current I_m is controlled by the PMOS transistor M_1. The PMOS transistor M_4 regulates the laser bias current, which has to be somewhat larger than I_{th} in order to allow fast laser modulation. By minimizing stored charge in the current steering transistors, fast rise and fall times with low jitter and skew can be achieved. The choice of the width of M_1, however, is a trade-off between low parasitic capacitance for small device width and a reduced output impedance for a large width. A device width of 500 μm has been chosen as a good compromise for $I_m = 15$–25 mA in [262].

The NMOS devices M_5 and M_6 together with the inverter between their gates perform a single-ended to differential conversion. The NMOS device M_3 controls the modulation current I_m. The PMOS transistor M_2 serves as a current mirror reference in the '1' state and is a self-biased inverter in the '0' state. In such a way U_{gs} of M_1 is not zero but approximately equal to the threshold voltage in the '0' state, when the laser is off, and a high modulation speed can be obtained. In the '1' state, I_2 is mirrored (amplified by a certain factor) through M_2 to M_1 and into the laser. With a 50 Ω resistor load instead of the laser and with $I_m = 25$ mA the rise and fall times were about 0.5 ns and the eye pattern was wide open at 622 Mb/s. When a chip with 8 CMOS drivers was incorporated into one package with a VCSEL array flip-chip mounted onto a BeO substrate, the bond-wire inductances somewhat

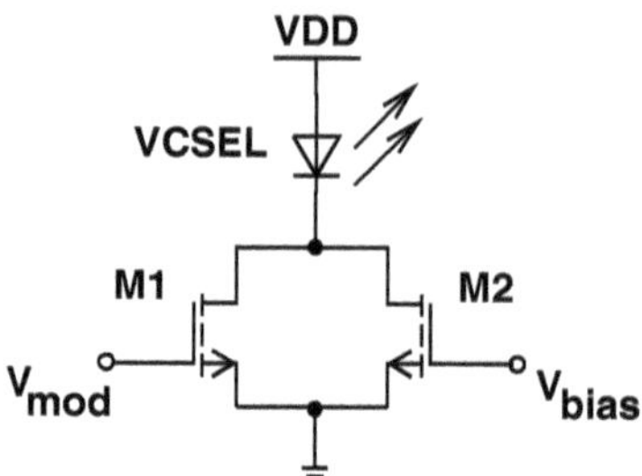

Fig. 6.11. Laser driver circuit with two NMOS transistors [263]

degenerated the slew rate [262]. A data rate of 622 Mb/s with a bit error rate of less than 10^{-9}, however, still has been obtained.

The cell size of one driver circuit with output pads was $175 \times 300\,\mu m^2$. The average power dissipation of one channel was 137 mW for an optical output power of 1 mW with $I_m = 20$ mA, where the laser consumed 75 mW (55 %), M_1 dissipated 45 mW (33 %), and its driver used 17 mW (12 %). For an array with many CMOS-VCSEL channels, this large power dissipation generates too much heat and better designs are necessary.

A laser driver circuit consisting of two N-channel MOS transistors (see Fig. 6.11) is advantageous compared to the shunt driver circuit in Fig. 6.9 with respect to power consumption. Here, M_2 is used to bias the laser above threshold ('0')and M_1 increases the laser current in order to obtain a high optical output power ('1'). Here, the current flowing continuously is lower than in the shunt driver circuit shown in Fig. 6.9. The circuit in Fig. 6.11 was implemented in [263] to modulate flip-chip bonded InGaAs quantum well VCSELs with an I_{th} of 6.4 mA.

The N-channel MOSFETs had a gate length of 0.8 µm and a gate width of 30 µm. With V_{mod} between 3.2 V and 5 V the optical output power could be controlled between 0.05 and 1.3 mW, when VDD = 8.5 V had been chosen. A possible modulation rate of 2 Gb/s for the circuit in Fig. 6.11 has been estimated [263].

6.4 Analog Circuits

In this section, a bipolar amplifier circuit for an optical flame detection system with a very high transimpedance of 1.1×10^9 V/A will be described. Another bipolar amplifier for audio CD systems is also described because of a so-called T-type feedback circuitry.

Two pixel circuits of a-Si:H CMOS image sensors with reduced cross-talk and a very high dynamic sensitivity range, respectively, an image sensor being adaptive to motion and light intensity, a self-calibrating logarithmic camera-on-a-chip (CoC), and an imager allowing read-out of regions-of-interest will follow. Furhermore, a fingerprint detector using lateral bipolar phototransistors is discussed. Innovative circuits for optical distance measurement and 3D cameras are described next. After these, smart pixel sensors are discussed. Then CMOS and BiCMOS circuits for optical storage systems are described.

Subsequently, continuous-mode fiber receiver circuits in bipolar SiGe, NMOS, BiCMOS, and CMOS technology are explained. New speed-enhancement techniques, plastic-optical-fiber (POF) receivers, as well as burst-mode and deep-sub-μm receivers then follow. A comparison of the performance of SOEIC receivers compactly represents the state of the art. Finally an opto-electronic PLL will be discussed.

6.4.1 Bipolar Circuits

The detector of a UV-sensitive OEIC [44] was already described in Sect. 2.1.3. The electronic circuit of the UV sensor system for flame detection where the photocurrent was only 20 pA to 1 nA will be explained here.

Due to the small photocurrents a large amplification with a large transimpedance of 10^9 V/A was necessary in order to obtain signal voltages of up to 1 V. Fortunately, the system did not require accurate control of the gain and, therefore, the current amplification factors of the NPN and PNP transistors of the complementary bipolar process could be fully exploited without a feedback circuitry. The circuit diagram of the bipolar amplifier is shown in Fig. 6.12.

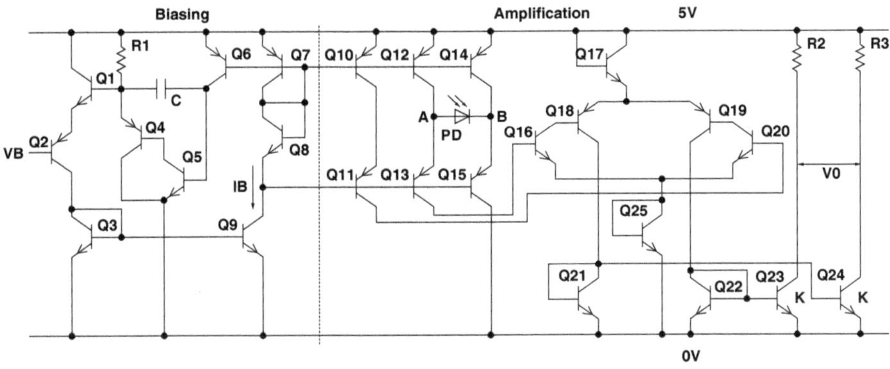

Fig. 6.12. Circuit diagram of a UV sensitive OEIC [44]

The transistors Q_{13} and Q_{15} both in common base configuration together with their current sources Q_{12} and Q_{14}, respectively, provide a low-impedance input for the photocurrent ($R_{in} = 1/g_{m,13}$). The UV photodiode is biased at zero volts. The anodic UV photocurrent I_{UV} (compare Fig. 2.6) flows into the emitter of Q_{13} and is injected into the base of Q_{16}. Transistor Q_{16} amplifies the photocurrent by its current gain factor β_{16}. This amplified current forms the base current of Q_{18}. At the collector of Q_{18}, the photocurrent is amplified by the value of the product $\beta_{16}\times\beta_{18}$. The collector current of Q_{18} is mirrored into Q_{24} by Q_{21} and finally transformed into a voltage by R_3.

The infrared-dependent cathodic current $I_{UV} + I_{IR}$ (see Sect. 2.1.3) is shunted to V_{CC} by Q_{14}. The transistors Q_{10} and Q_{11} supply a reference bias current I_B to Q_{20}, which is the input of the second branch of the differential amplifier with Q_{16}, Q_{18}, Q_{19}, and Q_{20}. A reference current amplified by β_{19} and β_{20} is mirrored via Q_{22} and Q_{23} to R_2. The voltage V_o across the output terminals of the circuit is, therefore, given by

$$V_o = \beta_{16}\beta_{18}KR_3(I_{UV} + I_B) - \beta_{19}\beta_{20}KR_2I_B. \tag{6.1}$$

When perfect device matching ($Q_{16} = Q_{20}$, $Q_{18} = Q_{19}$, and $R_2 = R_3$) can be assumed, the output voltage V_o does not depend on the bias current I_B:

$$V_o = \beta_{16}\beta_{18}KR_3I_{UV}. \tag{6.2}$$

For the complementary bipolar process, β is approximately 140. R_3 was designed to be 28 kΩ, and the mirror factor K was 2. This resulted in a transimpedance of 1.1×10^9 V/A.

Although the output voltage is independent of I_B, I_B has to be very well controlled for correct operation and it has to be very small, because of the large transimpedance value. The bias section Q_1 to Q_9 replicates the gain stage (Q_4 corresponds to Q_{18}, Q_5 corresponds to Q_{16}, and $R_1 = R_3$) and feeds back the proper bias current via Q_6, Q_{10}, and Q_{12}. I_B can be adjusted accurately in the nA range with the reference voltage V_B. For V_B ranging from 0 V to 3.5 V, I_B is adjustable between 15 nA and 0 nA. The compensation capacitor $C = 12$ pF is needed for the stability of the feedback loop.

The response time of the sensor of less than 100 ms was determined by the capacitance of the UV photodiode and the small photocurrents. The input-equivalent noise was smaller than 3.7 pA/$\sqrt{\text{Hz}}$ at 1 Hz. The power consumption of the UV-OEIC was 4 mW at 5 V. The total chip area was 4 mm^2 of which the UV photodiode occupied 1 mm^2.

After this example of an open loop very-high-gain amplifier, an amplifier with a T-type feedback network will be explained. Such a bipolar preamplifier OEIC for compact disk (CD) systems was described in [264]. The amplifier provides a high transimpedance gain, i. e. a low photocurrent is converted to a large voltage, and a relatively large bandwidth. In order to achieve these properties, the feedback via the T-type network shown in Fig. 6.13 is applied. The T-type network consists of the resistors R_{F1}, R_{F2}, and R_{F3}, whereby the values of R_{F1} and R_{F2} were chosen to be equal in [264]. The T-type network is an appropriate measure to simulate a high resistance R_F with low resistor values. The effective value R_F is

$$R_F = (R_{F1}R_{F2} + R_{F1}R_{F3} + R_{F2}R_{F3})/R_{F3} \tag{6.3}$$

and the output voltage V_o is obtained

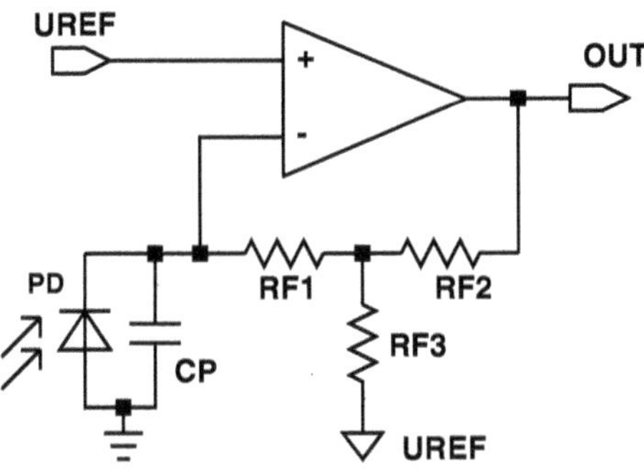

Fig. 6.13. T-type closed-loop configuration for high transimpedance gain [265]

$$V_o = -I_{ph}R_F. \tag{6.4}$$

In such a way it is possible to construct high-gain amplifiers without a high-resistivity polysilicon process module. Emitter polysilicon or gate polysilicon in a (Bi)CMOS process is sufficient. The T-type feedback can also be considered as a means to avoid large RC time constants of large resistance values, requiring a large polysilicon area with a large parasitic capacitance. Effective resistance values of 82 kΩ for a maximum photocurrent of 1.2 μA and of 328 kΩ for a photocurrent of 0.3 μA were realized in [264], whereby the true resistors had much lower values and the die area could be kept low at $400 \times 350\,\mu m^2$. It should be mentioned, however, that the input offset voltage of the operational amplifier is also amplified [265].

The circuit is shown in Fig. 6.14. Only NPN transistors are used in the OEIC for a compact disk (CD) system in order to achieve a large bandwidth. The transit frequency of the NPN transistors was 1.5 GHz. The operational amplifier has the voltage followers Q3 and Q4 in front of the common-emitter difference amplifier stage Q1/Q2 to obtain a low input current, whereby a large offset voltage due to the voltage drop across the T-type network caused by the base current of Q2 can be avoided.

The transistors Q2 and Q5 form a cascode stage to avoid a large Miller capacitance of Q2. High performance PNP transistors were not available and instead of a PNP current mirror load, the resistor R_C is used. Another voltage follower is used in order to obtain a low output impedance and to isolate the collector of Q5 from the load capacitance. The low-frequency open-loop gain was estimated to be $A_0 = 0.5g_{m2}R_C = 40$. A compensation network was reported to be necessary, however, it was not described in [264]. The photodiodes used in the CD-OEIC also were not described.

The complete CD-OEIC contained fast channels with a bandwidth of 16 MHz for a maximum photocurrent of 1.2 μA and slower channels for a photocurrent of 0.3 μA. For the fast channels, rise and fall times of 22 ns with a settling time to 0.1 % of 200 ns were reported. The systematic offset voltage due to the base current of Q4 was smaller than 4 mV. The power consumption for one fast channel was 9.8 mW at 5 V.

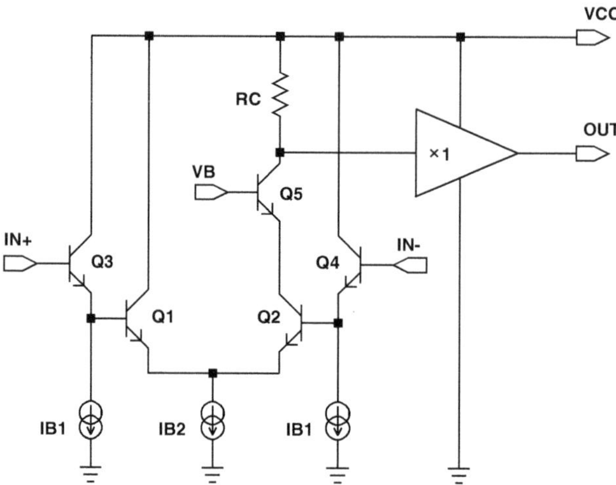

Fig. 6.14. Simplified circuit diagram of a bipolar OEIC for CD systems [264]

6.4.2 Two-dimensional CMOS Imagers

In Sect. 2.2.8, the concept of Thin Film on ASIC (TFA) was described. As an alternative to a self-structuring technology for the a-Si photodetectors in image sensors, an electronic circuit within each pixel was presented which reduces cross-talk between neighboring pixels [149]. The TFA sensor overcoming the coupling effect of neighboring photodetectors in an unstructured a-Si:H film by electronic means was called AIDA (Analog Image Detector Array). A circuit inside each pixel (in c-Si) provides here a constant rear electrode potential for the photodetectors, thereby eliminating lateral currents between neighboring photodetectors in the a-Si:H film. The photodetectors are used in a constant voltage mode. The circuit diagram of a pixel is given in Fig. 6.15. Each pixel consists of an a-Si:H photodetector, eight MOSFETs and one integration capacitance C_{int}. C_{int} is discharged by the photocurrent. The inverter M2, M3 and the source follower feedback M1 keep the cathode voltage of the detector constant. M4 limits the power consumption of the inverter. M5 restricts the minimum voltage across the integration capacitance to 1.2 V in order to always keep the constant voltage circuit working. The integrated voltage on C_{int} is read out via M7 in source follower configuration and via the switch M8. The reset operation is performed as M6 recharges C_{int} after readout. The effective integration time of the pixel is the time period between two reset pulses, because readout is nondestructive and is performed at the end of the integration period.

The integration time may be varied according to the brightness of the scene. By this means, the sensitivity is controlled for all pixels globally. The AIDA sensor consists of 128×128 pixels with a size of $25 \times 25\,\mu\text{m}^2$ each. The dynamic range of the sensor amounts to 60 dB for an integration time of 20 ms. The dynamic range can be extended significantly by means of the

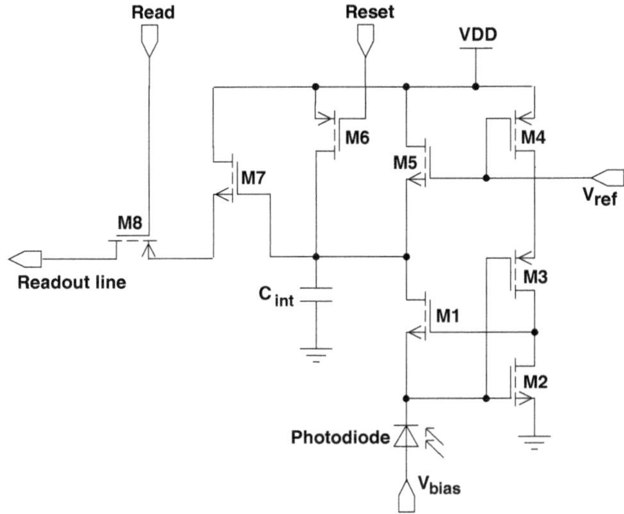

Fig. 6.15. Circuit diagram of a pixel in AIDA for operating the photodiode in a constant voltage mode [149]

sensitivity control. The sensor was tested for illumination levels as high as 80 000 Lux. No blooming effects or image lag were observed [149].

For applications of an image sensor in a vehicle guidance system, for instance, which requires a very high degree of safety, the global sensitivity control is not appropriate. A number of pixels with excessive illumination may be saturated, whereas only slightly illuminated pixels may generate signal voltages below the noise and dark current levels. As pixels with logarithmic output characteristics exhibited a seriously increased sensitivity to temperature changes and fixed pattern noise, a Locally Adaptive Sensor in TFA-technology (TFA-LAS) was suggested [149]. The TFA-LAS allows to control the integration time and, therefore, the sensitivity of each pixel individually [266]. Figure 6.16 gives the complete pixel circuitry realized in the c-Si below each a-Si:H pixel photodetector.

The photocurrent is integrated into the MOS capacitance C_{int} while the rear electrode of the photodiode (cathode) is kept at a constant potential (compare Fig. 6.15). The integration time value is calculated externally for each illumination period and programmed into the pixel as an analog voltage V_{prog}, which is stored on the capacitor C_{prog}. V_{prog} standing at *Dinout*, is switched via N13 to C_{prog} for this purpose. This voltage on C_{prog} is compared to a linear voltage ramp generated by the peripheral electronics during the integration phase. The comparator consisting of the transistors N1–N5 and P6–P8 starts the integration as soon as the voltage ramp rises above V_{prog} and stops it at the falling edge of the ramp. A standard 0.7 µm CMOS ASIC technology implementing the TFA-LAS pixel schematic of Fig. 6.16 enabled

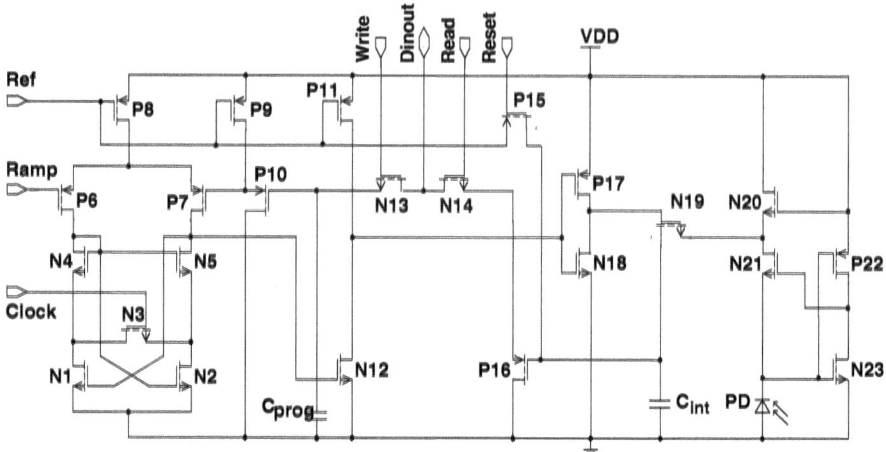

Fig. 6.16. Pixel circuit diagram of a locally adaptive image sensor [149]

a dynamic illumination range of more than 100 dB throughout the complete pixel array at any time. This corresponds to the outstanding dynamic range of c-Si PIN detectors, which is over 100 dB at an illumination intensity of 1000 Lux. For the TFA-LAS, the voltage range for the pixel signal amounts to 54 dB. The remaining dynamic range of 46 dB is included in the integration time and, therefore, in the programming voltage. The combination of both signal and programming voltage thus gives the information on the illumination level of a specific pixel. The TFA-LAS consisted of 64×64 pixels with a size of $40 \times 50\,\mu m^2$ each.

An image sensor in which each pixel adapts its integration time to motion and light intensity was proposed in [267]. The integratin time is shortened when motion is detected in the pixel or when pixel intensity becomes saturated. The adaptivity to motion and light significantly enhances temporal resolution and dynamic range of the image sensor. A scene containing both bright and dark regions can be captured by the pixels with shorter and longer integration times, respectively. Because of the adaptivity to motion, a higher temporal resolution is obtained in a moving area and a higher signal-to-noise ratio is achieved in a static area. The proposed sensor can be utilized for either high-speed imaging or wide dynamic range imaging, which is, for instance, beneficial to imaging systems for vehicles.

In order to control the integration time pixel by pixel, signal processing circuits are combined with the image sensor. The processing scheme of each pixel based on motion detection and saturation detection is shown in Fig. 6.17. Each pixel holds the integrated charge on the photodiode capacitance until motion or saturation is detected. Motion and saturation detection is processed every $t_{\triangle}$ seconds, which is the shortest integration interval. $t_{\triangle}$

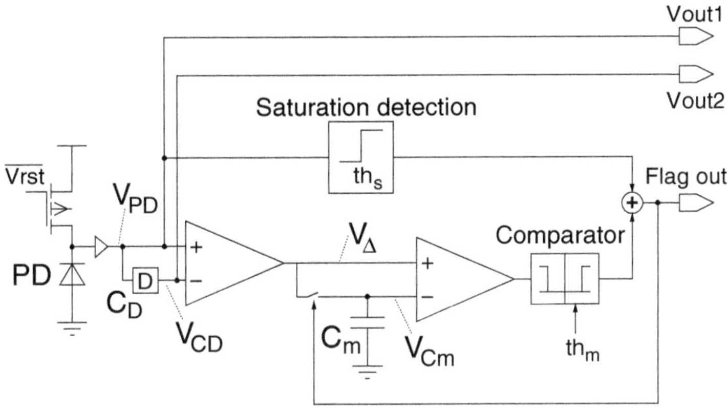

Fig. 6.17. Processing scheme for adaptive integration time in each pixel [267]

was set very small because the proposed sensor was intended to operate at a high frame rate [267].

The motion detection works as follows. First, $V_\triangle$ corresponding to the signal integrated during the shortest interval $t_\triangle$ is calculated by the difference between V_{PD} and V_{CD} (see Fig. 6.17). V_{PD} corresponds to the integration on the photodiode from the last reset to the present. V_{CD} is the voltage on the memory capacitance C_{D} which is delayed by $t_\triangle$ and stems from the previous integration. $V_\triangle$, therefore, can be considered as the change in pixel intensity during a $t_\triangle$ integration time. C_m is another memory capacitor that keeps the $V_\triangle$ of the last output. When the difference between $V_\triangle$ and V_{Cm} exceeds a certain threshold th_{m}, the pixel intensity is judged to have significantly changed and the pixel is considered to have observed motion. When this occures a flag signal is set and the pixel outputs V_{PD} and V_{CD}. Furthermore, the new $V_\triangle$ is stored on C_{m} and the photodiode is reset and integration is restarted, then.

The saturation is detected when V_{PD} exceeds the threshold th_s (Fig. 6.17). The flag signal is also activated and V_{PD} and V_{CD} are output, when this occures. V_{Cm} is also over-written by the corresponding $V_\triangle$, and the photodiode is reset and restarts integration. If the pixel detects no motion or saturation, the photodiode is not reset and keeps integrating without outputting V_{PD} and V_{CD}.

A prototype with 32×32 pixels of the adaptive image sensor was fabricated in a 1-poly 2-metal 1 µm CMOS technology [267]. The pixel size was $85 \times 85\,\mu\text{m}^2$. Together with memory and processing blocks, the total die size was $4.0 \times 6.1\,\text{mm}^2$. Each pixel contained 17 transistors for processing and 10 transistors for memory. Processing required 64 transistors per column in addition. A fill factor of 14 % was given. Power dissipation was 150 mW at 5 V. The minimum integration interval was 680 µs and the maximum integration interval was 16 ms in order to be able to display the output image on a monitor. A dynamic range of 56 dB, a readout noise of 3.1 mV RMS, a saturation charge level of 1.860.000 electrons, and a conversion gain of 1.1 μV per electron were achieved [267]. A good scalability for larger pixel arrays and low power consumption was claimed. In the case of 512 × 512 pixels, a power consumption of 2.4 W was estimated [267].

An interesting solution for a single-chip CMOS camera with logarithmic response was suggested [268]. The high-dynamic-range CMOS image sensor implements nonintegrating, continuously working photoreceptors with logarithmic response. The uniformity problem caused by device-to-device variations is greatly reduced by implementation of an analog self-calibration resulting in a fixed pattern noise (FPN) of 3.8 % (RMS) of an intensity decade at a uniform illumination of 1 W/m^2. The sensor provides a resolution of 384 × 288 pixels and a dynamic range of six decades in the intensity region from 3 mW/m^2 to 3 kW/m^2 spanning the range from moon light to sun and bright studio light. The image sensor contains all components required for operating as a camera-on-a-chip.

The method of correlated double sampling (CDS) is usually implemented in integrating image sensors [129]. The individual pixel offsets are corrected by taking the difference between the output signals before integration (reset state) and after integration. This concept, however, fails for continuously working photoreceptors, since there is no reset state. Off-chip methods with digital calibration using a frame memory for storing the individual pixel offsets have been established [269, 270]. Compared to another single-chip approach [271], the new solution achieves better correction due to a higher number of transistors and an additional capacitor.

Calibration is performed by a reference current instead of a photocurrent. All photoreceptors are stimulated by the same reference current during the calibration cycle. Without FPN, all pixel outputs would show an identical signal. In reality, however, there are device-to-device variations and these signals differ from each other. The self-calibration concept is shown in Fig. 6.18. An operational amplifier compares the individual output voltage to a reference voltage and adjusts each pixel until both voltages are equal. Then, the calibrated pixel state is stored on an analog memory (capacitor C) in every photoreceptor. Because the calibration amplifier is very area consuming, it is not included in the pixel itself, but is located at the border of the sensor array. One amplifier belongs to all pixels of one array column and calibrates them sequentially.

The pixel switches S_{p1} to S_{p5} are realized as single MOS transistors. PMOS switches in contrast to NMOS switches are drawn with a circle. The control signals CS, $\overline{CS}$, and $\overline{RS}$ are connected to all pixels of the same row, the lines I_{ref}, V_{out1}, V_{out2}, and V_{corr} belong to all pixels of the same column. During image acquisition the control line CS (calibration select) is set to LOW and the inverted line $\overline{CS}$ to HIGH. The switch S_{p2} consequently is closed whereas the switches S_{p1}, S_{p3}, and S_{p5} are open. The transistors M1

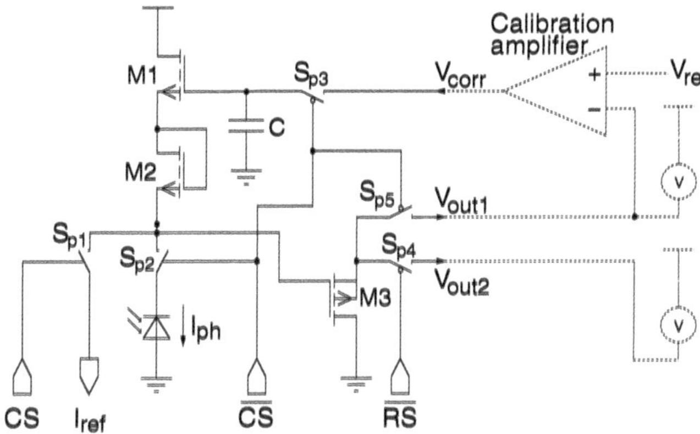

Fig. 6.18. Circuit diagram of self-calibrating logarithmic pixel [268]

and M2 working in subthreshold region for weak photocurrents convert the photocurrent I_{ph} into the logarithmic voltage V_{log}. The two transistors in series result in a doubled signal slope compared to a single transistor. The sensor signal V_{log} is buffered by the PMOS source follower M3. The current required for the operation of this buffer is provided by a current source located outside the pixel and shared with all pixels in a column. To select one pixel for readout, the control signal RS is made LOW switching on S_{p4} and connecting the buffer M3 to the V_{out2} line.

When CS is made HIGH and $\overline{CS}$ therefore becomes LOW, the switches S_{p1}, S_{p3}, and S_{p5} are closed and S_{p2} is opened. The image sensor changes to the calibration mode. The receptor circuits are no longer stimulated by the photocurrents but by the reference current I_{ref}. The sensor signal in calibration mode $V_{log,cal}$ is connected to the V_{out1} line guiding the sensor signal to the calibration amplifier. This operational amplifier compares V_{out1} to the reference voltage V_{ref} and delivers the correction voltage $V_{corr} = A(V_{ref} - V_{out1})$ where A is the gain of the calibration amplifier. A change of V_{corr} is transferred to $V_{log,cal}$ via M1 and M2. The receptor circuit in calibration mode, therefore, forms a feedback loop composed of the logarithmic photosensor and the operational amplifier.

When the pixel returns to the image acquisition mode for CS made LOW and $\overline{CS}$ made HIGH, S_{p3} is opened and V_{corr} is stored on the pixel capacitor C. Now, the photocurrent I_{ph} flows through M1 and M2. Due to the previous calibration, V_{out2} does not depend on the transistors M1, M2, and M3 and their actual properties. The effect of mismatch of these transistors in different pixels is removed by the calibration. Since mismatch of the photodiode responsivity or of charge injection of the switch S_{p3} is much lower, the originally dominating sources of mismatch, i. e. of FPN, are eliminated by the calibration. After calibration, however, mismatch of the photodiode

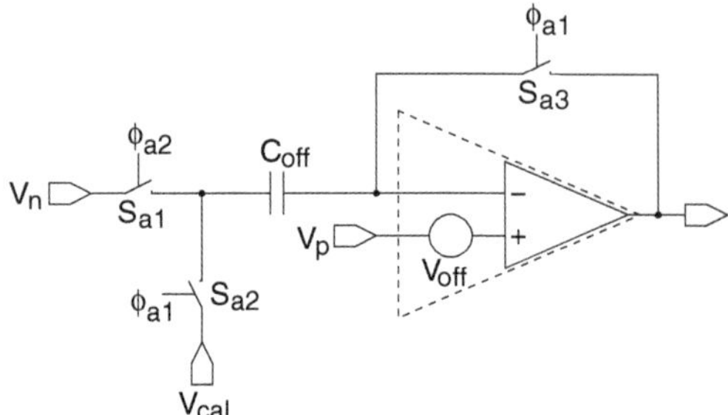

Fig. 6.19. Circuit diagram of autozeroing calibration amplifier [268]

responsivity and of charge injection of the switch S_{p3} are responsible for the remaining FPN.

To avoid variations between individual columns, the reference current sources and the calibration amplifier have to match precisely. The reference current sources are implemented as single MOS transistors biased with an adjustable gate voltage to adapt the reference current to the actual requirements. To achieve a good matching, each transistor was divided into five parts and spread over four adjacent columns. The drain current matching was found to be better than 3 % (RMS) for a reference current of 50 pA [268]. The input offset voltage of the calibration amplifier is reduced by implementing the autozeroing concept [272] depicted in Fig. 6.19. The feedback path through S_{a3} with Φ_{a1} HIGH allows to store the intrinsic amplifier offset V_{off} on the capacitor C_{off}. During normal operation Φ_{a1} is LOW and both inputs V_n and V_p are now shifted by the offset stored on C_{off} leading to a differential input offset of zero. In order to reduce the FPN strongly by the self-calibration mechanism, the amplifier has to provide a high gain [268].

The architecture of the complete image sensor implementing the described self-calibration mechanism is shown in Fig. 6.20. The sensor array contains 386×290 pixels, of which 384×288 can be selected for readout. The additional outer rows and columns are dummy structures to avoid variations in the outer visible pixels due to border effects. The autozeroing calibration operational amplifiers and the reference current mirrors are located above the sensor array. A clock circuit provides nonoverlapping clocks to control the calibration amplifiers. The reference current can be adapted with the help of a mirror circuit with adjustable gain. The bias generation block contains 19 digital-to-analog converters (DACs) each providing a voltage programmable in the range from 0–5 V. These voltages were required to bias and control the analog sensor circuits.

9-bit decoders and row drivers for selecting the calibration and readout row are placed to the left of the photosensor array. During readout the pixel signals are guided to the analog output multiplexer controlled by the 9-bit column decoder located below. CDS is applied in the readout amplifier to eliminate column FPN. Subsequently, a mixer with a sample-and-hold stage combines the sensor output with the video synchronization signal which is

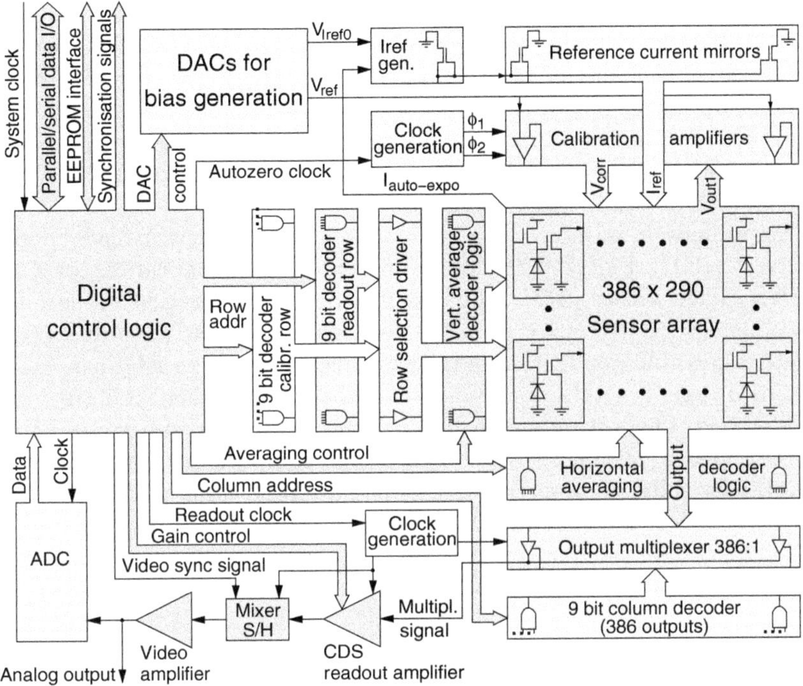

Fig. 6.20. Architecture of the complete self-calibrating logarithmic image sensor [268]

amplified by the video amplifier. An integrated analog-to-digital converter (ADC) generates a 10-bit digital value which can be read out via a parallel interface or a high-speed serial link. The digital logic block on the left side is responsible for the overall control of the image sensor chip. It enables operations like zooming or averaging and it selects all timing parameters. An EEPROM interface offers the possibility to save the chip parameter values in an external EEPROM. They can be automatically recovered after a system or power-on reset.

The self-calibrating logarithmic single-chip camera has been designed and fabricated in a 0.6 μm CMOS process [268]. Only one polysilicon layer was provided by this process. Therefore, all capacitors were implemented as metal-metal-poly sandwich structures or as MOS capacitors. The third metal layer has been used to protect all devices besides the photodiodes from the incident light. The photodiodes were N^+/P-substrate diodes with a quantum efficiency of approximately 50 % at 625 nm and a capacitance of 100 fF. The fill factor was about 30 %. The pixel pitch was quite large and amounted to 24 μm due to the high number of 10 transistors per pixel and due to the pixel capacitor with 150 fF realized as a MOS capacitor. The active die area was $11.5 \times 7.7\,\mathrm{mm}^2$. The photosensor array itself covered $9.2 \times 6.9\,\mathrm{mm}^2$ corresponding to somewhat more than the size of a 2/3-inch CCD sensor [268]. The slope of the logarithmic sensor could be adjusted between 130 and 720 mV per decade by changing the feedback capacitors in the switched-capacitor readout amplifier. With the help of reference structures it was found that the calibration mechanism improved the pixel nonuniformities by a factor of 10 or more. The power consumption of the image sensor was 150 mW at 5 V. The maximum pixel rate was 8 MHz [268].

Sometimes it is interesting to read out image areas called "regions of interest" (ROI). Figure 6.21 shows different ROI configurations of a 1k × 1k CMOS image sensor [273]. One possible readout operation mode allows to read out several ROIs (ROI_I ... ROI_{IV}) with different sizes and positions. Another possible application is target tracking were the ROI can follow the moving object. Both the start position and the size of each ROI are quantized at 32 pixels. The advantage of dividing the sensor into subgroups of 32 × 32 pixels is that a lot of chip area can be saved compared to a complete on-chip decoder for fully random access. It was stated that about 95 % of the area for such an decoder were saved by the ROI block technique [273].

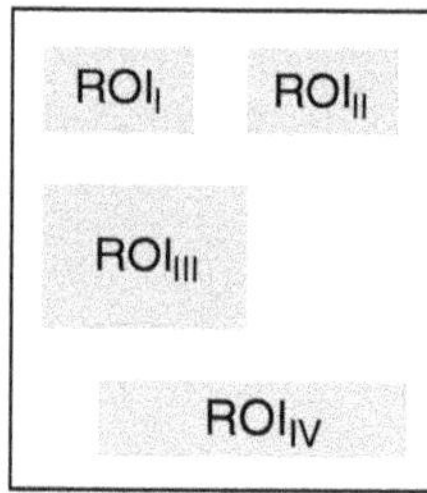

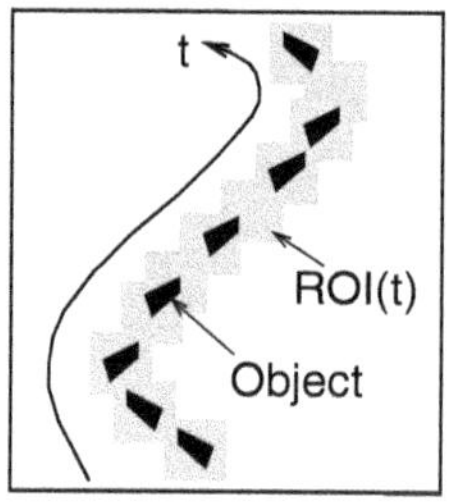

Fig. 6.21. Region-of-interest configurations (left: multiple regions of different size; right: target tracking) [273]

The architecture of the ROI imager is shown in Fig. 6.22. The CMOS image sensor in 0.5 µm technology contains a 1025 × 1024 pixel array with each pixel implementing the same pixel circuit as shown in Fig. 2.33. The row with number 1025 is used as dark row with light-independent pixels. These are used to compensate for dark current and channel mismatch. The digital circuitry at the left side of the pixel array contains shift registers for reset and select operation (exposure control) of the image sensor and the logic for ROI readout. Correlated double sampling (CDS) located at the bottom of each column is implemented to significantly reduce 1/f-noise and fixed pattern noise (FPN). A column parallel sample&hold stage and an analog four-channel multiplexer are implemented to provide a video frame rate at 512 × 512 pixel image resolution.

The chip area of the ROI imager realized with a standard N-well 0.5 µm CMOS process was 164 mm^2. The pixel pitch was 10 µm × 10 µm with a fill factor of 41 %. With a full frame resolution 15 frames per second were possible. With one integration time a dynamic range of 61 dB was reported.

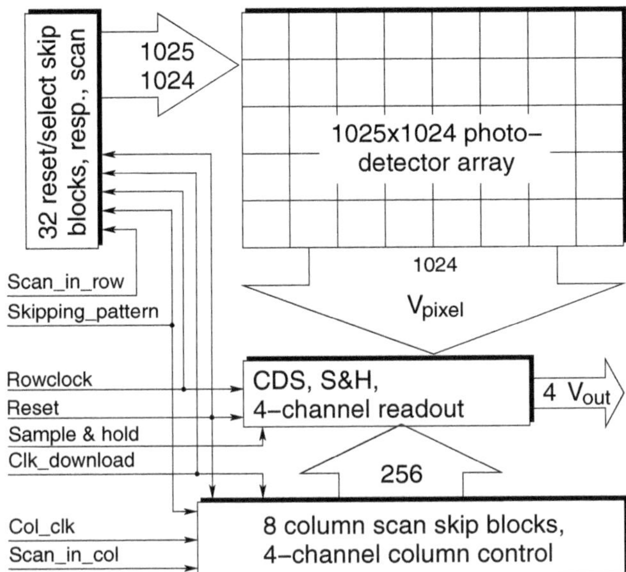

Fig. 6.22. Architecture of region-of-interest imager [273]

Using a multi-exposure technique the dynamic range could be extended to 92 dB [273]. The power dissipation was 300 mW at 3.3 V.

Another interesting CMOS image sensor in bulk silicon should be mentioned here. It combines a lateral bipolar phototransistor array and a current comparator for digitizing the image [154]. The lateral bipolar phototransistor described in Sect. 2.2.9, Fig. 2.37, was used as a photodetector in a fingerprint detection chip fabricated in standard CMOS technology. The fingerprint detection chip contained an 64×64 array of the lateral phototransistors.

The phototransistors in the array were scanned by connecting one at a time to a current comparator using MOS transistors as selection switches. The circuit of the current comparator is shown in Fig. 6.23.

A threshold must be established so that pixel current values may be compared in order to give a black/white image. The most convenient point to place the threshold is the average of all pixel currents in order to obtain a global average. It also may be advantageous to select only m pixels from the entire array to give a local average from a sub-array. The average of the pixel currents is then obtained by dividing the measured current by m. In the fingerprint detection system, for instance, four rows of the 64×64 array were used and a 256 : 1 current mirror was used to give the average pixel current. The current mirror gate voltage is stored on a 10 pF capacitor. From this capacitor the reference voltage V_{ref} is derived setting the reference current for the current comparator. During the following 4096 clock cycles, all pixels are selected one at a time and their photocurrents are digitized to '1' or '0' by the current comparator. During each clock cycle, one selection MOS switch is activated and the pixel current is pulled through the PMOS current mirror

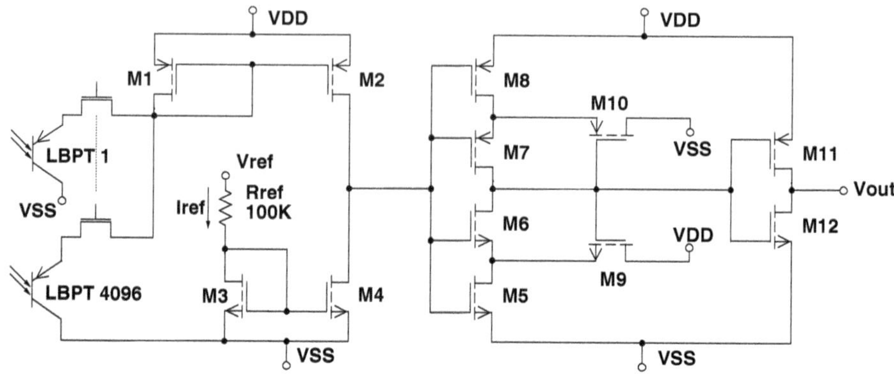

Fig. 6.23. Current comparator circuit of a fingerprint detector [154]

M1 and M2 (Fig. 6.23). If this pixel current is greater than the average current, which was set up through M3 and M4, V_{out} goes high; otherwise V_{out} goes low. The transistors M5–M12 compose a noninverting Schmitt trigger, which imposes a hysteresis and, in turn, stable digitization.

The field of application of such an image detection system is not limited to a fingerprint detection system, of course. Other types of pictures can also be digitized.

6.4.3 Optical Distance Measurement Circuits

The so-called time-of-flight (TOF) method is often used for optical distance measurement. The principle of TOF is shown in Fig. 6.24.

A light-emitting diode (LED) or a laser diode is used to send a light pulse. The light reflected at the object to which the distance has to be determined is detected by a receiver. The time difference t_{of} between emission of the light pulse and its reception at the receiver is directly proportional to the distance:

$$d_{obj} = \frac{c_0 t_{of}}{2} \tag{6.5}$$

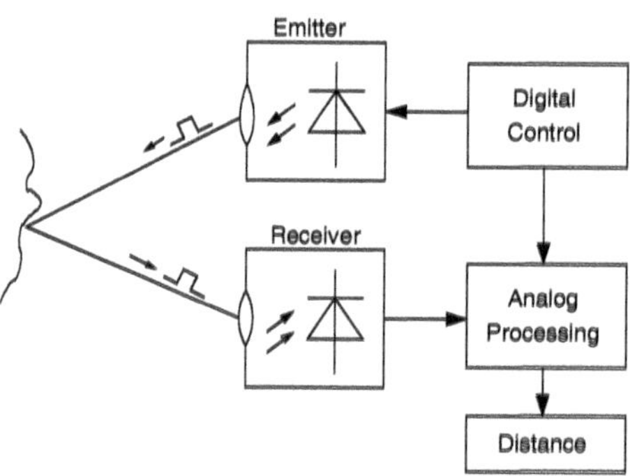

Fig. 6.24. Principle for optical distance measurement [274]

where c_0 is the velocity of light in vacuum or air. For short distances, where t_{of} is shorter than the light pulse, the phase shift from emitted pulse to detected signal can be used to determine the distance.

A time-of-flight optical distance measurement circuit can be advantageously based on the CMOS photonic mixer device (CMOS PMD) described in Sect. 2.2.10. The PMD combines the detection of the light and the correlation of this optical signal with the electrical signal used for modulation of the LED or laser diode [157, 275]. For short distances, the CMOS PMD, therefore, can be used as a detector for the phase difference between this light detected after reflection and this electrical signal. Figure 6.25 shows a circuit appropriate for optical distance measurement with a PMD.

The electrical signal that modulates the light source is applied to the two gates of the PMD in a differential form ($+u_m$ and $-u_m$). The photocurrents of the PMD i_a and i_b flow through the NMOS transistors M7 and M3, respectively. The inverters M9 and M10 with the source follower M7 as well as M1 and M2 with the source follower M3 keep the potential at the PMD source and drain largely independent from the size of the photocurrents by decoupling the PMD from the load capacitors C1 and C2. The PMD source/drain potential is kept at the inverter threshold which is at about $U_s/2 = 2.5\,\text{V}$. The photocurrents of the PMD i_a and i_b charge the capacitors C1 and C2. With a delay of the received light pulse compared to the electrical modulation signal due to the time-of-flight, a different charge is accumulated on the two capacitors resulting in a voltage difference. Simply speaking, the voltage difference $\triangle U_{ab}$ is proportional to the optical distance between the measure-

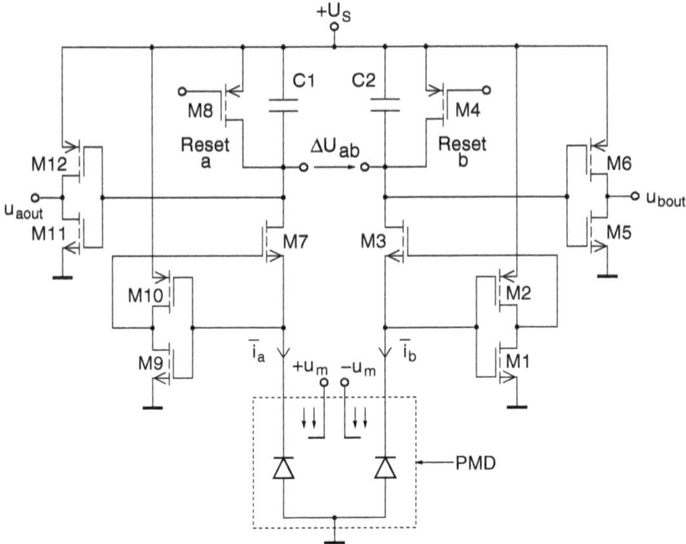

Fig. 6.25. Circuit for optical distance measurement with PMD [275]

ment unit and the target. Actually the sum of both charges also has to be determined to eliminate the influence of the target reflectance, since the distance is proportional to $\triangle U_{\mathrm{ab}}/(\triangle U_{\mathrm{a}} + \triangle U_{\mathrm{b}})$. The inverters M5 and M6 as well as M11 and M12 amplify the voltage difference $\triangle U_{\mathrm{ab}}$ to $U_{\mathrm{aout}} - U_{\mathrm{bout}}$. A good matching between these inverters is essential. Background light is integrated on both capacitors in the same way. Without precautions like compensation of background light it can drive the inverters into saturation. In practice, measurements with at least three different phase differences between electrical laser modulation signal and u_{m} are performed to obtain an unequivocal distance result [275].

When no such a PMD shall be used, conventional photodiodes can be used as a light detector. Background light can be measured without any laser pulses in the cycle W_1 (Fig. 6.26). The following cycles are used to determine the time-of-flight of the laser echo. A number of $N-1$ consecutive integration windows is used to reduce the influence of noise on the distance measurement. The basic idea of these consecutive integration windows is to accumulate the small signal caused by each pulse, adding at each pulse the voltage signal of the last packet to the previous ones for a suitable number of pulses. A better signal-to-noise ratio is obtained by averaging the noise contributions on many samples.

Figure 6.27 shows the block diagram of an implemented pixel architecture. The photodiode read-out stage is a voltage amplifier. A self-biasing cascode amplifier resulting in a good gain, low input-referred noise, immunity to transistor mismatch, and good pixel-to-pixel uniformity was chosen [274].

The operation of the first stage is characterized by three major steps [274]: (i) the photodiode capacitance is reset to V_{pol} by switching on transistor S_{res}. (ii) The cascode amplifier is properly biased by switching S_{bias} on; the two transistors Mn1 and Mp2 are forced into their saturation region, biasing the amplifier at its highest gain operating point regardless of transistor mismatch. (iii) The switch S_{bias} is finally opened. A voltage value suitable to keep the amplifier in the high-gain region is sampled on C_{bias} and the amplifier works as a standard cascode amplifier. The gain of the amplifier implementation was 45 with an input-referred noise of $13\,\mu\mathrm{V}_{\mathrm{rms}}$.

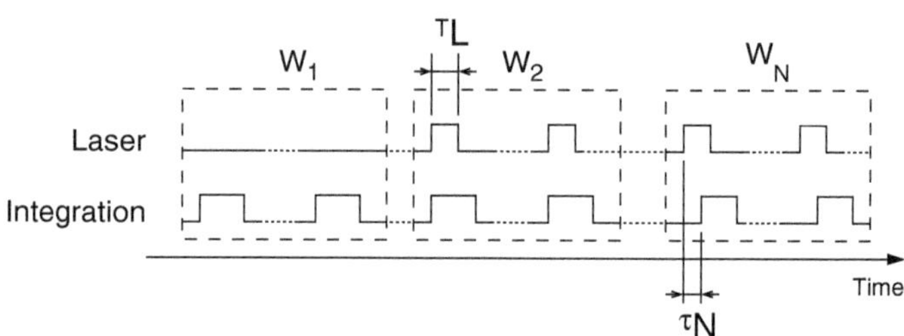

Fig. 6.26. Successive integration with N windows for each distance measurement [274]

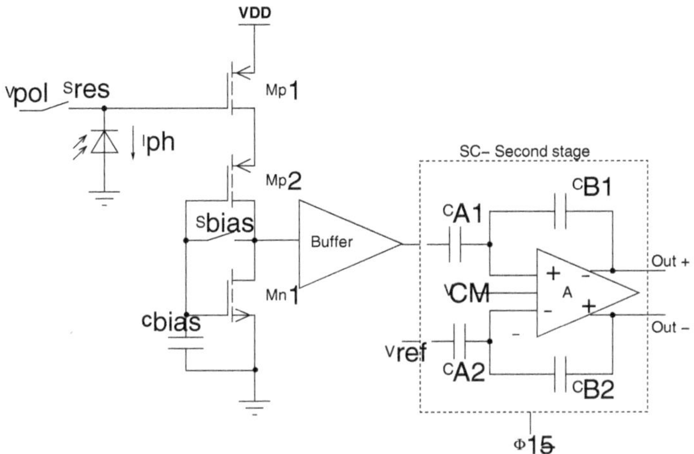

Fig. 6.27. Block diagram of circuit for optical distance measurement [274]

The second stage (Fig. 6.27) performs the correlated double sampling (CDS) filtering and the charge accumulation. The greatest parts of kT/C and flicker noise contributions are removed. A fully differential topology was chosen to reduce switch charge injection effects and "environmental" noise because of its good common-mode signal rejection ratio. Switched-capacitor (SC) circuits with their digital noisy power supply lines, therefore, become possible. A folded cascode architecture was employed in the SC-amplifier resulting in high gain, small area occupation, low noise, and low power consumption.

The operation of the second stage is like this [274]: The accumulation capacitances $C_{\mathrm{B1,2}}$ are reset first while the OTA is kept in a voltage follower configuration and its input offset is sampled onto the feedback capacitances. Then the first stage reset value (including low-frequency noise contribution) is sampled onto C_{A1}, while the reference value Vref is sampled onto C_{A2}. Then the charge stored onto $C_{\mathrm{A1,2}}$ is discharged onto the feedback capacitors $C_{\mathrm{B1,2}}$ and the amplifier outputs change to the reset value $\triangle VOut_1 = Out + -Out- = Vres$. The signal value Vsig is subsequently sampled onto C_{A2} (while Vref is sampled onto C_{A1}) and the stored charge is then fed to $C_{\mathrm{B1,2}}$. After the five cycle procedure, the amplifier output gives $\triangle VOut_1 = Vres - Vsig$. The five-cycle procedure is repeated at each pulse, thereby accumulating the charge purged by the reset and noise equivalent charge. The output signal after the i-th packet is $\triangle VOut_i = \triangle VOut_{i-1} + (Vres - Vsig)_i$.

A test chip fabricated in a 0.6 µm 3M-2P CMOS technology had a good accuracy for distances from 7.5 to 10 m at pulse numbers $N = 64$ and $N = 128$. At shorter distances the distance error of the test chip was up to about 20 % [274]. The test chip allowed a variation of the pulse number N between

4 and 1024. The area of a pixel was $180 \times 160\,\mu m^2$ with a fill factor of 14 %. The circuitry was screened by metal 3 against light incidence.

6.4.4 Three-dimensional CMOS Cameras

A CMOS imager chip for three-dimensional (3D) imaging applications containing a 32×2 photodiode array, fast synchronous electronic shutter, analog switched-capacitor (SC) readout, multiple double short time integration (MDSI), clocking, pulse synchronization, and control was presented [276]. These features allow optical time-of-flight (TOF) measurements of laser pulses reflected from a target. The circuit diagram of each pixel in this 3D imager is shown in Fig. 6.28.

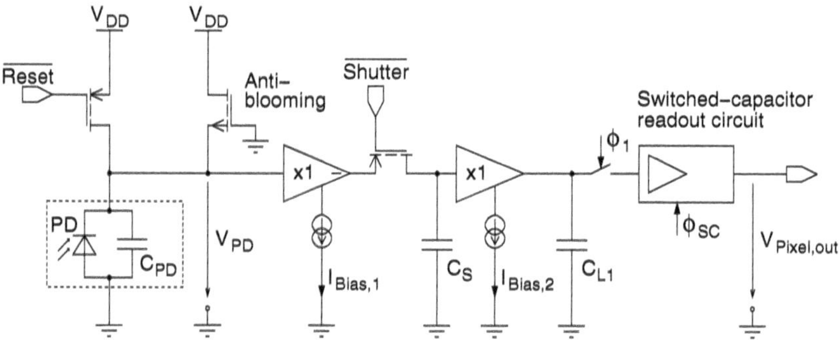

Fig. 6.28. Pixel circuit of 3D-imager [276]

To achieve a high speed the photodiode voltage VPD is buffered by a source follower. The voltage stored on C_S is then transferred using a second buffer to a switched-capacitor readout circuit. This SC readout circuit performs several tasks: sample&hold, correlated double sampling (CDS), and analog accumulation if laser pulse bursts are used [276]. The electronic shutter is activated by switching the reset transistor off and the shutter transistor on.

The measurement principle of the 3D TOF imager is illustrated in Fig. 6.29. A light pulse of a few ns duration generated by an IR laser diode illuminates the field of view by using a simple optical diffuser to widespread the laser beam. Simultaneously or after a fixed delay time, the electronic shutter is activated. The photocurrent of the reflected light including background illumination of the scene in the field of view is integrated at the photodiode array PD. The amount of the generated charge depends on synchronous timing of the laser diode and electronic shutter as well as on the propagation delay of the reflected pulse, i. e. on the distance. After a few ns (T_1), the electronic shutter is switched off and the corresponding voltage V_1(ill) of each pixel is stored on the pixel storage capacitor C_S. This measurement is repeated with the same shutter time and without active illumination (laser

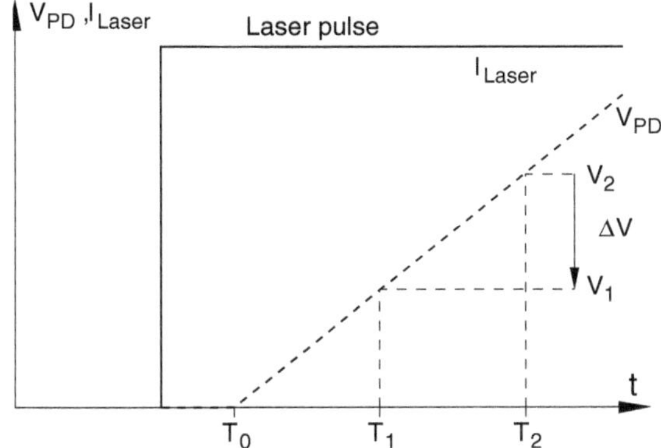

Fig. 6.29. Principle of measurement used in the 3D-imager [276]

is off). The resulting voltage $V_1(dark)$ is then subtracted from the previous measurement resulting in $V_1 = V_1(ill) - V_1(dark)$ at the SC readout circuit. A second measurement cycle with a different integration time, i. e. the shutter is closed at T_2 (see Fig. 6.29), results in a second voltage V_2. With these two cycles the time-of-flight T_0 can be extrapolated and the distance d can be obtained as [276]:

$$d = \frac{1}{2} c_0 \frac{V_2 T_1 - V_1 T_2}{\triangle V}, \tag{6.6}$$

where c_0 is the velocity of light. With this approach the influence of the background light and the reflectivity of the object are cancelled.

Optionally to increase the accuracy or the distance range, each of the two measurement cycles may be repeated m times applying laser pulse bursts thus yielding MDSI [276]. The resulting voltage differences are accumulated in the SC readout circuit. An improvement of the signal-to-noise ratio (SNR) by a factor $\sqrt{m}$ results extending the range of the distance sensor.

Several important aspects have to be mentioned. The CDS operation reduces not only offsets and low frequency noise of the buffers but also of the SC readout circuit. Shutter clocking is very important since shutter phase noise, i. e. a jitter or time offset between the different shutters in the array, results in a distance inaccuracy. A shutter phase noise of 1 ps, for instance, corresponds to a 0.15 mm distance error [276]. To obtain best synchronous shutter clocking, synchronization circuits ensure minimum skew (see Fig. 6.30).

The 3D CMOS imager was fabricated in a 0.5 µm three-metal-layer, one-polysilicon (3M1P) CMOS process. Two linear photodiode arrays with 32 pixels each were integrated on the 3D imager chip. N-well/P-substrate photodiodes with an area of 260 µm^2 were implemented. The die had an area of 42 mm^2 including all circuitry and bondpads. By employing subtraction of readings with laser ON and OFF and by using short pulses and shutter times, insensitivity to background illumination even with direct sun light was achieved [276]. The light falling on the chip circuitry was blocked by

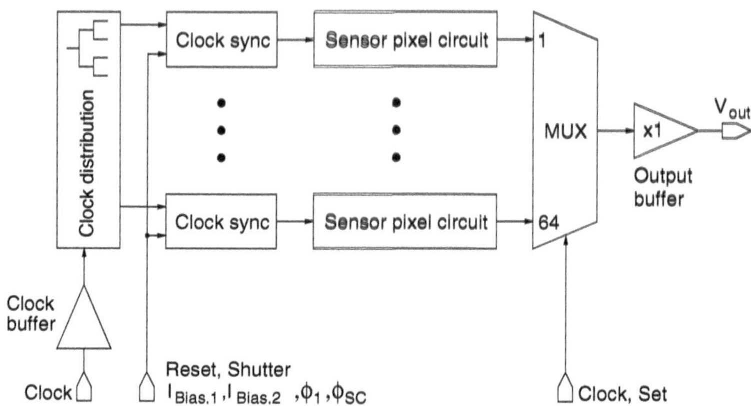

Fig. 6.30. Block diagram of 3D-imager chip [276]

a metal3 shielding. Another trick was necessary to combine low-noise and high-speed in analog CMOS circuits. High bias currents are required during operation. To avoid thermal problems dynamic power management using switched bias currents I_{Bias} (see Figs. 6.28 and 6.31) was employed. The power dissipation was 330 mW at a single 3.3 V supply. The minimum shutter period was 30 ns. Measurements verified an insensitivity to background light up to 10 kLux [276]. Due to noise optimization the measured noise-equivalent-power (NEP) was 5.2 W/mm^2 at 30 ns shutter time. Calibrated distance measurement showed a linearity error of less than 5 % for different pulse counts. For a 3D image an accuracy of ± 1 cm was reported [276]. The range resolution of less than 5 cm for one pulse and for a measurement range of more than 5 m was given. The range resolution became less than 1 cm for burst operation with 100 pulses. A maximum frame rate of 20.000 frames/s reducing to 4.000 frames/s in 100-pulses burst operation for 3D images at a

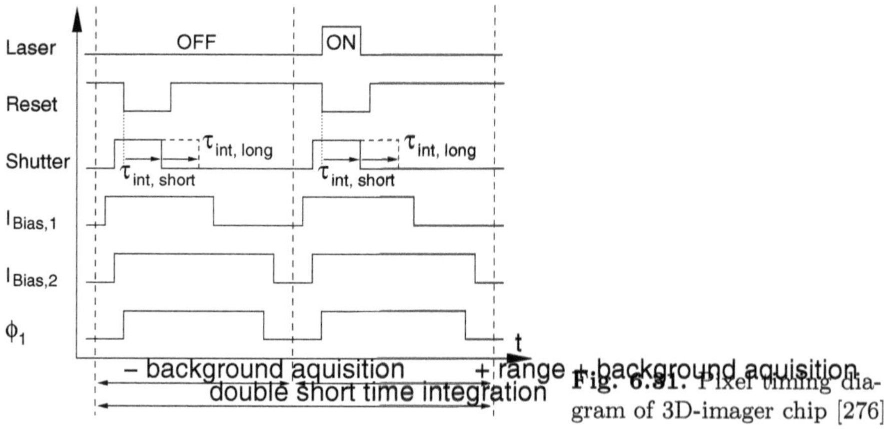

Fig. 6.31. Pixel timing diagram of 3D-imager chip [276]

66 MHz clock was reported. Because the sensor array implemented only two pixel points in one lateral dimension, a linear mechanical scan was required for 3D image recording. The measurement range was up to 10 m depending on the number of pulses [276].

6.4.5 Smart Pixel Sensors

State-of-the-art CMOS technology allows to integrate photodetectors, analog, and digital electronic functionality on the same "smart image sensor" chip or even in so-called "smart pixels". In such a way pixels with a very high functionality can be realized with properties which are difficult to achieve or which are not possible to achieve with conventional photosensors. Some examples out of the whole variety of smart pixel sensors like linear transforming smart pixels, optically adressable gate arrays, smart pixel sensors for real-time optical tomography for medical and technical applications as well as sensors for basic machine vision tasks with the ultimate goal of creating "seeing chips" will be addressed in the following.

Let us consider linear transforming smart pixels first. The fact that the shape of photodiodes can be chosen almost arbitrarily allows an easily realizable way to a huge amount of photosignal processing capability. We want to assume an one-dimensional light intensity distribution $I(x)$ that falls onto a photodiode whose photosensitive width $w(x)$ is varying with the distance x. Because the responsivity of photodiodes can be linear over more than 10 orders of magnitude [277], the output signal s_{pd}

$$s_{pd} = \int_a^b I(x)w(x)dx \tag{6.7}$$

of the photodiode results. Therefore, parametrized linear transforms of incident light distributions can be determined without additional processing elements, when the photodiode shape is chosen properly. A triangular shape, for instance, allows to determine the centroid of an one-dimensional light distribution [278]. An amplitude-modulated sine shape comprises a Fourier-transform pixel which calculates the weighted sine and cosine part of a signal, with which the magnitude and phase of an incident light pattern could be obtained accurately [278]. This principle was implemented in several optical metrology and encoder products with an accuracy of below 0.1 μm for linear and rotational distance measurements [279].

An improved Segmented Fourier Phase Detector (SFPD) was presented in [279]. The construction principle of this SFPD is shown in Fig. 6.32. A sine and cosine function are superimposed within one period. In a CMOS chip 12 periods of 80 μm each were implemented. Six photosensitive segments, P1–P6, were present. The photocurrents of these segments can be converted to six proportional voltages by transimpedance amplifiers. By the proper addition of these voltages U_{P1}–U_{P6}, purely sinusoidally modulated signal voltages, U_{PD1}–U_{PD4}, are obtained [279]:

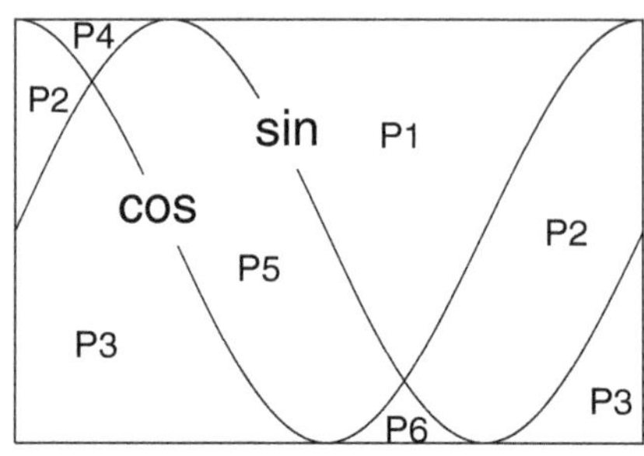

Fig. 6.32. Construction principle of SFPD [279]

$$\begin{aligned} U_{\mathrm{PD1}} &\propto U_{\mathrm{P1}} + U_{\mathrm{P4}} + U_{\mathrm{P5}}\,, \quad & U_{\mathrm{PD2}} &\propto U_{\mathrm{P2}} + U_{\mathrm{P3}} + U_{\mathrm{P6}}\,, \\ U_{\mathrm{PD3}} &\propto U_{\mathrm{P1}} + U_{\mathrm{P4}} + U_{\mathrm{P2}}\,, \quad & U_{\mathrm{PD4}} &\propto U_{\mathrm{P5}} + U_{\mathrm{P3}} + U_{\mathrm{P6}}\,. \end{aligned} \tag{6.8}$$

The desired quadrature signals U_{sin} and U_{cos} result as differences in both signal pairs:

$$U_{sin} = U_{\mathrm{PD1}} - U_{\mathrm{PD2}}\,, U_{cos} = U_{\mathrm{PD3}} - U_{\mathrm{PD4}}\,. \tag{6.9}$$

The phase angle Φ can be obtained from

$$\Phi = arctan\frac{U_{sin}}{U_{cos}}\,. \tag{6.10}$$

The relative position x_Φ within a grating period p is proportional to the phase angle Φ: $x_\Phi = p\Phi/2\pi$. The detector design and signal processing can be simplified, since both small photodetectors P4 and P6 appear only in combination with the large photodetectors P1 and P3, respectively. Therefore, P4 and P1 as well as P6 and P3 can be connected to form the new photodetectors P1' and P3': P1' = P1+P4 and P3' = P3+P6. In such a way only four transimpedance amplifiers are necessary and the photocurrents of the small detectors P4 and P6 do not have to be amplified separately [279].

Another very interesting form of smart pixels are optically programmable gate arrays. Field programmable gate arrays (FPGAs) are today widely used to develop electronic systems cost and time efficiently. FPGAs contain a large number of digital configurable logic blocks (CLBs) with which many logical functions can be realized. Furthermore programmable signal routing components between CLBs and input-output blocks are implemented in FPGAs. The FPGA's configuration data is often stored in a program memory which is read in from an external storage device at power-up or when a new configuration has to be programmed. Since despite of clock periods in the nanosecond range the configuration times of FPGAs can be in the millisecond range, optically programmable gate arrays (OPGAs) [280] may further increase the computational speed of advanced microelectronic systems. The central element of an OPGA is a combination of a smart pixel and a CLB, resulting in a so-called optically configurable block (OCB). A few hundred photons are sufficient to change the logic function of an OCB. Therefore, smart pixels can contribute to a much closer connection between photonics and electronics especially since computation seems to be carried out electronically and data seem to be transferred optically and to be stored optically for a long period in the future.

Optical low-coherence tomography (OCT) [281] is a rather new and sophisticated technique for acquiring three-dimensional images with micrometer resolution of many biological and technical objects which are diffusely reflecting making it difficult to distinguish details in their bulk by conventional methods. In the beginning of OCT, the image acquisition had to be done by mechanical scanning with a single-channel (point) OCT detector, due to the complexity and necessary performance of the signal processing circuitry. Figure 6.33 shows the block diagram of such a circuitry. A smart pixel for OCT applications has to implement photodetector, offset cancellation (done with the block FB by changing the current in the PMOS current source CS, because the modulation signal is typically −60 dB or far below the DC signal), low-noise preamplifier, rectification, programmable low-pass filtering and random access readout.

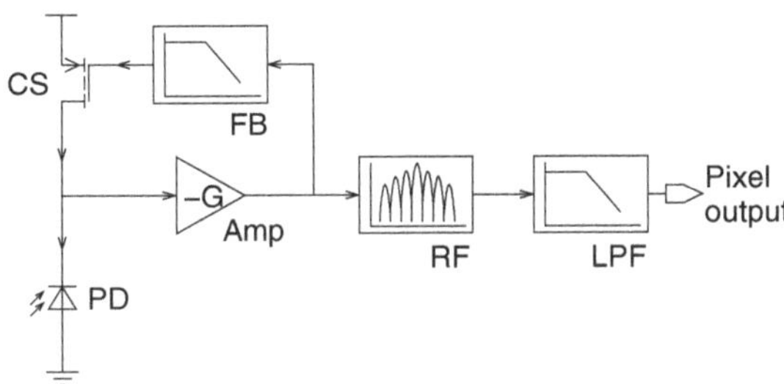

Fig. 6.33. Principle of pixel for optical low-coherence tomography [278]

All these functions were realized in one smart OCT pixel with 20 MOSFETs [282]. An OCT chip in 2 μm CMOS technology with a bipolar transistor option implementing 58×58 smart OCT pixels for simultaneous imaging of parallel OCT channels was demonstrated. The total die area of the OCT chip was $7.2 \times 7.2\,\mathrm{mm}^2$ with a pixel pitch of 110 μm. The photodiodes had a size of $35 \times 35\,\mu\mathrm{m}^2$. A sensitivity of −58 dB, i. e. a minimum detectable optical power of 6.3 fW was observed compared to the shot-noise limit of 1.2 fW. The detection limit thus was 7.2 dB above the shot-noise limit probably due to electronic noise in the signal processing circuitry [282]. This OCT chip represents an interesting step towards a new type of inexpensive real-time imaging without ionizing radiation with many possible applications in medicine, biology, and technical surface topography [278].

Sensors for basic machine vision are a large field of current investigations. The increasing demand for fast and efficient low-power and low-cost vision systems is also being led by the needs of surveillance, automotive and mobile systems market. This market asks for the capabilities of segmentation, recognition as well as classification of characters, faces, and postures, for instance. The traditional way to fulfil these tasks is: (i) acquisition of the image on a low-cost CCD camera and (ii) software processing of the raw image information on a digital platform (personal computer, digital signal processor (DSP), or application specific integrated circuit (ASIC)). Although the computational capabilities of digital platforms improve every year, they are far from low-cost and portability, as needed for the visual applications targeting the grand public [283]. A possible solution is to shift part of the computation into the sensor itself, where in each pixel analog preprocessing can be performed [284–289]. In this way the camera becomes a more sophisticated device, i. e. a so-called retina. In many cases the pictures taken are preprocessed electronically, involving algorithms that perform preprocessing in parallel for or at all pixels. Each pixel has to be equipped with its own analog or digital pre-processor which processes its own photosignal and these of its neighbors independently of each other at the same time. Here only one recent example for such a massively parallel approach can be included [283]. In this example, each pixel exploits the local image surroundings to calculate local gradients. Similar to biological vision systems which can focus attention on the most conspicuous portion of a scene, the smart pixels described in [283] implement a most interesting concept to read out the gradient signals: the gradient information is temporally ordered from the largest to the smallest gradient's magnitude, resulting in address-event coded output.

The main idea for computation of the spatial gradients is based on the definition of the derivative of a 2-dimensional function performed in the vector direction $\mathbf{r}$

$$\frac{\partial I(x,y)}{\partial \mathbf{r}} = \frac{I(x,y)}{\partial x} cos\alpha + \frac{I(x,y)}{\partial y} sin\alpha \tag{6.11}$$

where α is the vector's angle. This angle can be swept in time using steering functions and the computation of the gradient on a discrete lattice square at the pixel level can be performed by

$$\frac{\partial I_{\mathrm{i,j}}}{\partial \mathbf{r}(\mathbf{t})} = \frac{I_{\mathrm{i+1,j}} - I_{\mathrm{i-1,j}}}{2} cos\omega t + \frac{I_{\mathrm{i,j+1}} - I_{\mathrm{i,j-1}}}{2} sin\omega t \tag{6.12}$$

where $I_{\mathrm{i,j}}$ represents the local luminance at pixel (i,j) as dedected by the photodiode and ω is the angular frequency of the steering functions. In such a way, the local derivative in the direction of vector $\mathbf{r}$, i. e. the angle ωt, is continuously determined as a linear combination of two basis functions (the derivatives in x and y directions). This is the principle of steerable filters [290]. A function is called steerable under a transformation (translation, rotation, scaling, etc.) when all transformed versions of this function can be expressed using linear combinations of a fixed, finite set of basis functions. Here the sine and cosine functions are interpolating functions. The first nearest neighbor pixels in x and y directions are involved here in the compution of the gradient. Due to the interpolating functions, sub-pixel precision is possible, i. e. derivatives in directions between discrete pixel locations can be calculated. The spatial gradient is by definition the vector $\mathbf{r(t)}$ for which (6.12) reaches its maximum. If the image, i. e. its luminance fluctuates substantially slower in time than $1/\omega$, (6.12) represents a sine wave in time, which is called *local steering function* in the following, for which amplitude and phase are related to the magnitude and direction of the gradient, respectively.

As a consequence, the coding of information is straight forward. After the first revolution of the interpolating functions, each local steering function went through a maximum, which can be captured by a simple peak-detection mechanism and stored locally on small (parasitic) capacitors. Then, at the next period of the interpolating functions, a monotonic decreasing threshold function (at best with a constant slope) is triggered (see Fig. 6.34). When the decreasing threshold equals the stored maximum value, address-event pulses, encoding the location of the corresponding pixel, are generated and transmitted on an asynchronous bus [291]. In this scheme, the time of appearance of an event on the bus is directly related to the gradient's magnitude, and the sending of gradient information off-chip is ordered from high to low values. The gradient's direction similarly contained in the phase of the local steering functions, is also pulse-encoded in time by detecting the zero-crossing.

Because all communications are completely asynchronous, i. e. each pixel can access the bus independently from the others, collisions have to be avoided. Therefore, each pixel is allowed to access the bus only once during a frame acquisition resulting in a slightly complicated logical structure of the pixel. Another powerful measure to avoid collisions is to dynamically control the decreasing threshold function. The threshold can be decremented faster in the first part of the acquisition, when information is typically sparse, and slower in the latter part of the acquisition. Still another collision avoidance mechansim is used: a pixel momentarily raises the global threshold when emitting a pulse to prevent other pixels with similar maximum values from firing.

The chip architecture is characterized by five main blocks (Fig. 6.35): a serial peripheral interface (SPI) block, a digital control block, an analog bias block, communication circuitry (one for each bus), and the pixel array with photodiodes and analog computation circuitry in each pixel. Let us now have a look on these blocks in more detail.

Two low-level computations, global normalization and low-pass spatial filtering, are performed on the image before extraction of the spatial gradients. The normalization is done by regulating the total current. It is needed to guarrantee proper operation under different illumination. Low-pass filtering was necessary to reduce high-frequency noise introduced by the derivative operation and for obtaining subpixel accuracy [283]. For proper gradient extraction and minimization of events generated onto the two buses for magnitude and

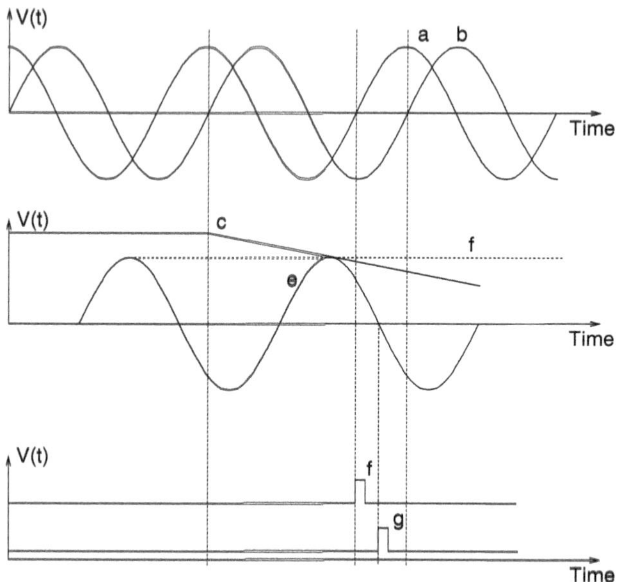

Fig. 6.34. Signal activity during frame acquisition. After completion of the first period of the interpolation functions a and b, the dynamic threshold c starts decreasing. When the threshold equals the stored maximum d of the local steering function, a gradient's magnitude pulse f is generated. A zero-crossing pulse g is emitted during the first descending flank of e following the magnitude pulse f [283]

direction information, a state machine was implemented in the pulse generator block of each pixel. This state machine starts acquisition of a frame after receiving a *sync* signal provided by the user. Prior to an acquisition, the old stored maximum values are cleared. The local steering function then starts to be calculated. During its first period, the maximum values are detected and stored locally. The decreasing threshold is launched and the pixels wait for it to match their locally stored maximum value. A communication pulse is generated on the "magnitude" bus when matching eventually occurs. Then, the pixel switches to the second state where it waits for the zero crossing on the descending flank of the local steering function. When this happens, a second pulse is emitted; this time onto the "direction" bus. Now, the pixel enters the third state where it is in stand-by and cannot generate any pulses until a new *sync* signal is received and the whole scheme starts again.

The serial peripheral interface (SPI in Fig. 6.35) serves for programming of the chip. With its help, the normalization current, the width of the spatial low-pass filter, the analog biases, the gain of the photo-receptors, and the frame rate can be changed. The digital control block generates the threshold function as a monotonic decreasing ramp as well as the sine and cosine interpolating functions. Discrete values are computed at fixed time intervals and then converted into analog reference currents by the analog bias block. This

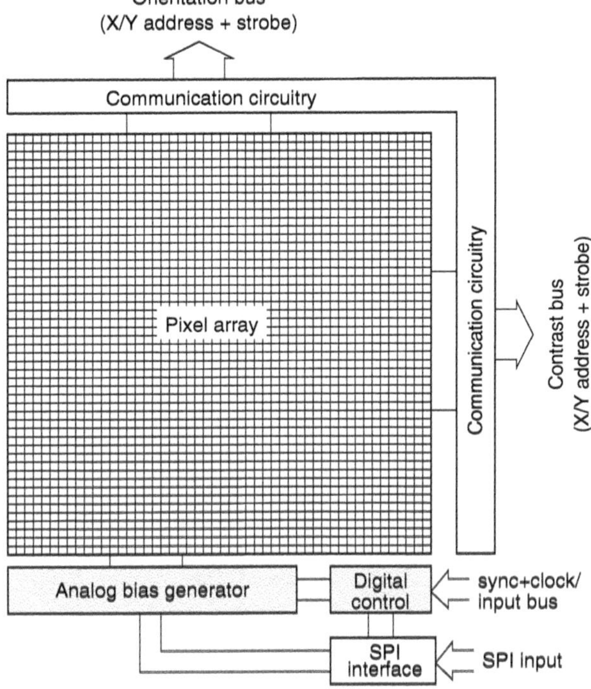

Fig. 6.35. Architecture of gradient extraction chip [283]

block supplies fixed global bias voltages or currents, which are needed by the pixels as references. The communication circuitry consists of two 14-bit buses. Each bus contains 7 bits for the pixel x- and y-address each. When a pulse is generated in a pixel, the communication circuitry encodes its (x,y) address in a digital word. To indicate whether the code is valid or not, a strobe signal is also provided.

In the following, the circuits will be described in more detail. Figure 6.36 shows the overall pixel structure. The photodiode measures the local luminance. The pixel is connected to its closest neighbors to perform low-pass filtering and to compute the local spatial gradient. A multiplier (see below) performs the multiplication of basis function and interpolating functions in current mode. The result of current summation from the nearest neighbors at the summation node represents the local steering function (e in Fig. 6.34 and 6.36). A detection block samples its maximum value, compares it with the dynamic threshold, and determines its zero crossing. The two remaining blocks in the pixel structure (Fig. 6.36) encode the gradient's information onto the buses and guarantee the correct sequencing of the state machine. The multiplier and detection block covered about 60 % of the pixel area [283]. The pulse generators took about 20 % and the photodiode about 15 % of the

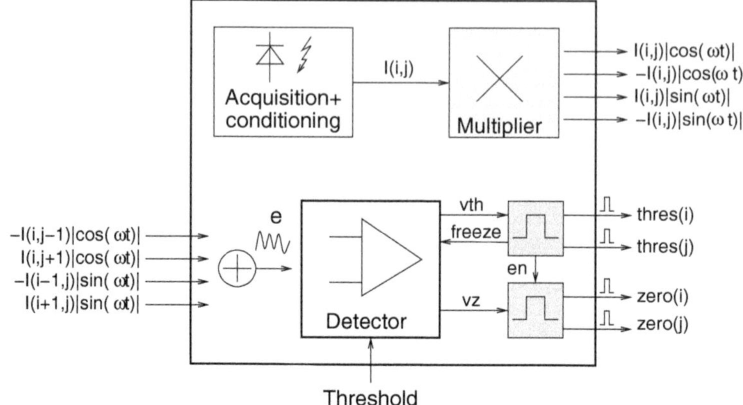

Fig. 6.36. Pixel block diagram of gradient extraction chip [283]

total pixel area. The connections and the signal conditioning circuitry occupied the rest of the pixel area.

Figure 6.37 shows the current mode circuit for light-to-current conversion and signal conditioning. An N-well/P-substrate photodiode was used. For normal illumination conditions, the photocurrent ranged between a few pA for dark current and about 10 nA for full light incidence. The current normalization is done by a weighted current mirror. The weights self-adjust in order to keep the total output current fixed to a certain programmable value. Another advantage of the current mode is that the photocurrents are directly available without integration. The low-pass filtering is done in a dif-

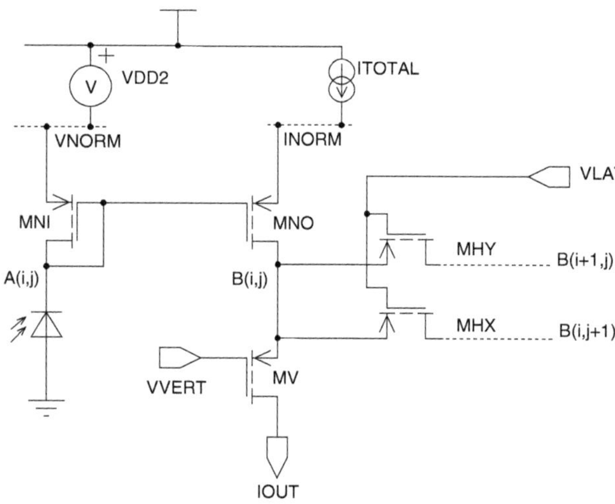

Fig. 6.37. Circuit for light-to-current conversion and signal conditioning [283]

fusion network formed by pseudo-conductances laterally connecting the south $(B(i+1,j))$ and east $(B(i,j+1))$ nearest neighbor pixels. This network of pseudo-conductances is achieved by MOS transistors operating in weak inversion [292]. The lateral and vertical pseudo-conductances in the diffusion network can be controlled by VLAT and VVERT.

The multiplier depicted in Fig. 6.38 calculates the local steering functions. One-quadrant multipliers were chosen to simplify the pixel circuit. This was possible because the sine and cosine interpolating functions change sign at well defined instants. Therefore, it was possible to generate a positive and a negative replica of the desired product with an error of less than 1 % [283]. These replicas were switched at appropriate times to distribute the right signal to the proper neighbor. In such a way the silicon area could be kept smaller than for a four-quadrant multiplier.

The core of the multiplier consists of matched transistors MIN and MCOS biased in weak inversion. The output current, therefore, is given by

$$I_{\mathrm{COS}} = I_{\mathrm{IN}} exp \frac{V_{\mathrm{REF}} - V_{\mathrm{COS}}(t)}{U_T} \tag{6.13}$$

where I_{IN} is the low-pass filtered normalized photodiode output current IOUT of Fig. 6.37. The voltages V_{REF} and V_{COS} have to be set so that

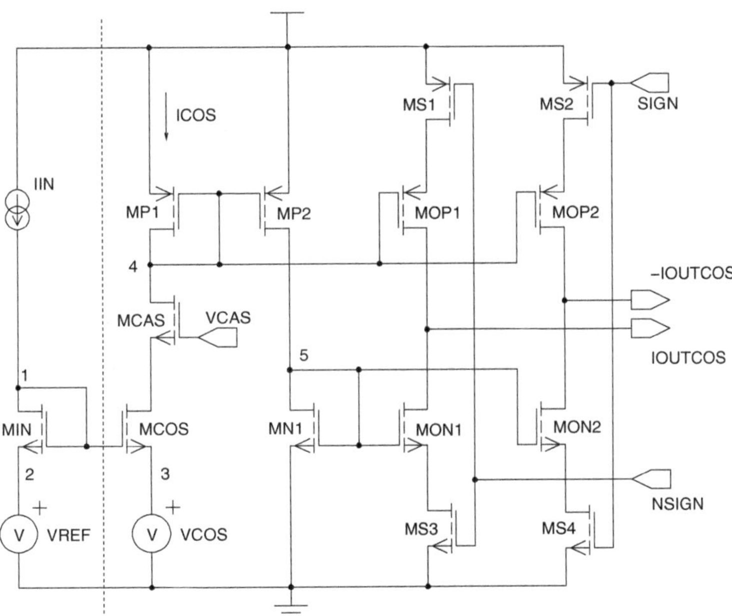

Fig. 6.38. Circuit of the multiplier for the cosine product. The right part is replicated to obtain the sine product, while the left part is common to the two half circuits [283]

$$exp\frac{V_{\text{REF}} - V_{\text{COS}}(t)}{U_{\text{T}}} = |cos\omega t| \tag{6.14}$$

in order to obtain the desired product. Since the output is a current, the most accurate way to generate voltages is with reference currents. Two reference transistors in the bias block being identical to MIN and MCOS supply voltages V_{REF} and $V_{\text{COS}}(t)$ on the basis of currents so that

$$I_{\text{COS}}(t) = I^*_{\text{COS}}(t)\frac{I_{\text{IN}}}{I_{\text{REF}}} \tag{6.15}$$

where $I^*_{COS}(t)$ is a rectified sinusoidal reference current delivered from the bias block. In such a way the multiplications do not depend on physical process parameters. MIN in MCOS have to match to each other of course. The parameter I_{REF} allows multiplier amplification.

Four-quadrant multiplication is achieved in the following way: transistors MP1, MP2, and MN1 replica the cosine current, while transistors MOP1, MOP2, MON1, and MON2 form the current mirrors output, which are connected to the neighbor pixels. Transistor switches MS1/MS3 and MS2/MS4 select the right branch according to the input levels SIGN and NSIGN. For a high SIGN (NSIGN is then low), a positive current flows from IOUTCOS and a negative one from −IOUTCOS. The sign of these currents changes for a low SIGN (NSIGN = high). The silicon area of the multiplier block was 1920 μm^2 [283].

The detection block extracts the magnitude and direction of the local gradients. This unit detects the zero crossing of the local steering function, samples and stores its maximum value, and determines the time when the threshold function matches the stored maximum value. The circuit diagram of the detection module is shown in Fig. 6.39. At first, the detection block determines the sign of the local steering function denoted IIN in Fig. 6.39. This task is performed by the comparator consisting of transistors MZ1, MZ2, MC1, MC2, MC5, and MC6. The output signal OUTZERO of this comparator is fully digital. When input current flows into the circuit (through MZ1 to ground), OUTZERO is low. When IIN flows out from the input (sourced by MZ2 from VDD), OUTZERO is high. The logical level of OUTZERO, therefore, indicates the sign of IIN.

For a negative input current IIN, it is drawn from the maximum detection circuit formed by transistors MSH, MSPL1, MSPL2, MDIS, and MSW. The maximum detection works only when IIN is negative, i. e. when MZ2 is conducting. In this case, transistor MSW is also conducting and MDIS acts on the gate of MSPL1 to make its drain current equal to the input current. The transistor MDIS, however, is only active as long as the magnitude of IIN increases. MDIS is cut off when the magnitude of the input current starts to decrease. When this occurs, the potentials at nodes 1 and 2 (see Fig. 6.39) rise and turn off MSW. Therefore, the output transistor MSPL2 is isolated and it now holds the maximum current reached by the input. The parasitic capacitance of node 4 including the gate-source capacitance of MSPL2 is exploited

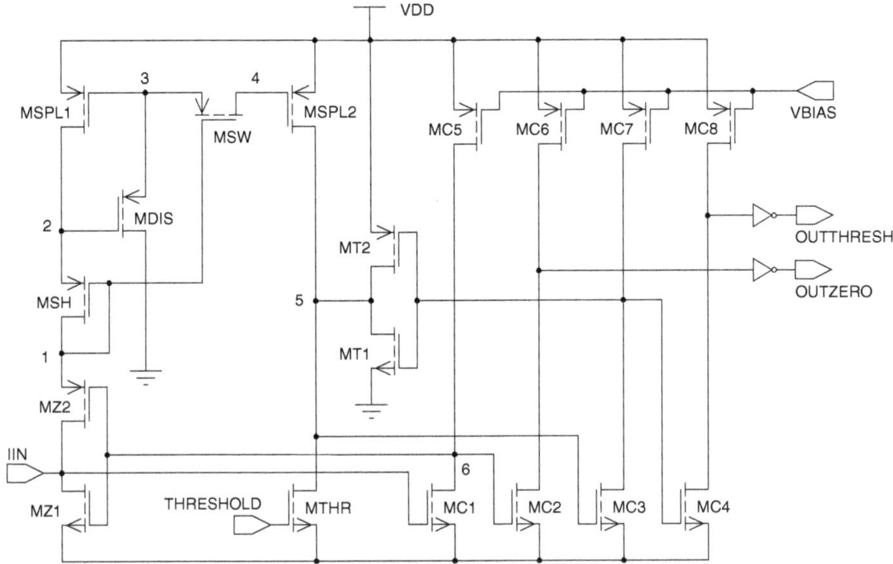

Fig. 6.39. Circuit of the detection block [283]

to hold the maximum value. The isolation device MSW is necessary to avoid coupling to the hold node 4 via the gate-drain capacitance of MSPL1. This capacitance desaturates quickly when the input current starts to decrease. Without the isolation transistor MSW, the fast variation of the drain-gate voltage of MSPL1 would induce a charge on node 4, which in turn would change the stored maximum value too quickly. The isolation transistor MSW reduces this effect but cannot eliminate it completely [283]. The relative error is larger for low currents. Therefore, the current from the photodiode is amplified in the multiplier block to obtain a local steering current largely unaffected by this coupling.

When the maximum value is available on transistor MSPL2, the comparison with the dynamic threshold can start. The current difference I_D(MSPL2) $- I_D$(MTHR) is compared to zero by the comparator formed by the transistors MT1, MT2, MC3, MC4, MC7, and MC8. This comparator is identical to the one described above. The digital value of the output voltage OUTTHRESH indicates whether the stored maximum is larger or lower than the threshold. The transition of this digital value constitutes an event. Reset transistors are required to clear the stored maximum values before a new frame acquisition starts. The reset transistors are not included in Fig. 6.39 [283]. The output signals OUTTHRESH and OUTZERO were buffered with an inverter each before they are connected to the next stage. Behind these inverters, threshold crossings are represented by high-to-low transitions and zero crossings are represented by low-to-high transitions [283].

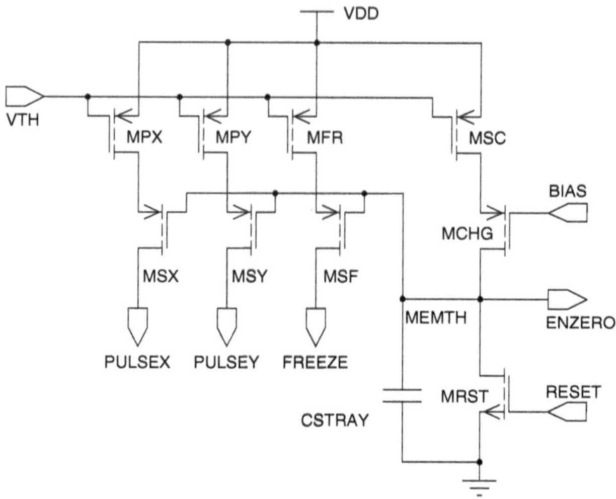

Fig. 6.40. Circuit of the threshold pulse generator [283]

The next stage is the pulse generator block. The two circuits shown in Figs. 6.40 and 6.41 were derived to generate and communicate the threshold-crossing and zero-crossing events, for which a current pulse has to be generated only once for each pixel during one frame acquisition. For this purpose, the pixel communication circuitry has to keep track of the latest history of the pixel in two one-bit registers included in each pulse generator (MEMTH in Fig. 6.40 and MEMZ in Fig. 6.41). These registers also exploit parasitic stray capacitances to save chip area.

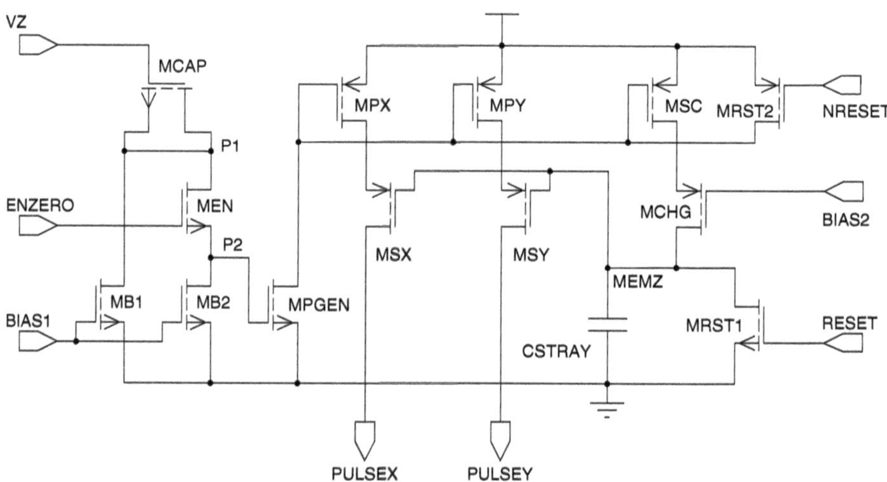

Fig. 6.41. Circuit of the zero-crossing pulse generator [283]

A *sync* signal starts each frame acquisition and brings all pixels into state 1. A reset is applied to the zero-crossing pulse generator making its inputs RESET and NRESET (Fig. 6.41) high and low, respectively. To realize that any zero crossing in state 1 has to be disrecarded before a threshold crossing has been communicated, the memory nodes MEMTH and MEMZ of the pulse generators are set to VSS each time when a *sync* signal is applied. In this state threshold pulses are possible because MSX and MSY (Fig. 6.40) are on and zero-crossing pulses are disabled, because switch MEN in Fig. 6.41 is off. For a threshold matching, a high-to-low transition happens on input signal VTH and transistors MPX, MPY, and MFR are made conducting. At the same time transistor MSC is turned on and the charge current of MCHG increases the potential of the memory node, bringing MEMTH into the high logical state. This new state causes that switches MSX and MSY are turned off. In such a way two current pulses out of the nodes PULSEX and PULSEY result. The length of these pulses is set by a bias current via a programmable BIAS signal.

The zero-crossing generator is enabled when transistor MEN is switched on by a logical high value at MEMTH. MEMTH = 1 and MEMZ = 0 correspond to state 2 of the state machine, in which the threshold generator is disabled. When a zero crossing is detected on the descending flank the local steering current produces a low-to-high transition of signal VZ and node P2 is brought high through the coupling capacitor MCAP and switch transistor MEN. This potential change at node P2 causes MPGEN to turn on transistors MPX and MPY resulting in X and Y current pulses from PULSEX and PULSEY of Fig. 6.41. Transistor MSC is also on and delivers charging current through MCHG to the memory node MEMZ, which is therefore pulled high. With a time delay determined by BIAS2, the transistors MSX and MSY and the pulses are turned off. Now MEMTH = 1 and MEMZ = 1 hold, the state machine is in state 3, and both generators are disabled.

The transistors MB1 and MB2 in Fig. 6.41 are necessary to bring down nodes P1 and P2 after a zero-crossing event, when the pulse generators are in the initial state, i. e. before a threshold event. When the input signal goes high in this state (corresponding to detection of a zero crossing), node P1 is also drawn high, but no pulse is generated due to node P2 being isolated from node P1 by MEN, which is off. The potential on P1 would remain high without MB1 and as soon as MEN would be switched on by the threshold pulse generator. P2 would follow P1 to the high state and a pulse would be generated belonging to a threshold detection instead to a zero-crossing detection.

The column (x-axis) and row (y-axis) address is encoded by two pulses from PULSEX and PULSEY. A third FREEZE pulse (see Fig. 6.40) is needed to avoid collisions. A freeze pulse being present momentarily increases the threshold current by slightly charging the stray capacitance of node THRESHOLD in the detection block (see Fig. 6.39). In such a way, other pixels with maximum values close to the threshold are prevented from emitting pulses during the time needed by the currently emitting pixel to send its information onto the bus.

The communication circuitry generates the address code for firing pixels. It also produces a STROBE signal, which can be used off-chip to latch the actual value of the x- and y-axes addresses. These addresses determine the location of the active pixel. The time of address appearance on the bus characterizes either the magnitude or direction of the spatial gradient depending on which pulse generator has been active [283].

The chip with an array size of 100×100 was implemented in a $0.5\,\mu m$ analog CMOS technology with a total silicon area of $86\,mm^2$ [283]. The size of each pixel was $80 \times 80\,\mu m^2$ with a fill factor of 15 %. The total number of transistors in the chip was 850.000, whereby the number of transistors per pixel was 84. The static power consumption was 50 mW. A frame rate of 1000 Hz was possible and the sensitivity of the retina spanned from $0.3\,mW/m^2$ to $10\,W/m^2$. The key to such fast processing is the application of the theory of steerable filters in an analog VLSI implementation to compute the magnitude and direction of the gradients. An irradiance of $2\,mW/m^2$ was necessary to enable reliable edge detection. The standard deviation of the frequency distribution of the errors on the direction of the gradients was 7.7° [283].

Let us summarize. The very sophisticated signal processing circuitry in each pixel truely justifies to classify the silicon retina chip described in [283] as a smart pixel chip. While this "silicon retina" delivers only preprocessed data, the interpretation of the actual image content still has to be done off chip. The required interpretation logic, however, presumably can be integrated one day on the same chip which will then be a single-chip machine vision system, a so-called "seeing chip".

6.4.6 CMOS Optical Sensors

Here a low-level light sensor and preprocessor OEIC for a disposable medical probe for chemoluminant diagnostic applications shall be introduced. A 10-Hz 5–$10\,\mu W/cm^2$ (corresponding to a maximum photocurrent of $4\,\mu A/cm^2$) input optical signal in the visible spectrum has to be analyzed by this OEIC, which contains the sensor, analog circuits, an analog-to-digital converter (ADC), digital circuits and an output modulator (see Fig. 6.42).

One of the advantages for integrating the optical sensor in a disposable probe is the elimination of cleaning the probe between measurements. The disposable light sensor must be cheap and the number of I/O pins must be small to allow reliable contacting of the detachable parts. Usually a serial bus interface is used leading to a number of three pins for the two power lines and the data pin. The number of pins, however, can be reduced down to only two, since the output data can be modulated on the power lines as shown in Fig. 6.42.

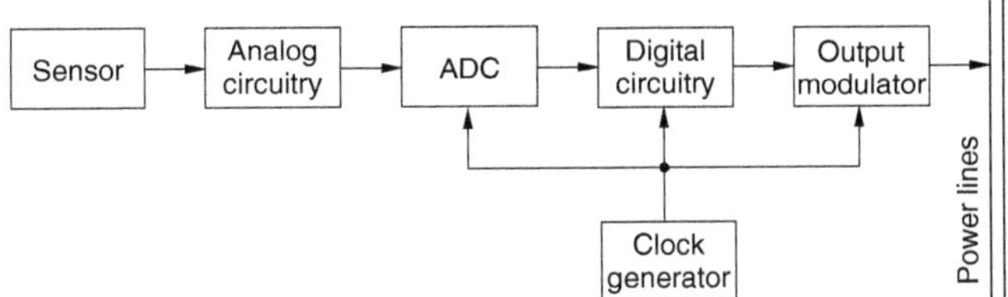

Fig. 6.42. Smart sensor system for medical applications [293]

Usually current-to-frequency conversion [294, 295] thanks to its simplicity and high precision is a good candidate for analog-to-digital conversion, when no fast measurements are required. Operation of those current-to-frequency converters is based on counting the number of pulses produced by a current-controlled oscillator (CCO) over a constant period of time to obtain a digital presentation of the incident light intensity. A simple first-order sigma-delta modulator for analog-to-digital conversion was successfully used in an area image sensor [296]. The signal-to-noise ratio (SNR) of a sigma-delta modulator theoretically improves by 9 dB for each doubling of the oversampling ratio. Doubling the measurement time in the CCO scheme, which corresponds to doubling the oversampling ratio, in contrast, improves the SNR only by 3 dB [293]. A first-order sigma-delta ADC, therefore, can achieve potentially a higher SNR than a CCO, whereby the first-order sigma-delta ADC is only slghtly more complicated. A single-bit first-order sigma-delta modulator ADC was chosen due to its inherent linearity, robustness against process deviations, and simplicity. The sensor architecture is shown in Fig. 6.43.

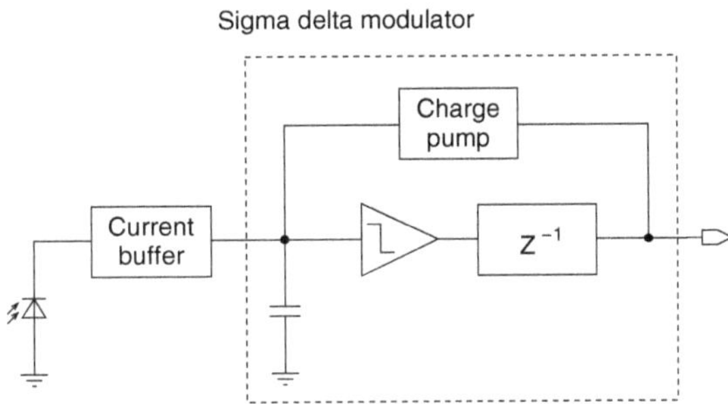

Fig. 6.43. Architecture of sensor for medical applications [293]

The sensor reported in [293] consists of an integrated photodiode with an area of $500 \times 500\,\mu m^2$, a current buffer, and the sigma-delta modulator. The photocurrent from the photodiode is buffered by the current buffer into an integrating capacitor. The photocurrent is integrated on an external integrating capacitor rather than on the diode junction capacitance itself to improve the photocurrent-to-voltage conversion efficiency [293]. This is advantageous because the integrating capacitor is much smaller than the diode junction capacitance achieving a larger output voltage for the same charge. The integrator voltage is compared to the threshold level by sampling the comparator. As soon as the integrator voltage drops below the threshold, a constant amount of charge is pumped into the integrator. In such a way a first-order sigma-delta modulation loop is formed. The output is a one-bit digital signal, modulated by the input photocurrent.

Let us examine the circuit of the current buffer before the first-order sigma-delta modulator. Due to the small area of the integrated photodiodes compared to a cm^2, the current buffer has to provide a sufficient bandwidth for input currents as low as tens of picoamps. For such a purpose, the usual common gate circuit shown in Fig. 6.44a does not perform satisfactory. The input pole of this circuit is determined by the transistor transconductance and the photodiode junction capacitance [293]:

$$\frac{i_{\text{out}}}{i_{\text{ph}}}\|_{C.G.} = \frac{g_{\text{m,N2}}}{sC_{\text{D}} + g_{\text{m,N2}}} \tag{6.16}$$

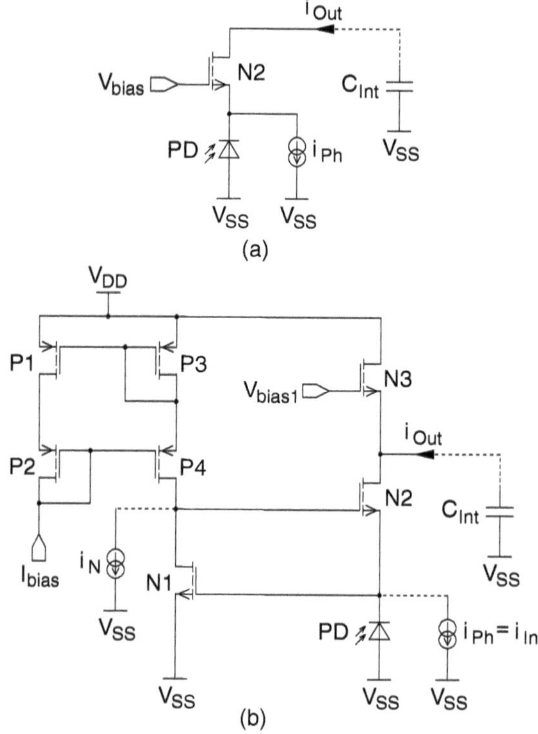

Fig. 6.44. Common gate **(a)** and boosted current buffers **(b)** [293]

The maximum photocurrent for the photodiode area of $0.25\,\text{mm}^2$ is $4\,\mu\text{A/cm}^2 \times 0.25\,\text{mm}^2 = 10\,\text{nA}$. At this very low current, transistor N2 is in subthreshold with

$$g_{\text{m,N2}} = \frac{qI_{\text{ph}}}{k_{\text{B}}T} \tag{6.17}$$

For a photocurrent of 10 pA, the bandwidth can be calculated to be only about 0.6 Hz, which is much too low for the diagnostic application. The transconductance does not depend on the transistor dimensions in subthreshold operation, thus this property cannot be improved by transistor sizing. Another circuit has to be chosen. The circuit illustrated in Fig. 6.44b possesses a better performance.

The boosted configuration [297] with feedback effectively decreases the input impedance by the loop transfer and increases the output impedance by the same factor. There is a feedback loop via $c_{gs,N2} + c_{gd,N1}$. The loop transfer function is [293]:

$$LT(s) = \frac{c_{\text{f}}^2}{C_{\text{D}}(c_{\text{f}} + c_{\text{gs,N2}})} \times \frac{(s + \frac{g_{\text{mN2}}}{c_{\text{f}}})(s - \frac{g_{\text{mN1}}}{c_{\text{f}}})}{(s + \frac{g_{\text{mN2}}}{C_{\text{D}}})(s + \frac{g_{\text{ds,N1}}}{c_{\text{f}} + c_{\text{gs,N2}}})} \tag{6.18}$$

where $c_{\text{f}} = c_{\text{gs,N2}} + c_{\text{gd,N1}} \approx c_{\text{gs,N2}}$. For $z_1 < p_2$, i.e. the frequency of the first zero is smaller than that of the second pole, the possibility of complex poles is eliminated. Increasing the bias current enlarges g_{ds1}, and z_1 becomes smaller than p_1 for sufficiently large bias currents around $1\,\mu\text{A}$ [293].

The block diagram of the first-order single-bit sigma-delta modulator and its implementation are shown in Fig. 6.45 (a) and (b), respectively. The comparator is an one-bit quantizer formed by two inverters in series. The comparator threshold does not affect the modulator performance [298], thus the circuit is not sensitive to transistor threshold voltage variations. The transmission gate is the sampling delay element. The comparator activates the charge pump (P5, P6, and C_p), which is a single-bit digital-to-analog converter, via the NAND gate. The operation is synchronized by two-phase 1-kHz clocks Φ_1 and Φ_2 (see Fig. 6.46).

At a low transition of Φ_1 (event I), P6 is cut off. At a low transition of Φ_2 (event II), the comparator transmission gate opens and the result propagates to the NAND gate. P5 turns on simultaneously and charges C_{P} to the supply voltage. At event III (high transition of Φ_2), the result is latched on C_{Store} and P5 is turned off. Subsequently, on event IV (high transition of Φ_1), if $V_{C\text{Int}} < V_{\text{Th}}$, the V_{bias2} voltage is set on the P6 gate, pumping a constant amount of charge Q_{P} into C_{Int}, and the cycle repeats.

The amount of charge pumped into the integrator is $Q_{\text{P}} = C_{\text{P}}(V_{\text{DD}} - V_{\text{bias2}} - V_{\text{Th}})$. The exact amount of Q_{P} is, of course, subject to V_{Th} variations, but there is no need to control it precisely. The gain and offset of the modulator response are affected by charge level variations, but the response stays linear [298]. As a result, the circuit can be calibrated for fixed gain and offset variations by a single calibration measurement [293].

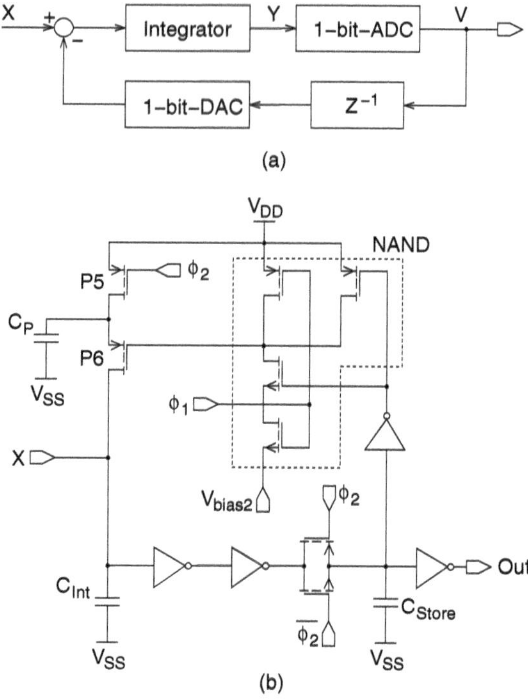

Fig. 6.45. Single-bit first-order sigma-delta modulator **(a)** and its implementation **(b)** [293]

The maximum pulse rate is the 1-kHz sampling rate of the modulator. The input current, i. e. the photocurrent, can be obtained from the modulator pulse rate as follows. The integration capacitor C_{Int} receives the same amount of charge at each modulator pulse, resulting in $\Delta V = 0.5\,\mathrm{V}$ on the capacitor. The input current charging, assuming that the modulator does not saturate, is

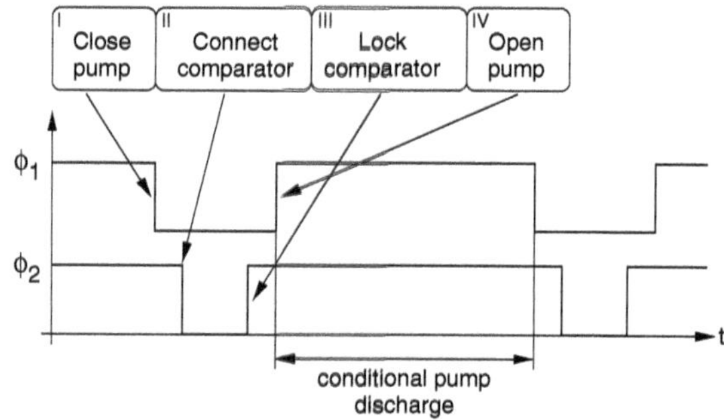

Fig. 6.46. Clock diagram [293]

$$I_{\text{in}} = \frac{N_{\text{pulses}}}{T_{\text{measurement}}} \times \Delta V \times C_{\text{Int}} \tag{6.19}$$

where $N_{\text{pulses}}/T_{\text{measurement}}$ is the output pulse rate. The output pulse rate is proportional to I_{in}.

The OEIC chip for diagnostic applications was realized in a 0.5 µm 1P3M CMOS technology. The circuit consisted of 31 transistors and occupied an area of $160 \times 110\,\mu\text{m}^2$. The power supply was a single 3.3 V source. The signal bandwidth of 10 Hz was verified [293]. The leakage current of the photodiode of $80\,\text{pA/cm}^2$ limited the dynamic range (ratio between the maximum photocurrent and the leakage current) to 10^4 or 80 dB. In order to avoid rather rare metastability problems, the authors suggested to introduce a positive feedback (flip-flop) into the modulator loop instead of the transmission gate [293].

Let us have a look onto the SNR of this medical sensor OEIC finally. The noise is determined by three major contributors: photodiode shot noise, current buffer noise, and the reset noise of the charge pump. Transistor N1 is the dominant noise source in the current buffer. The current buffer noise can be represented as an equivalent current noise source i_{n} at the drain of N1 (see Fig. 6.44). The noise transfer function was computed as [293]:

$$A_{\text{noise}}(s) = -\frac{(sc_{\text{gd,N2}} - g_{\text{m,N2}})\cdot(sC_{\text{D}} - g_{\text{ds,N2}})}{(s - p_1)\cdot(s - p_2)} \tag{6.20}$$

This equation possesses a low-frequency zero $g_{\text{ds,N2}}/C_{\text{D}}$ in the noise transfer, which effectively suppresses the 1/f-noise of the feedback amplifier that would have dominated otherwise.

Simulations showed that the dominant noise source is the photodiode shot noise for frequencies below 100 Hz [293]. The output current of the current buffer is integrated on C_{Int} for one sampling period. C_{Int} accumulates some noise charge in addition to the signal charge. The charge pump reset noise ($k_{\text{B}}TC$ noise) is added to the integrator charge noise and the total charge noise power is [293]:

$$\overline{q_{\text{n}}^2} = \int_0^\infty S_{\text{i,in}}(f)\left(\frac{sin(\pi f T_{\text{s}})}{\pi f}\right)^2 df + k_{\text{B}}TC_{\text{P}} \tag{6.21}$$

where $S_{\text{i,in}}(f)$ is the noise spectral density of the current buffer output signal and T_{s} is the sampling time. in the digital frequency domain, the entire noise power spectrum is folded into the range from 0 to π, whereas the signal occupies only a fraction of this range due to oversampling. For a 10-Hz input bandwidth and a sampling frequency of 1 kHz, the signal occupies the range from 0 to $\pi/50$. The out-of-band components are cut off by the decimation low-pass filter. The true noise power after decimation, therefore, is 50 times lower than after (6.21). Referring the output noise to the input gave $i_{\text{n,in,rms}} = 0.82\,\text{pA}$, while the rms photodiode shot noise alone was about 0.81 pA [293].

A first-order sigma-delta modulator implies quantization noise [298]:

$$\overline{n^2} = \frac{\pi^2 \triangle^2}{36} R_{\mathrm{ov}}^{-3} \tag{6.22}$$

where $\triangle$ is the distance between the quantization levels. In [293] $\triangle$ is the maximum effective current pumped by the charge pump: $\triangle = Q_{\mathrm{P}}/T_{\mathrm{S}}$. R_{ov} is the oversampling ratio defined as the ratio of the sampling frequency f_s and the input signal Nyquist frequency $2f_0$:

$$R_{\mathrm{ov}} = \frac{f_s}{2f_0}. \tag{6.23}$$

From (6.22) a 9-dB noise decrease follows for each doubling of the sampling frequency. With the power of a sine wave with amplitude $\triangle/2$ as a reference level, the signal-to-noise ratio (SNR) is [293]:

$$SNR = \frac{(\frac{\triangle}{2})^2}{2} \frac{36}{\pi^2 \triangle^2} R_{\mathrm{ov}}^3 = \frac{9}{2\pi^2} R_{\mathrm{ov}}^3. \tag{6.24}$$

According to this equation the SNR is about 51 dB for an oversampling ratio R_{ov} of 64 and about 60 dB for $R_{\mathrm{ov}} = 128$ [293]. It is necessary to reduce the quantization noise to the noise floor of photodiode shot noise and current buffer noise. This can be achieved for a high enough sampling rate. The noise floor was reached for a sampling time of approximately 200 µs corrsponding to an oversampling ratio R_{ov} of about 256 [293]. The SNR for this oversampling ratio is 69 dB.

An analog ASIC facilitating the focal point position detection in a Hartmann-Shack wavefront sensor for optical wavefront measurements was described [299]. The phase information of the light beam can be obtained by detecting the lateral deviation of each focal point of the Hartmann-Shack sensor's lens array caused by the partial distortion of the aberrated wavefront within the according lens. As a future aim, measuring optical distortions of the human eye was mentioned in [299]. Therefore, the ASIC was optimized to process an optical incident power of 1 nW per focal point, because the incident power is below 0.01 W/m^2 due to security limitations when directly penetrating the eye's retina by the measuring beam. To prevent eye movements from disturbing the measurements, a frame repetition rate of several hundred hertz was necessary.

A single-path measurement scheme often applied in an optical wavefront detection system is referred to as the Hartmann-Shack wavefront detection scheme [299]. A Hartmann-Shack sensor (HSS) consists of an array of optical lenses arranged within the optical path to divide the incident light beam into a matrix of subapertures (Fig. 6.47a).

Each lens focuses into a focal point located within the lens array's focal plane. When a local tilt is present in the wavefront within the aperture of the

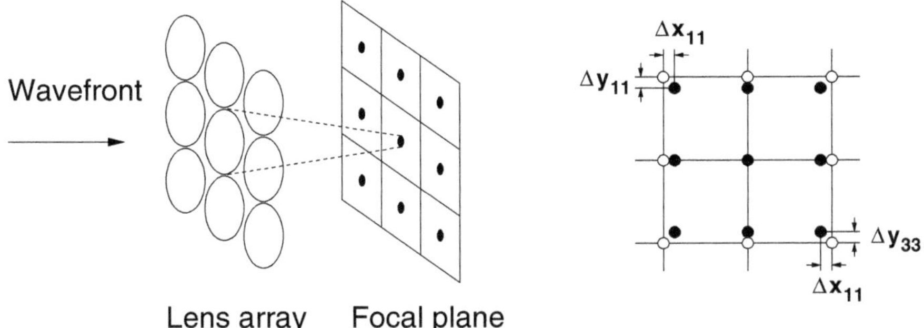

Fig. 6.47. (Left) Principle of Hartmann-Shack wavefront detection. (Right) Focal point pattern of a reference wavefront (open circles) and of a distorted wavefront (filled circles) [299]

lens, the focal point is deviated perpendicular to the optical axis (Fig. 6.47b). The magnitude of the deviation is in a first order proportional to the magnitude of the local tilt in the wavefront. The tilt can therefore be quantified. The light beams's total wavefront can be reconstructed by means of a least-square-fit of the measured data of all focal point deviations. The mathematical description of the optical path difference (OPD) between the calculated wavefront and an arbitrary reference wavefront thereby is commonly based on high-order two-dimensional Zernike polynomials [300, 301].

A lens array is mounted in parallel to the photodetector matrix in a distance of the focal length. When the partial wavefront's average is tilted within the aperture of a lens, the corresponding focal point is defected (Fig. 6.48a).

Assuming paraxial optics, i. e. $sin(x) = x$, the defection $\triangle x$ can be calculated from [299]

$$\frac{dW}{dx} = \frac{\triangle x}{f_{\mathrm{L}}}, \tag{6.25}$$

where dW/dx is the wavefront's tilt in x-direction and f_{L} is the focal length of the lenses in the lens array. For $f_{\mathrm{L}} = 53\,\mathrm{mm}$ as an example, a wavefront tilt of $dW/dx = 200\,\mathrm{nm}/400\,\mu\mathrm{m}$ causes a focal point deflection of $\triangle x = 26.5\,\mu\mathrm{m}$.

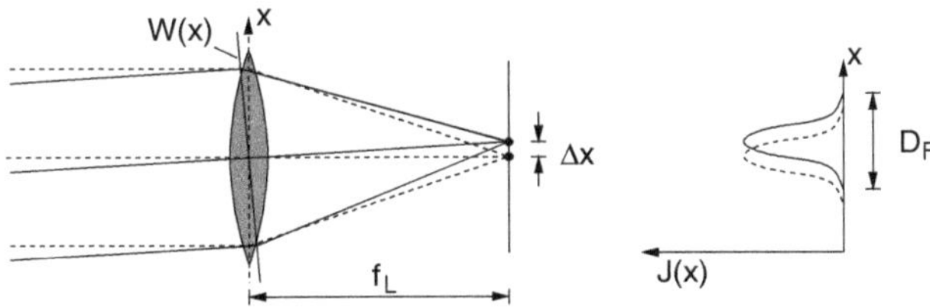

Fig. 6.48. (Left) Deviation of the focal point due to a tilt in the wavefront within the lens. (Right) Intensity distribution of a focal point [299].

The lateral intensity distribution of a focal point can be approximated by a gaussian distribution (Fig. 6.48b). According to Fourier optics, the intensity distribution is circular with minima and maxima of decreasing intensity. The diameter D_{F} of the first minimum-intensity circle (often referred to as the diameter of the *Airy disc*) is [299]

$$D_{\mathrm{F}} = \frac{1.22\lambda f_{\mathrm{L}}}{D_{\mathrm{L}}} \tag{6.26}$$

with the lens diameter D_{L}. For $\lambda = 780\,\mathrm{nm}$ and $D_{\mathrm{L}} = 400\,\mu\mathrm{m}$, $D_F = 126\,\mu\mathrm{m}$ is obtained which is much larger than $\Delta x = 26.5\,\mu\mathrm{m}$. Thus the signal processing circuitry has to determine the location of the maximum intensity by distinguishing only slightly differing values.

The Hartmann-Shack sensor (HSS) ASIC described in [299] combines photosensitive devices for image detection and electrical circuits for focal point position detection (Fig. 6.49). For easy readability, only 4×4 lenses and focal point detectors are drawn whereas the ASIC actually included 16×16 focal point detectors.

Analog circuitry is implemented to evaluate focal point intensity distributions for the maximum by use of the winner-take-all (WTA) algorithm. Parallelism is inherent in this approach as originally presented by Lazzaro [302]. Each focal point detector (called cluster in Fig. 6.50) contains an array of photodiodes and surrounding analog signal processing circuitry to find the location of the focal point's intensity maximum. After detection, the position data is transferred to digital circuitry, where reformatting and compression is performed. The position detection clusters are linked together via shift registers to push the position data to the compression unit (Fig. 6.51). Finally data are sent to the output for a least-square-fit calculation of the Zernike polynomial coefficients.

The lateral dimensions of a cluster were $400 \times 400\,\mu\mathrm{m}^2$ to fit the lens diameter. The chip area of the signal processing circuitry had to be kept as small as possible to enlarge the photosensitive area. The photodiodes are placed in a special pattern within the photodetector array of one cluster. The array

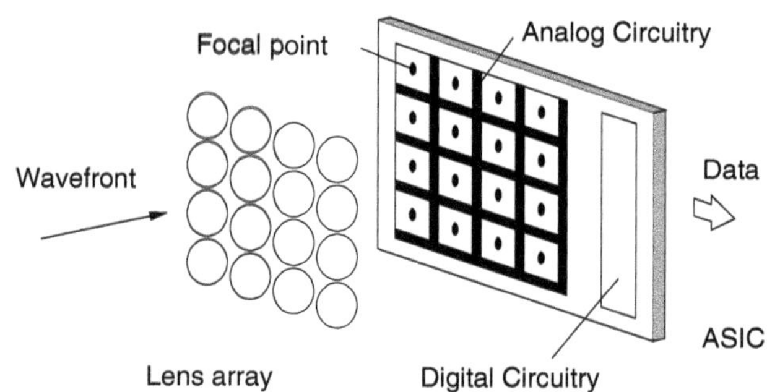

Fig. 6.49. Architecture of the HSS system [299]

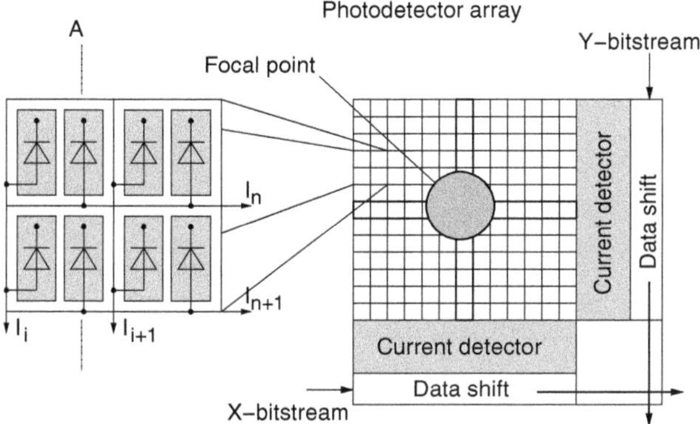

Fig. 6.50. (Left) Zoomed illustration of the photodetectors within one cluster. (Right) Global architecture of one cluster [299]

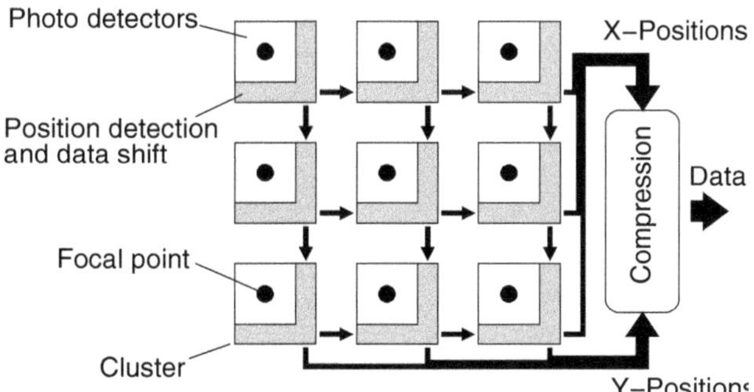

Fig. 6.51. Data flow in the HSS [299]

contains a number of 19×19 photodetector pixels, each consisting of two adjacent photodiodes covering an area of $p^2 = 17.6 \times 17.6\,\mu m^2$ [299]. The left diode is connected to its column neighbors and the right diode is connected to its row neighbors (Fig. 6.50b). In such a way, a number of 19 stripes in each direction is realized, allowing to read out the stripe currents of the x- and y-direction simultaneously. Figure 6.52 shows a cut through the chip along line A in Fig. 6.50 with the intensity distribution of a focal point. Each photodiode stripe is illuminated by a fraction of the incident power within the length p and the row currents I_n are generated.

Let us now have a look onto the signal processing circuitry in the HSS. The most important part of the ASIC's signal processing circuit is the current detector for determining the maximum current among the 19 generated photocurrents. MOSFETs are used for that purpose. The winner-take-all (WTA) algorithm proposed in [302] fits best to the task to detect locally a maximum of several photocurrents. The basic circuitry of the WTA topology for position detection as presented in [302] is shown in Fig. 6.53. Three cells are drawn

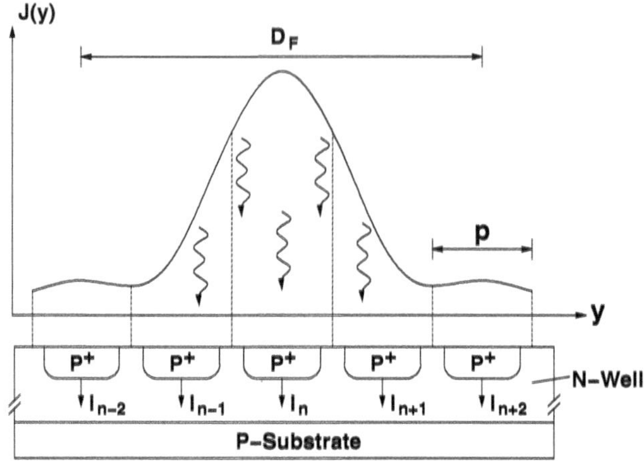

Fig. 6.52. Intensity distribution of a focal point (not to scale) [299]

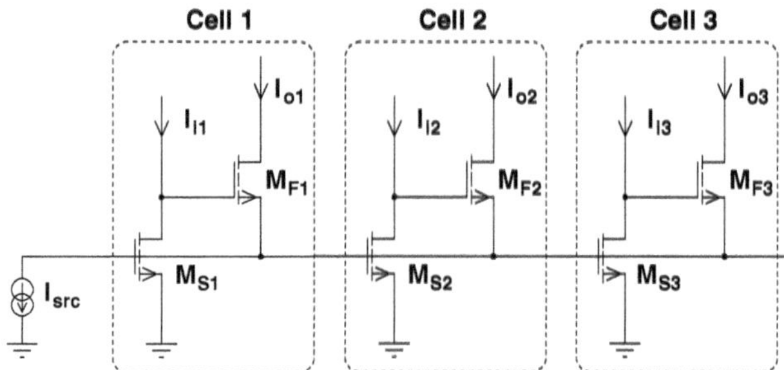

Fig. 6.53. WTA topology illustrated with three cells implementing n-channel MOSFETs [299]

for the detection of three currents. Each cell contains the two MOSFETS M_S and M_F. M_S senses the input current I_i and generates a V_{DS}, which is connected to the gate of M_F. The gates of all transistors M_S are connected, and owing to the same transfer characteristics ($I_D(V_{DS})$), one of them generates the highest drain potential due to its highest $I_D(I_i)$ and V_{GS} of the according M_F is the highest of all transistors M_F. Most of the current of the current source I_{src} is sunk by this M_F and guided to I_o, inhibiting current flow in the other transistors M_F. Since speed properties of this algorithm are essential to the HSS, the transient response has to be investigated in detail.

The transient response of a cell is mainly limited by the capacitance at the drain node C_Ψ of M_{Sn} (Fig. 6.54). When the input current I_{ph2} becomes larger than a constant I_{ph1}, three time steps occur until the output currents change significantly [299]. In the first step, the photocurrent I_{ph2} changes to a higher magnitude, but the gate potential of M_{S2} keeps the same since it is still defined by M_{S1} and M_{F1}. The parasitic capacitor $C_{\Psi2}$ is charged

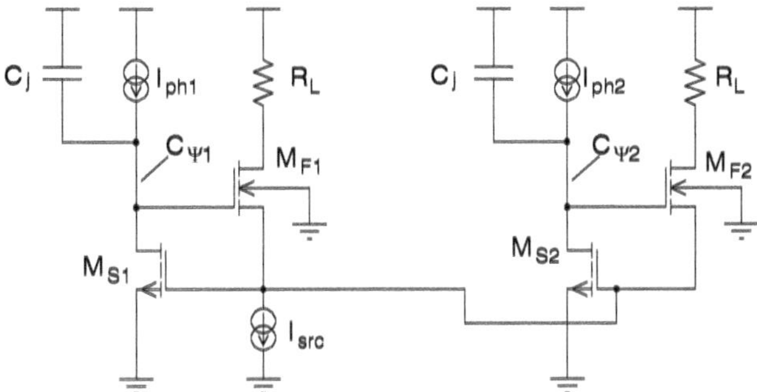

Fig. 6.54. Photodiode junction capacitance and parasitic capacitors $C_{\Psi 1}$ and $C_{\Psi 2}$ [299]

by $\triangle I2 = I_{\mathrm{ph2}} - I_{\mathrm{D,S2}}$ to raise V_{DS} of M_{S2}. In the second step, $V_{\mathrm{DS}}(M_{\mathrm{S2}})$ is high enough to significantly affect the current through M_{F2}. Now, M_{F2} inhibits the current flow through M_{F1} by increasing its drain current to almost I_{src}, and the gate potential of M_{S1} and M_{S2} changes to a slightly higher value. In the third step, the capacitor $C_{\Psi 1}$ is discharged, reducing the drain potential of M_{S1} to its equilibrium value. The first time step of recharging the parasitic capacitor $C_{\Psi 2}$ clearly dominates the response time, because of the small amount of $\triangle I$ for charging.

The junction capacitance of the photodiode stripe with 19 P^{+}/N-well diodes is about 2.3 pF at a reverse voltage of 2 V. According to circuit simulations for a change of the photocurrent I_{ph2} from 46 to 54 pA at a constant I_{ph1} of 50 pA, it takes about 0.6 s to raise V_{DS} of M_{S2} by a value of 1 V from the initial value to change the output current conditions of the two cells significantly [299]. To reduce this unacceptable high value of response time, therefore, different extensions to the basic WTA topology have been introduced.

The largest influence to the response time of a WTA cell stems from the photodiode capacitance C_{j}. A regulated cascode configuration with transistors M_{C} and M_{R} as illustrated in Fig. 6.55 can be applied to reduce the capacitance value appearing at the drain node of the transistors M_{Sn}. The resulting parasitic capacitor at the drain nodes of the transistors M_{Sn} is then dominated by the drain junction capacitor of M_{S} and the gate capacitor of M_{F}. It was stated that it takes about 13 ms with the cascode configuration to change the output current conditions significantly [299].

A further improvement of response speed can be achieved by setting the node potential of the drains of transistors M_{Sn} to a defined value, which bypasses the first time step. For that purpose, the transistors M_{In} are introduced at the nodes Ψ_{n} (Fig. 6.56). With them a start-up value of for instance $V_{\mathrm{init}} = 0.9\,\mathrm{V}$ can be set. Since all M_{Fn} conduct the same drain current I_{D} with $(\Sigma I_{\mathrm{D}}(M_{\mathrm{Fn}}) = I_{\mathrm{src}})$, the gate-source voltages V_{GS} of transistors M_{Sn} are forced to a low value almost shutting off transistors M_{Sn}. When the nodes

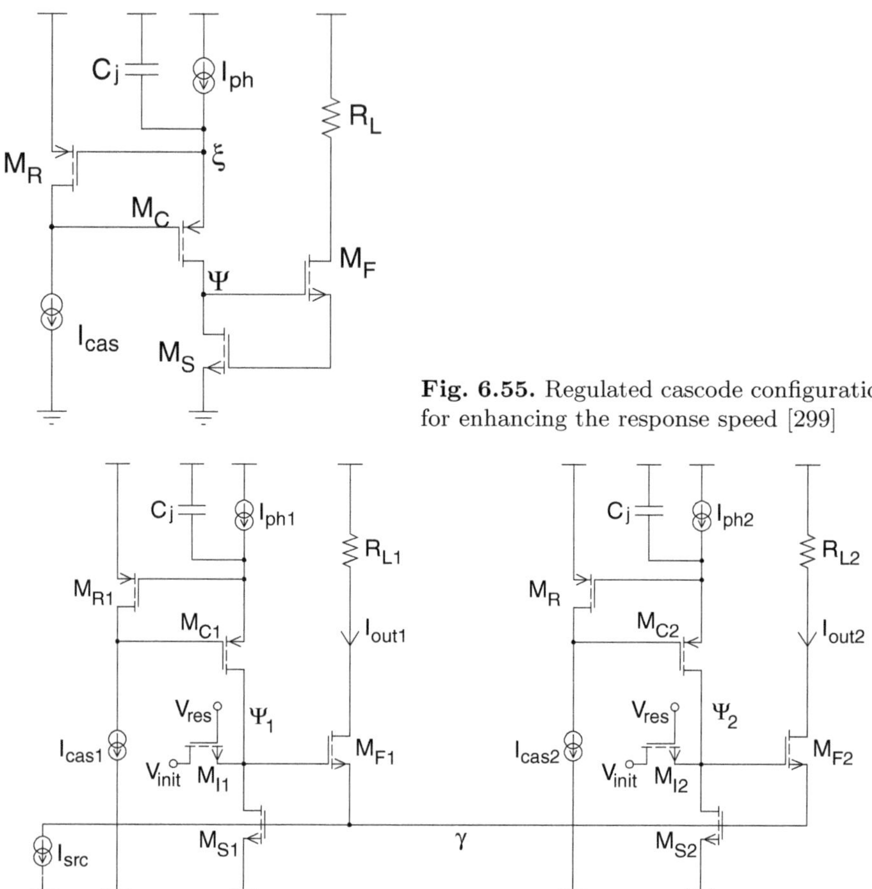

Fig. 6.55. Regulated cascode configuration for enhancing the response speed [299]

Fig. 6.56. Introduction of an initial condition for the drain potentials of M_S [299]

Ψ_n are released for an integration time t_{int}, the capacitors $C_{\Psi n}$ are recharged by the complete photocurrents, as long as $V_{GS}(M_{Sn})$ still prevents significant current flow in transistors M_{Sn}. For the highest photocurrent I_{phi}, M_{Fi} sinks most of I_{src}, inhibiting current flow in the other transistors M_{Fn}, and the cell wins. Due to recharging the capacitors $C_{\Psi n}$ with the complete photocurrent instead of the difference of photocurrents, the response speed is accelerated by about an order of magnitude [299].

In addition to these two measures, local excitory feedback is applied as suggested by Starzyk [303], which feeds back a portion of the output current to the input current of each cell, thereby boosting the effect of charging the capacitors $C_{\Psi n}$. To adapt the large output current in the range of nanoamperes to the small input photocurrent in the range of picoamperes, a new feedback topology has been implemented. A PMOS current mirror is added, where the feedback attenuation is controlled with the body effect of the output MOSFETs (Fig. 6.57). The amount of current, which is fed back to the input node, is determined by considering the subthreshold drain current characteristics proposed by Vittoz [153] for the body effect, resulting from applying different potentials to the source and bulk terminals of M_{FB}. The ratio $I_{\mathrm{out}}/I_{\mathrm{fb}}$ is given by

$$\frac{I_{\mathrm{out}}}{I_{\mathrm{fb}}} = e^{V_{\mathrm{SB}}(n-1)/nV_{\mathrm{T}}}, \tag{6.27}$$

where $n \approx 1.5$ and $V_{\mathrm{T}} = k_{\mathrm{B}}T/q$. According to simulations, a further increase of response speed can be achieved, when the feedback current is well adapted [299].

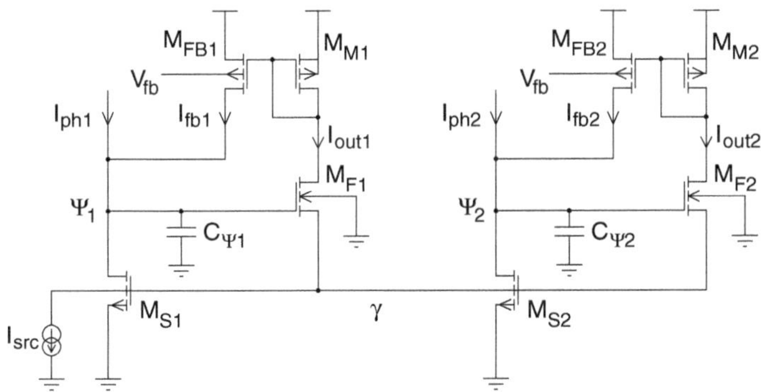

Fig. 6.57. Feedback configuration of a portion of the output currents to the input nodes with the current mirrors $M_{\mathrm{Mn}}/M_{\mathrm{FBn}}$ [299]

Implementing all three described topologies in the WTA configuration, the transient response time could be reduced from 0.6 s to below 1 ms [299]. Noise and mismatch, however, increase due to these measures. These issues will be discussed next.

In the picoampere range, noise generally is important and has to be considered. The gate potential of transistors M_{Fn} is very sensitive to noise and mismatch. Two voltage sources therefore are placed at their gates (Fig. 6.58). The first one is taken to model the noise voltage $\overline{v_{\mathrm{n}}^2}$, the second one to consider the effect of mismatch with an offset voltage v_{os}.

Different noise sources contribute to $\overline{v_{\mathrm{n}}^2}$. Poisson's distribution of uncertainty in counting single events has to be applied. The uncertainty n_{n} of the number of electrons being accumulated at the capacitor C_{Ψ}, thus, can be calculated from the charge Q on the capacitor [299]:

$$n_{\mathrm{n}} = \sqrt{N} = \sqrt{\frac{Q}{q}}. \tag{6.28}$$

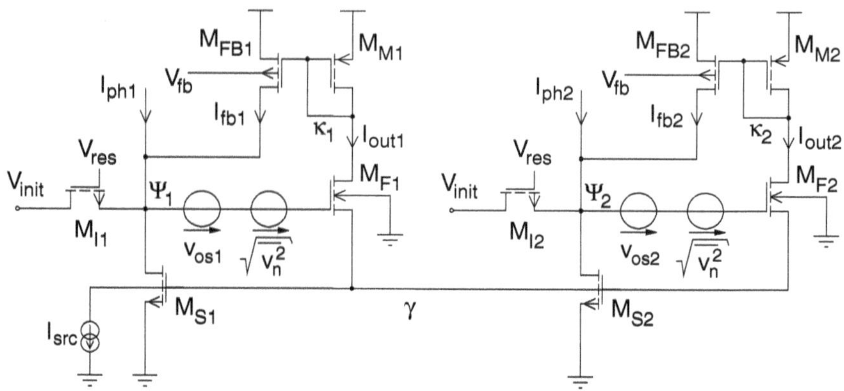

Fig. 6.58. Noise and mismatch considerations of a WTA cell [299]

The first noise contribution to $\overline{v_{\mathrm{n}}^2}$ is caused by the photocurrent I_{ph}, the leakage current of the photodiode I_{L}, and the feedback current I_{fb} [299]:

$$\overline{v_{\mathrm{n1}}^2} = (qn_{\mathrm{n}}\frac{1}{C_\Psi})^2 = q\frac{Q}{C_\Psi^2} = \frac{q}{C_\Psi^2}\int_0^t I_{\mathrm{ph}} + I_{\mathrm{L}} + I_{\mathrm{fb}} d\tau. \qquad (6.29)$$

The second noise contribution stems from reset noise by setting the capacitor C_Ψ to a defined value before the integration period. During this period, the capacitor is charged to a voltage with an uncertainty resulting from thermal resistor noise. This noise can be obtained by considering the spectral density of current noise and impedance at the node Ψ [299]:

$$\overline{v_{\mathrm{n2}}^2} = \int_0^\infty \frac{2q(I_{\mathrm{ph}} + I_{\mathrm{L}} + I_{\mathrm{fb}})}{|\mathrm{i}2\pi f C_\Psi + g_{\mathrm{ds}}(MI) + g_{\mathrm{ds}}(MS)|^2} df. \qquad (6.30)$$

The third noise contribution is caused when transistor M_{I} is shut off. Several electrons are pushed to the node Ψ due to charge feedthrough from the gate of M_{I}. The number of these electrons is [299]

$$n_{\mathrm{i,ft}} = \frac{1}{q} C_\Psi \Delta V. \qquad (6.31)$$

With Poisson's statistics, the third noise contribution is obtained [299]:

$$\overline{v_{\mathrm{n3}}^2} = \frac{1}{C_\Psi^2}(\sqrt{n_{\mathrm{i,ft}}}q)^2 = \frac{1}{C_\Psi} q\Delta V. \qquad (6.32)$$

Flicker noise (1/f-noise) causes the fourth contribution to noise [299]:

$$\overline{v_{\mathrm{n4}}^2} = \int_0^{f0} \frac{KF}{2fC_{\mathrm{ox}}WLK'} df. \qquad (6.33)$$

Thermal noise of the channel resistance of M_{F} transforms to its gate and forms the fifth contribution to noise [299]:

$$\overline{v_{\mathrm{n5}}^2} = \int_0^{f0} \frac{8k_{\mathrm{B}}T}{3g_{\mathrm{m}}} df. \tag{6.34}$$

The bandwidth of the transfer function of transistor M_{F} can be calculated with a small-signal analysis of I_{src}, M_{F}, M_{M}, and the parasitic capacitors C_{κ} of the node κ [299]:

$$f_0 = 1.57 f_1 = 1.57 \frac{1}{2\pi} \frac{g_{\mathrm{m}}(M_{\mathrm{M}}) + g_{\mathrm{ds}}(M_{\mathrm{F}})}{C_{\kappa}} df, \tag{6.35}$$

where the relation $f_0 = 1.57 f_1$ considers a transfer function with a single pole at the frequency f_1.

The different noise contributions are independent and the equivalent noise of the circuit shown in Fig. 6.58 is therefore obtained as [299]

$$\overline{v_{\mathrm{n}}^2} = \Sigma_{i=1}^{5} \overline{v_{\mathrm{ni}}^2} = \overline{v_{\mathrm{np}}^2}(I_{\mathrm{ph}}) + \overline{v_{\mathrm{nc}}^2}, \tag{6.36}$$

where $\overline{v_{\mathrm{np}}^2}$ represents the noise depending on the amount of charge on the capacitor C_{Ψ} accumulated by the photocurrent I_{ph} and where $\overline{v_{\mathrm{nc}}^2}$ represents the noise floor.

When a photocurrent of 10 pA is accumulated on C_{Ψ} during an integration period t_{int} of 1 ms, the uncertainty of the voltage across the capacitor C_{Ψ} was calculated in [299] from (6.36):

$$\frac{\sqrt{\overline{v_{\mathrm{n}}^2}}}{\triangle V_{\Psi}} = \frac{2.7\,\mathrm{mV_{rms}}}{200\,\mathrm{mV}} \approx 1.3\%. \tag{6.37}$$

When a significant current I_{fb} is flowing, the noise properties of the circuit depend strongly on the ratio of $I_{\mathrm{ph}}/I_{\mathrm{fb}}$, which is a nonlinear function of $\overline{v_{\mathrm{n}}^2}$ [299]. For that case, a simple analytical expression could not be derived [299]. Measurement of focal point positions revealed an error due to noise. In feedback mode with an incident optical power of 1 nW ($\lambda = 680$ nm) per focal point and with an integration time of 1 ms, the detection of the focal point position suffered from an average noise level of 1.89 pixels for the x-direction and 1.81 pixels for the y-direction, with a pixel size of 17.6 µm [299].

Mismatch between the WTA cells is also critical for the functionality of the WTA scheme. The accuracy of distinguishing input currents for differences is based on comparing node potentials of several identically built circuit blocks. To estimate consequences of fixed pattern noise (FPN), the voltage source v_{os} has been added in the WTA cell (Fig. 6.58). Neglecting mismatch of the photodiode stripes, the most important mismatch results from V_{GS} deviations of transistors M_{S}, M_{F}, and M_{FB}, from different charge injection values of M_{I}, and from differing values of the capacitors C_{Ψ} [299].

Noise and mismatch contribute to the total uncertainty of $V_{\mathrm{GS}}(M_{\mathrm{F}})$ [299]:

$$\sigma(V_{\mathrm{GS}}(M_{\mathrm{F}})) = \sqrt{v_{\mathrm{n}}^2 + v_{\mathrm{os}}^2}. \tag{6.38}$$

The accuracy of the WTA function can be estimated by forming the ratio of this uncertainty to the capacitor voltage difference $\triangle V_{\Psi}$ due to the accumulated photocurrent during the integration period. The difference of the photocurrent of two neighboring photodiode stripes is determined by the light spot diameter and the stripe width and is therefore at most 8 % [299]. Thus, a wrong decision of the WTA circuit happens when $\sigma(V_{\mathrm{GS}})/\triangle V_{\Psi} > 8\,\%$.

Optical glasses of spectacles were measured, to demonstrate the main issue of the HSS ASIC to quantify wavefront aberrations. For that purpose the optical glasses were placed between the laser source and the lens array. The focal point positions captured by the ASIC were fed into a personal computer to perform a least-square-fit for calculating the Zernike polynomial coefficients, from which spherical and cylindrical aberrations can be derived [300]. For integration times of 1, 2, and 4 ms and frame repetition rates of 500, 333, and 200 Hz, respectively, the observed relative errors for the spherical and cylindrical aberrations of the optical glasses were 7.5, 6, and 4.5 %, i. e. 0.15, 0.12, and 0.09 diopters [299].

In order to prevent illumination and photogeneration in the sensitive analog circuitry, the ASIC was shielded completely by a metal layer, except for the photosensitive areas. The ASIC was fabricated in a 0.6 µm N-well CMOS process with single poly and triple metal. The chip dimensions were $7.7 \times 8.2\,\mathrm{mm}^2$ [299].

6.4.7 BiCMOS Optical Sensors

There is a huge variety of optical sensors. It is not possible to describe all of them here. In the following a movement detection system working with infrared light consisting of a low-noise input amplifier, a low-pass filter, and a 10-bit tracking analog-to-digital converter as well as a laser driver with a low current noise is discussed [178]. This system was integrated on a chip in 0.5 µm standard BiCMOS technology. The photodiode is shown in Fig. 2.52.

The low-noise input amplifier will be introduced next. The system works with a high level of dc background light and a very weak level of ac signal light in the frequency range 10 Hz to 50 kHz. Since the spectral density of the shot noise of a photodiode is proportional to the total photocurrent (high dc part and weak ac part), whereas the ac-signal-to-noise-ratio is important in the system, the only way to reduce the total integrated shot noise is to limit the noise bandwidth. This has to be done properly, because signals in the operating bandwidth have to stay undamped. The circuit principle of the input stage implemented in the system is illustrated in Fig. 6.59.

The frequency characteristics is programmable for minimizing the noise bandwidth and for avoiding peaks in the noise transfer characteristics. The

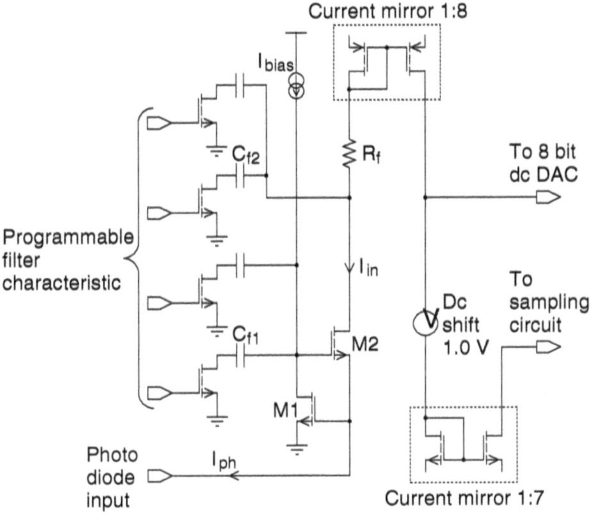

Fig. 6.59. Principle of low-noise input amplifier [178]

photodiode is biased reversely with a constant voltage due to NMOS transistor M2 with its transconductance g_{m2}. The PIN photodiode in standard BiCMOS technology shown in Fig. 2.52 was integrated together with the low-noise input amplifier depicted in Fig. 6.59. The photocurrent is filtered by the programmable low-pass filter of second order. The transfer function of the input stage is [178]

$$\frac{I_{\text{in}}}{I_{\text{ph}}}(s) = \frac{1 + s\frac{C_{f1}}{g_{m1}}}{s^2 \frac{C_{f1} C_D}{g_{m1} g_{m2}} + s\frac{C_{f1}}{g_{m1}} + 1} \tag{6.39}$$

with the photodiodes capacitance C_D, the sum of the programmable filter capacitors C_{f1}, and the transconductances g_{m1} and g_{m2} of the input transistors. R_f and C_{f2} form an additional low-pass filter for I_{in} and complete the filter of second order. Current mirrors realize the required amplification for the photocurrent. The compensation of the dc part of the photocurrent is done between the two current mirrors. For a reliable operation of the 8-bit DAC for the dc compensation an additional dc voltage shift is necessary as included in Fig. 6.59 [178].

As typical values for the ac part of the photocurrent 0.5 nA and 0.5 µA for its dc part were mentioned in [178]. The dc part multiplied by the amplification of the first current mirror is compensated by the dc DAC being controlled by a digital algorithm.

The application of the system required a 10-bit word for the ac part every 10 µs. A clock frequency of 1 MHz was used to transmit the 10-bit word via a serial interface. A tracking ADC (Fig. 6.60) to save chip area could be

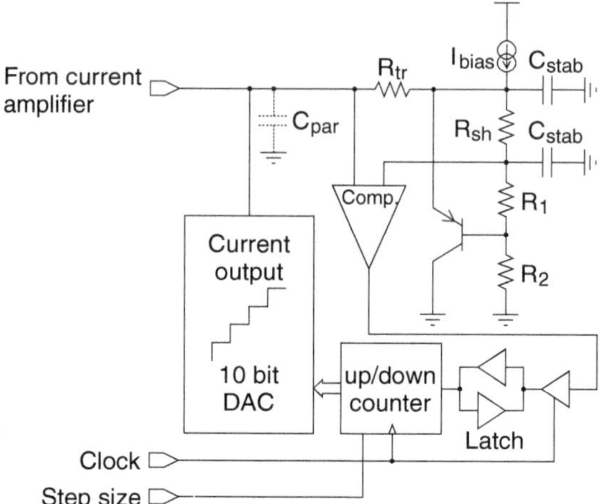

Fig. 6.60. Tracking analog-to-digital converter as sampling system [178]

implemented for the 10-bit ADC assuming a more or less continuous input signal.

In the tracking ADC, the current difference between the input current and the current of the 10-bit DAC produces a voltage drop across the trans-impedance resistor R_{tr}. This voltage drop is limited by the least-significant-bit (LSB) current of the 10-bit DAC. The comparator decides every clock cycle, whether the counter should count up or down. To synchronize the comparator output it has to be latched before the counter changes its output. The sampling speed of the system was limited by the maximum allowed current consumption. The design of the comparator in the tracking ADC, therefore, had to be done carefully by optimizing the speed and limiting the current consumption of this circuit module. The parasitic input capacitance C_{par} of the tracking ADC had to be minimized by a carefully performed chip layout, because the charge/decharge current of the parasitic capacitance at the input node is represented by the LSB DAC current of 50 nA and therefore puts a speed limit for the tracking ADC [178].

The voltage divider formed by R_{sh}, R_1, and R_2 serves as a bias structure for the comparator input. Adaptive count steps are used in the 10-bit up/down counter in order to follow faster changing signals with a higher slew rate [178].

The system discussed here is a light transmitting and receiving system. A laser diode with a wavelength of 780 nm has to be driven in this system. In order to achieve a photocurrent of 1 μA in the photodiodes integrated in the movement detection chip, the laser diode needs a maximum current of about 11 mA [178]. Figure 6.61 depicts the low-noise laser driver circuit also implemented in the BiCMOS optical sensor chip.

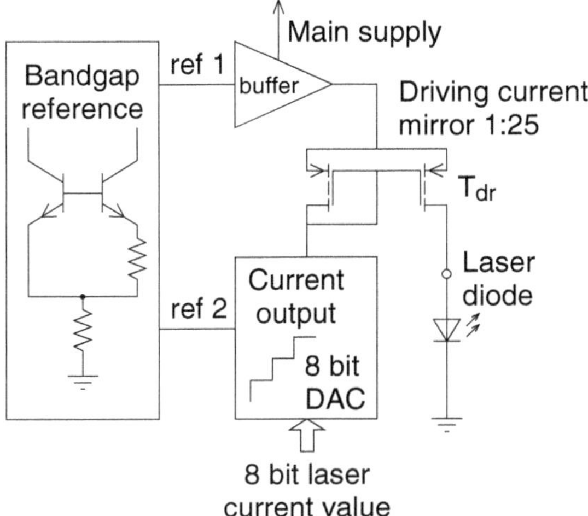

Fig. 6.61. Block diagram of low-noise laser driver circuit [178]

Special measures are necessary to ensure that the current through T_{dr} is free of any disturbances from the power supply. For that purpose the bandgap reference supplied voltage ref1 in Fig. 6.61 is amplified in the buffer in such a way that the minimum voltage across the driver transistor T_{dr} is at least 300 mV for the highest laser diode current of 11 mA. The reference current ref2 (Fig. 6.61) in addition has to be provided by a buffered bandgap reference voltage and a temperature compensated resistor network in order to become independent from the main supply [178].

It was very important for the application that the emitted light was free of any noise, because noise in the emitted light would reduce the signal-to-noise ratio of the receiving part in the system. Since above the laser threshold current the emitted light is approximately proportional to the laser diode current, the driver output current had to be free of current noise. With the configuration shown in Fig. 6.61 an integrated RMS noise current of only 0.5 µA up to 300 kHz at a driving current of 11 mA was achieved [178].

Due to the various tricks described, the innovative light detection chip achieved low-noise amplification, high sampling speed, and a low over-all current consumption of 4 mA in a voltage operating range from 3.0 V to 5.5 V whereby the laser diode current was not included. The proposed system consisting of six receiving channels, the laser driver unit, voltage regulation, and a digital interface was realized within a chip area of about 18 mm^2. The integration of such an analog-digital system also including the optical detectors on the mixed-signal chip posed a challenge. The coupling of clock signals from the digital part of the system to the sensors consuming a large area and generating only a very weak signal, however, could be avoided by placing the digital part near the chip corner and locating digital signal lines as well as digital pads as far away as possible from the sensor array and the input amplifiers [178].

6.4.8 CMOS Circuits for Optical Storage Systems

Compact disk (CD), CD-ROM, and Digital-Video-Disk/Digital-Versatile-Disk (DVD) are optical storage (OS) systems with a storage capacity of the order of 1 to 10 GByte [304]. The stored information is read with a focused laser beam. Depending on the stored state of 0 or 1, more or less light intensity is reflected into the read circuit, which we will call OS-OEIC. Special arrangements of 6 to 8 photodiodes are implemented in OS-OEICs in order to obtain the signals for tracking and for focusing in addition to the RF signal, which contains the stored information (see Fig. 6.62). Digital-video-recording (DVR) systems allow also writing of information by the focused laser beam. In contrast to the name *digital*-video-disc-system, the DVD-OEIC is a purely *analog* front-end circuit, which is a key-device for the whole DVD-system. OS-OEICs for CD and CD-ROM are also analog circuits. The data rate and therefore the speed of the optical storage system are determined by the OS-OEIC.

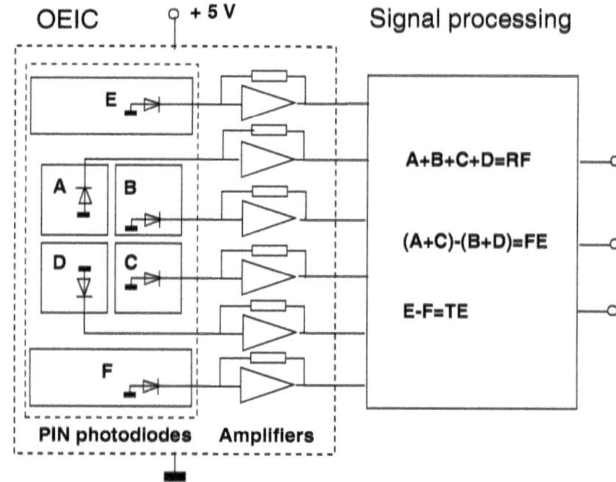

Fig. 6.62. OEIC for compact disc, CD-ROM and DVD applications

In OS systems, accordingly, the demand for fast OEICs is steadily increasing, especially in the red spectral range ($\lambda \approx 635$–$650\,\text{nm}$). The integration of both the optical devices and the electronic circuits on the same chip leads to a smaller die area, to lower manufacturing costs and to faster systems. Furthermore, the reliability and the immunity against electromagnetic interference of OEICs are enhanced compared to a two-package solution with a photodiode package and an amplifier package. In comparison to a two-chip solution with a photodetector chip being wire-bonded to an amplifier chip in one package, the die area consumption of a monolithic OS-OEIC is smaller, because the area of a photodiode in an OS system is smaller than the area of a bondpad.

For fast OS systems, integrated PN photodiodes are not sufficient, because only a bandwidth of 10–15 MHz is achievable [62,64]. Although a −3dB bandwidth of 32 MHz can be derived from the frequency response of a P^+ to N-substrate photodiode shown in Fig. 2.13, the photocurrent already begins to decrease at a frequency of 2 MHz due to slow diffusion of photogenerated carriers. Hence, PIN photodiodes are required. It was shown that for the integration of a PIN photodiode in a standard twin-well CMOS process (1 μm), which uses epitaxial wafers, little modification is necessary. Compared to the published approaches in [35] and [305] with standard-buried-collector (SBC) based bipolar OEICs, less additional process complexity is required [76] as was already pointed out in Sects. 2.1.3 and 2.2.5. The cross section of a CMOS-OEIC was already shown in Fig. 2.19.

In order to obtain fast integrated PIN photodiodes, the standard doping concentration of the epitaxial layer of $1 \times 10^{15}\,\mathrm{cm}^{-3}$ has to be reduced to approximately $5 \times 10^{13}\,\mathrm{cm}^{-3}$ [76]. This reduction of the doping level does not influence the electrical parameters of the CMOS devices, since the MOSFETs are placed in wells and, therefore, the model parameters for circuit simulation did not have to be modified for the design of the OEICs [84]. ESD (electrostatic discharge) and latch-up immunity could be obtained by appropriate layouts.

OS-OEICs have to fulfill a stringent requirement concerning the output offset voltage. A fast and low-area CMOS optoelectronic integrated circuit (OEIC) for optical storage (OS) systems containing photodiodes and transimpedance amplifiers with a low offset voltage was presented [306]. The photodiodes and the amplifiers were monolithically integrated in a 0.6 μm CMOS process. A replica transimpedance amplifier and a low-offset operational amplifier were implemented for offset reduction. An expression for the standard deviation of the output offset voltage was derived analytically [306]. A -3dB frequency of 147 MHz and a sensitivity of 24 mV/μW at 670 nm were measured, as well as an offset voltage deviation of 4 mV. With this performance, an innovative low-cost CMOS OEIC was suggested being a true alternative to BiCMOS OEICs developed and used in the years 2002 and 2003.

Optical storage systems such as Audio CD, CD-ROM (compact disk read only memory), DVD (digital video disk or digital versatile disk), or DVR (digital video recording) require OEICs with a high bandwidth, a high sensitivity, and a low output offset voltage compared to an external reference voltage V_{Ref}. To achieve high bandwidths and low offset voltages bipolar or BiCMOS processes are preferred [176, 181]. In [182] an OS OEIC exploiting complementary bipolar current mirrors with a bandwidth of 250 MHz in BiCMOS technology was described. Apart from the absent description of the essential idea, fast PNP transistors were needed, which are only available in costly BiCMOS processes. The OEIC proposed in [306], however, was implemented in a low-cost 0.6μm CMOS process.

Usually operational amplifiers in transimpedance configuration perform the photocurrent to voltage conversion [176, 307, 308]. But low-offset CMOS operational amplifiers need large-area input transistors. Therefore it is difficult to achieve high-speed CMOS OS OEICs [307, 308]. Fiber receivers often use inverters in transimpedance amplifiers as for instance in a 155 Mb/s-receiver [309] or in a 1 Gb/s-receiver [310]. To obtain a high bandwidth in the CMOS OS OEIC, the transimpedance amplifier of the OEIC proposed in [306] consists of a one-stage inverter with a polysilicon feedback resistor and two source followers, which reaches a bandwidth of 147 MHz and a sensitivity of 24 mV/μW. In the operating temperature range and due to process variations, the dark-level output voltage of the CMOS OS OEIC would vary from 2.43 V to 2.64 V. To reduce this output offset voltage, a replica offset cancellation (ROC) circuit suggested for BiCMOS OS OEICs in [181] was used. Alternatively, this could be done by subtracting the output voltage of the replica amplifier from the output voltage of the transimpedance amplifier [177, 311], but this method requires a fast subtracting circuit. This disadvantage can be avoided, if a rather slow low-offset operational amplifier is used to adjust the output bias voltage of both the inverter-transimpedance and the replica inverter-amplifier to be equal to the reference voltage V_{Ref}. With this method we can achieve an output offset voltage of much less than 15 mV.

To be able to judge the over-all performance of an OS OEIC a figure of merit FOM, which combines the bandwidth $f_{3\mathrm{dB}}$, the sensitivity S and the maximum output offset voltage V_{offset} of an OS OEIC, can be defined [306]:

$$FOM = \frac{f_{3\mathrm{dB}} \cdot S}{V_{\mathrm{offset}}} \qquad (6.40)$$

Combining the results mentioned above, this CMOS OS OEIC described here featured an outstanding figure of merit of 287 MHz/μW.

Besides the use of a low-cost CMOS process, a complete OS OEIC, using the proposed circuit, would only occupy a chip area of about $1\mathrm{mm}^2$ [306]. Common comparable OEICs need chip areas of $2\,\mathrm{mm}^2$ [307] to $3.25\mathrm{mm}^2$ [176, 181].

The simplified schematic of the OEIC is shown in Fig. 6.63. The transimpedance amplifier consists of a single-stage inverter N1/P1 and a feedback resistor Rfb. Two source-followers N2/N3 and P2/P3 are appended to drive a load capacitance of up to 10 pF. VB1 and VB2 are bias voltages. The replica transimpedance amplifier has the same structure and exactly the same layout as the transimpedance amplifier. A very simple method to vary the DC level of the output voltage was implemented by the inverters N7/P7 respectively N8/P8. Because of the use of long-channel transistors, their output resistance is rather high (approximately 500kΩ). So they act as bipolar current sources pulling the output voltages of the inverters N1/P1 and N4/P4 up or down

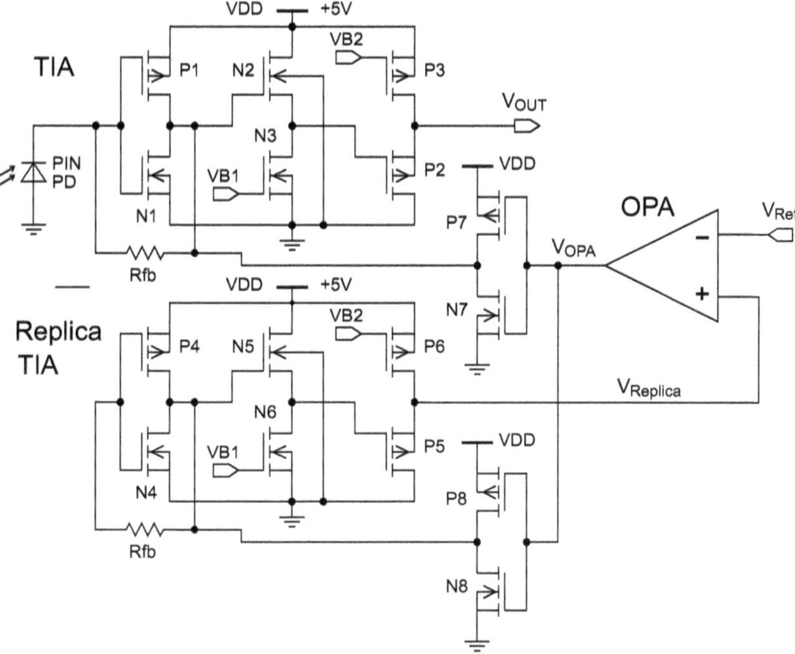

Fig. 6.63. Simplified schematic of low-offset CMOS DVD OEIC [306]

by ±0.34 V. A low-offset operational amplifier is used to regulate the output voltage of the replica transimpedance amplifier $V_{\rm Replica}$ to the same level as the reference voltage $V_{\rm Ref}$. The operational amplifier uses a standard differential input stage with a complementary output stage. A similar structure is described in [155]. Because the used CMOS operational amplifier has no systematic offset and a gain being sufficiently large, the output voltage should be equal to the reference voltage $V_{\rm Ref}$ if no light is coupled into the photodiode. But as a result of non ideal matching of the transistor pairs N1/N4, P1/P4, N2/N5, P2/P5, N7/N8, P7/P8 and of the input transistors of the operational amplifier, an output offset voltage occurs. We assume that the values of the threshold voltages $V_{\rm T}$ of the MOS transistors are distributed normally with the variances $\sigma^2_{\rm VT}$. The variance $\sigma^2_{\rm VT}$ can be derived by [306]

$$\sigma^2_{\rm VTi} = \frac{AVTOL^2}{W_{\rm i}L_{\rm i}} , \tag{6.41}$$

where $L_{\rm i}$ is the length and $W_{\rm i}$ is the width of the transistor i. AVTOL is a process dependent matching parameter. Table 6.1 lists the $\sigma^2_{\rm VT}$.

To determine the standard deviation of the output offset voltage, an equation describing the dependence of the output voltage on the transistor threshold voltages $V_{\rm Ti}$ and the input offset voltage $V_{\rm IO}$ of the CMOS operational amplifier is derived [306]. At first we determine the output voltage of the in-

Table 6.1. Variances and their weights needed in (6.56) [306]

Parameter	a_j	σ_j^2 (mV^2)
VTN1	0.4866	21.45
VTP1	-0.5134	12.89
VTN2	-1	5.005
VTP2	1	0.669
VTN4	-0.4866	21.45
VTP4	0.5134	12.89
VTN5	1	5.005
VTP5	-1	0.669
VTN7	0.1301	20.02
VTP7	-0.1032	12.03
VTN8	-0.1301	20.02
VTP8	0.1032	12.03
$V_{\mathrm{offset,OPA}}$	-1	0.858

verter N1/P1, which is also the output voltage of the inverter N7/P7. Equation (6.42) shows the relation of their drain currents.

$$I_{\mathrm{DS,N1}} + I_{\mathrm{DS,N7}} = I_{\mathrm{DS,P1}} + I_{\mathrm{DS,P7}} \,. \tag{6.42}$$

Using a simple large-signal MOSFET model, (6.43) follows, where V_{OPA} is the output voltage of the operational amplifier.

$$\begin{aligned} &\frac{KPN}{2}\frac{W_{\mathrm{N1}}}{L_{\mathrm{N1}}}\left(V_{\mathrm{DS,N1}} - V_{\mathrm{TN1}}\right)^2 + \\ &\frac{KPN}{2}\frac{W_{\mathrm{N7}}}{L_{\mathrm{N7}}}\left(V_{\mathrm{OPA}} - V_{\mathrm{TN7}}\right)^2 = \\ = &\frac{KPP}{2}\frac{W_{\mathrm{P1}}}{L_{\mathrm{P1}}}\left(V_{\mathrm{DD}} - V_{\mathrm{DS,N1}} - V_{\mathrm{TP1}}\right)^2 + \\ &\frac{KPP}{2}\frac{W_{\mathrm{P7}}}{L_{\mathrm{P7}}}\left(V_{\mathrm{DD}} - V_{\mathrm{OPA}} - V_{\mathrm{TP7}}\right)^2 \,. \end{aligned} \tag{6.43}$$

The gain factor of the NMOS transistors KPN is higher than the gain factor of the PMOS transistors KPP. To compensate this imbalance, the width of the PMOS transistors W_{Pi} of all inverters are chosen by the relation (6.44):

$$\frac{W_{\mathrm{Pi}}}{W_{\mathrm{Ni}}} = \frac{KPN}{KPP} \,. \tag{6.44}$$

To minimize the length of the following equations the definitions (6.45), (6.46) and (6.47) are established:

$$k = \frac{W_{\mathrm{N1}} L_{\mathrm{N7}}}{W_{\mathrm{N7}} L_{\mathrm{N1}}} = \frac{W_{\mathrm{N4}} L_{\mathrm{N8}}}{W_{\mathrm{N8}} L_{\mathrm{N4}}} \tag{6.45}$$

$$V_{iA} = V_{DD} - (V_{TPi} + V_{TNi}) \tag{6.46}$$

$$V_{iS} = V_{DD} - (V_{TPi} - V_{TNi}) \, . \tag{6.47}$$

Now (6.43) is solved for the drain-source voltage $V_{DS,N1}$, which is equal to the output voltage of the inverter N1/P1

$$V_{DS,N1} = \frac{V_{1S}}{2} + \frac{V_{7A}V_{7S} - 2V_{OPA}V_{7A}}{2kV_{1A}} \, . \tag{6.48}$$

The output voltage V_{OUT} of the transimpedance amplifier can be derived by equation (6.49)

$$V_{OUT} = V_{DS,N1} + V_{GS,P2} - V_{GS,N2} \, . \tag{6.49}$$

Inserting (6.48) into (6.49), equation (6.50) follows:

$$V_{OUT} = V_{GS,P2} - V_{GS,N2} + \frac{V_{1S}}{2} + + \frac{V_{7A}}{2kV_{1A}} (V_{7S} - 2V_{OPA}) \, . \tag{6.50}$$

In an equivalent way equation (6.51), which describes the output voltage of the replica transimpedance amplifier $V_{Replica}$, can be obtained:

$$V_{Replica} = V_{GS,P5} - V_{GS,N5} + \frac{V_{4S}}{2} + + \frac{V_{8A}}{2kV_{4A}} (V_{8S} - 2V_{OPA}) \, . \tag{6.51}$$

Equation (6.52) applies, if the voltage gain of the operational amplifier is adequately large. V_{IO} is the input offset voltage of the CMOS operational amplifier:

$$V_{Replica} = V_{Ref} - V_{IO} \, . \tag{6.52}$$

If equations (6.51) and (6.52) are combined, the output voltage of the operational amplifier V_{OPA} can be derived:

$$V_{OPA} = \frac{kV_{4A}(V_{IO} - V_{Ref} + V_{GS,P5} - V_{GS,N5})}{V_{8A}} + + \frac{kV_{4A}V_{4S} + V_{8A}V_{8S}}{2V_{8A}} \, . \tag{6.53}$$

If this result is inserted into (6.50), equation (6.54) follows:

$$V_{OUT} = V_{GS,P2} - V_{GS,N2} + \frac{V_{1S}}{2} + \frac{V_{7A}}{2kV_{1A}V_{8A}} \bullet \bullet \Big[2kV_{4A}(V_{Ref} - V_{IO} + V_{GS,N5} - V_{GS,P5}) + + V_{7S}V_{8A} - V_{8A}V_{8S} - kV_{4A}V_{4S}\Big] \, . \tag{6.54}$$

In (6.55), the output offset voltage V_{offset} is defined:

$$V_{\text{offset}} = V_{\text{OUT}} - V_{\text{Ref}} \ . \tag{6.55}$$

The expected value of V_{OUT} is equal to V_{Ref}, so the expected value of V_{offset} is zero. The variance of the output voltage σ^2_{OUT}, which is equal to the variance of the offset voltage, can be obtained by:

$$\sigma^2_{\text{OUT}} = \sum_{\text{j}} a_{\text{j}}^2 \sigma_{\text{j}}^2 \tag{6.56}$$

$$\text{with} \quad a_{\text{j}} = \frac{\partial V_{\text{OUT}}}{\partial V_{\text{j}}} \ . \tag{6.57}$$

V_{j} stands for the threshold voltages of the transistors N1-8 and P1-8 as well as the input offset voltage V_{IO} of the operational amplifier. In the calculation of the derivations of V_{OUT}, it is possible to assume that the variances of the gate-source voltages are equal to the variances of the threshold voltages. Because of their rather large area it is not necessary to consider the threshold voltage variances of the transistors N3, P3, N5, and P5. The values of a_{j} are derived from (6.57) and (6.54). The variances σ_{j}^2 are derived from (6.41). The values of a_{j} and σ_{j}^2 are listed in Table 6.1. In consideration of equation (6.56) from these values the standard deviation of the output voltage $\sigma_{\text{OUT}} = 5.486\,\text{mV}$ follows. This means that 99.4 % of the produced OEICs have a dark-level output voltage of $V_{\text{Ref}} \pm 15\,\text{mV}$ [306].

A PIN photodiode, similar to that described in [77], was applied on a P^+ substrate. The anode of the photodiode was formed by the P^+ substrate and was connected to ground. The cathode of the PIN photodiode was formed by an N^+ source/drain region. An antireflection coating (ARC) was used to achieve a high photodiode quantum efficiency $\eta = 96.5\,\%$ at $\lambda = 670\,\text{nm}$. Rise times of 660 ps and fall times of 620 ps were measured at the photodiode.

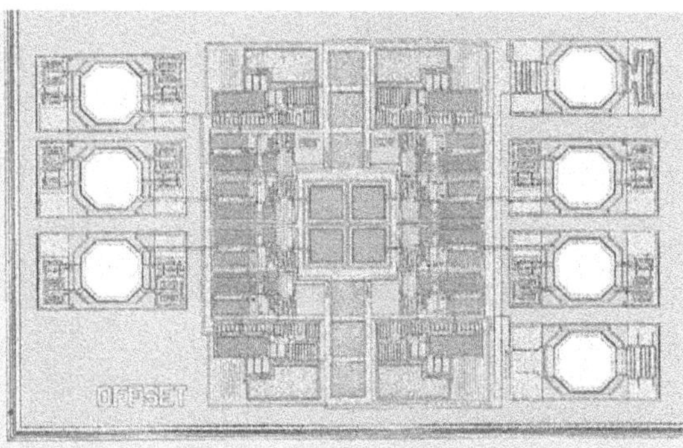

Fig. 6.64. Chipfoto of low-offset CMOS DVD OEIC [306]

To exploit the available chip area, four identical OS receivers were implemented on one chip. The chip was fabricated using a 0.6 μm CMOS process. Figure 6.64 shows a microphotograph of the four receivers each with an associated photodiode. The whole chip including the bondpads occupies an area of 930 µm times 555 µm. Figure 6.65 shows one of the four receivers in detail, consisting of the PIN photodiode (PD), the transimpedance amplifier (TIA), the replica transimpedance amplifier (Replica), and the operational amplifier (OPA).

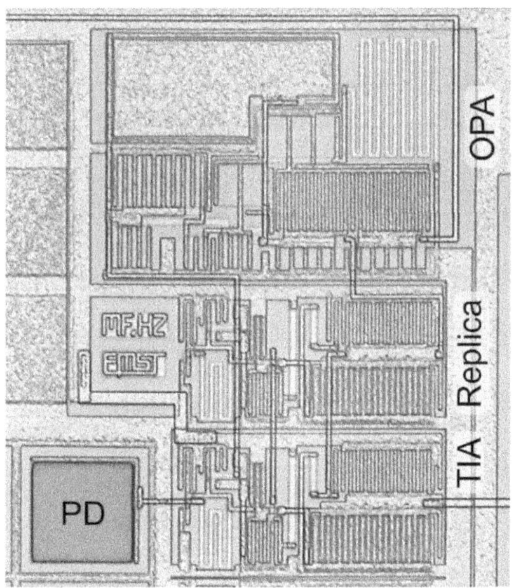

Fig. 6.65. Chipfoto of one channel in low-offset CMOS DVD OEIC

The OS OEIC was measured on wafer level. The results of one receiver typical for all four receivers are shown in Table 6.2.

For all measurements the supply voltage $V_{\rm DD}$ was 5 V and the reference voltage $V_{\rm Ref}$ was 2.5 V. The light was coupled into the photodiode via a single-mode optical fiber adjusted on a wafer prober. The light source was a laser diode with an optical wavelength of $\lambda = 670$ nm modulated via a bias-tee. The output voltage was measured with a picoprobe (input capacitance 0.1 pF, bandwidth 3 GHz). The transient response was measured with a digital storage oscilloscope Tektronix CSA8000. Rise times of 2.4 ns and fall times of 2.45 ns were measured. The frequency response was measured with a network analyzer Hewlett Packard 8753E. The frequency response of the OS receiver, displayed in Figure 6.66, shows a flat response with a bandwidth of 147 MHz. The bandwidth as well as the supply current are lower than the simulated values (see Table 6.2), which is caused by process tolerances. The

Table 6.2. Summary of results for the low-offset CMOS OEIC [306]

	SpectreS simulation	Measured results
Supply current	13.6 mA	10.6 mA
Sensitivity	29.2 mV/μW	23.9 mV/μW
Transimpedance	58.3 kΩ	46.0 kΩ
Bandwidth	191 MHz	147 MHz
Rise time	1.85 ns	2.40 ns
Fall time	1.87 ns	2.45 ns
Noise voltage density (at 100 MHz)	$36\,\mathrm{nV}/\sqrt{Hz}$	-

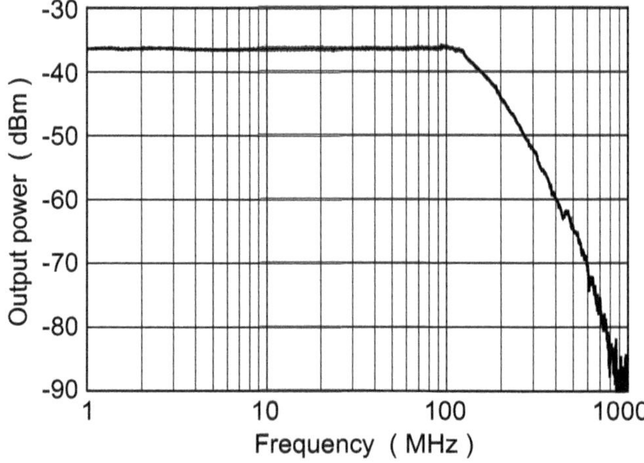

Fig. 6.66. Frequency response of low-offset CMOS DVD OEIC [306]

lower sensitivity could be explained by tolerances of the feedback resistor or of the antireflection coating of the photodiode [306].

The results of the DC output offset voltage measurements are shown in Figure 6.67. A number of 42 chips was measured along 2 rows across a whole wafer. The mean values of the output offset voltages of the outputs 1 to 4 are: −11.47 mV, −11.02 mV, −9.07 mV and −17.41 mV [306]. The associated standard deviations are: 3.19 mV, 4.12 mV, 4.51 mV, and 3.55 mV. The mean value of the offset voltages of all 168 channels shown in Fig. 6.67 is −12.24 mV, the standard deviation is 4.04 mV [306]. Though the standard deviation is less than the calculated value of 5.486 mV, the mean values are definitely deviating from 0 V for the output offset voltage. The source of these aberrations could be traced back. Because of the rather high supply currents (Table 6.2), voltage drops along the ground conductors appear, which cause voltage differences between the transimpedance amplifiers and their associ-

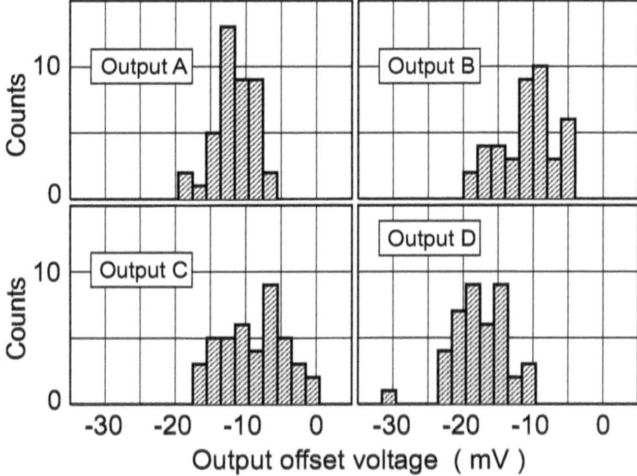

Fig. 6.67. Distribution of output offset voltage in low-offset CMOS DVD OEIC [306]

ated replica TIAs [306]. In a redesign this drawback can be avoided by a proper designed layout.

The OEIC presented in [306], nevertheless, featured an outstanding figure of merit $FOM = 287\,\mathrm{MHz}/\mu\mathrm{W}$. Other OS OEICs achieved figure of merits from 32.9 MHz/μW [307] to 83.5 MHz/μW [176].

Three suggestions for improvement could be made [306]: First, the single-stage inverters (N1/P1, N4/P4) could be replaced by 3-stage inverters. The higher gain of this topology results in a higher bandwidth of the trans-impedance amplifiers. Secondly, the simple method of shifting the DC level by the inverters N7/P7 and N8/P8 has an influence on the value of the trans-impedance and therefore on the sensitivity of the OEIC. This effect can be minimized by level-shifting within the proposed 3-state inverter. Finally, it has to be mentioned, that an improved layout can minimize the output offset voltage.

An optical receiver for optical storage systems implemented in a low-cost 0.6 μm CMOS process was described. In a first attempt, a bandwidth of 147 MHz and a sensitivity of 23.9 mV/μW could be achieved. The output voltage offset deviation of 4.04 mV was smaller than the calculated value. This OS OEIC achieved an outstanding figure of merit of 287 MHz/μW with a potential of improvement. It was shown that four independent receivers including the photodiodes and bondpads can be placed within a chip area of $0.51\,\mathrm{mm}^2$. A complete 8-channel DVD OEIC should therefore only occupy a total chip area of approximately $1\,\mathrm{mm}^2$ [306].

6.4.9 BiCMOS Circuits for Optical Storage Systems

OEICs for optical storage (OS) systems with an enhanced data rate contain four fast channels (A–D) for data extraction and focus control plus four slower channels (E–H) for tracking control with a ten times larger sensitivity (Fig. 6.68).

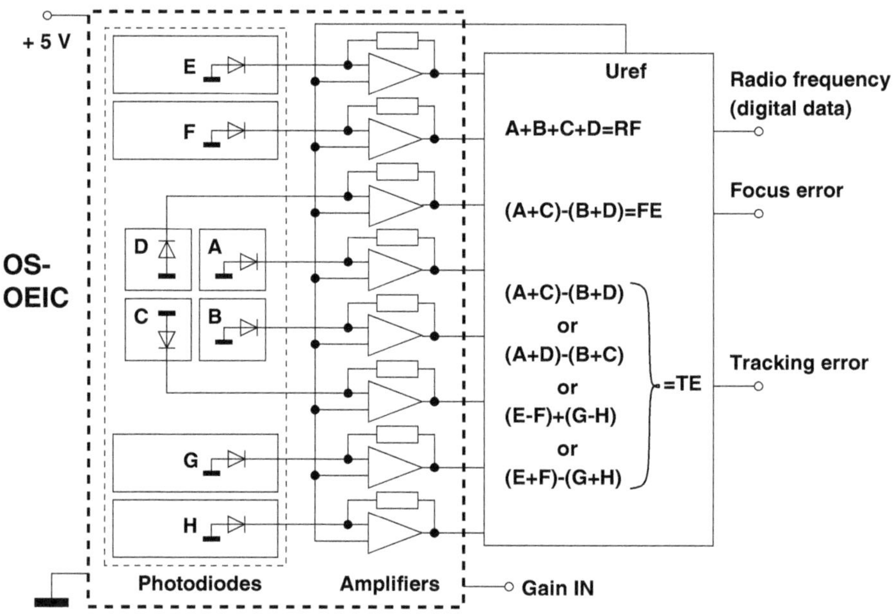

Fig. 6.68. Block diagram of an OS-BiCMOS-OEIC

No process modifications were necessary in the BiCMOS process implementing a double photodiode shown in Fig. 2.47 [173]. The schematic of one fast channel of an eight-channel OS-BiCMOS-OEIC is shown in Fig. 6.69 [173]. Polysilicon-polysilicon capacitors are available in the 0.8 µm BiCMOS process and a transimpedance amplifier is realized here. The gain is switchable between high (H), medium (M), and low (L). Only NPN transistors are used in the signal path and the resistor loads R1 and R2 are implemented in the difference amplifier in order to achieve a high −3dB bandwidth. The value of 44 MHz was confirmed by measurements for the high gain [174]. The emitter follower Q5 reduces the output impedance of the operational amplifier.

The base currents of Q1 and Q2 in the input stage of the operational amplifier are approximately 5 µA. Such a large input current of the operational amplifier, flowing through the transimpedance resistor R3, leads to a systematic output offset voltage of approximately 0.1 V, when no special

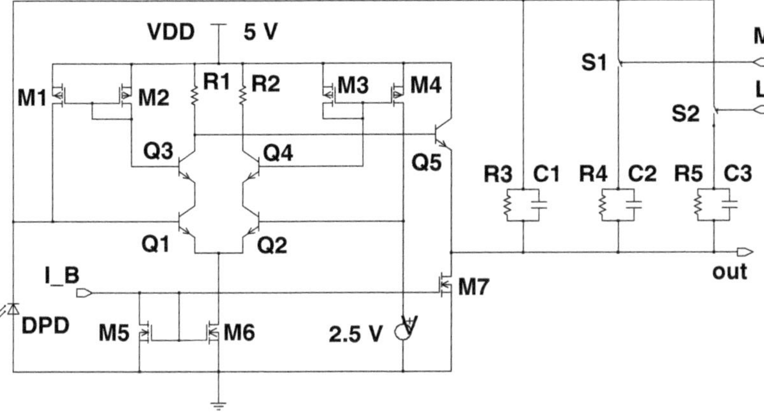

Fig. 6.69. Schematic of the fast channels A–D in an OS-BiCMOS-OEIC [173]

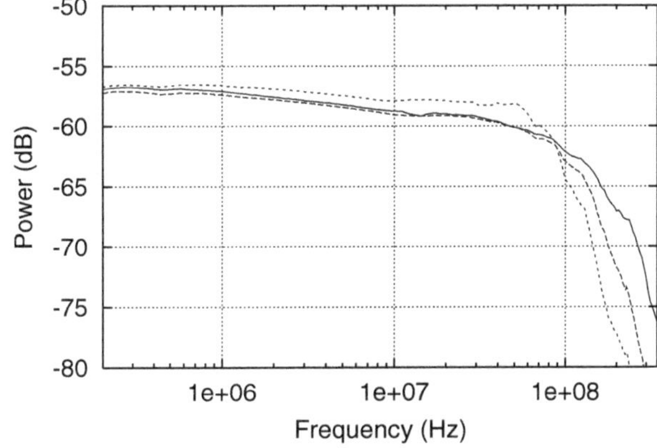

Fig. 6.70. Frequency response of an BiCMOS OEIC for optical storage systems for three different optical input powers, i. e. three different gain factors [173]

measures are taken. These special measures are: The transistors Q3 and Q4 are used to sense the base currents of Q1 and Q2, respectively. The current mirrors M1/M2 and M3/M4 mirror the base currents of Q3 and Q4 into the bases of Q1 and Q2, respectively [312]. With this bias current cancellation, the output offset voltage could be reduced to less than 9 mV [173].

The frequency responses for the three different gain factors of the OS-BiCMOS-OEIC are shown in Fig. 6.70. The −3dB bandwidth for the high gain is 44 MHz, for the medium gain it is 59 MHz, and for the low gain it is 64 MHz. The slightly decreasing frequency responses between 1 MHz and 40 MHz are due to parasitic capacitors in the amplifier and not due to slow carrier diffusion of photogenerated carriers in the double photodiode (DPD).

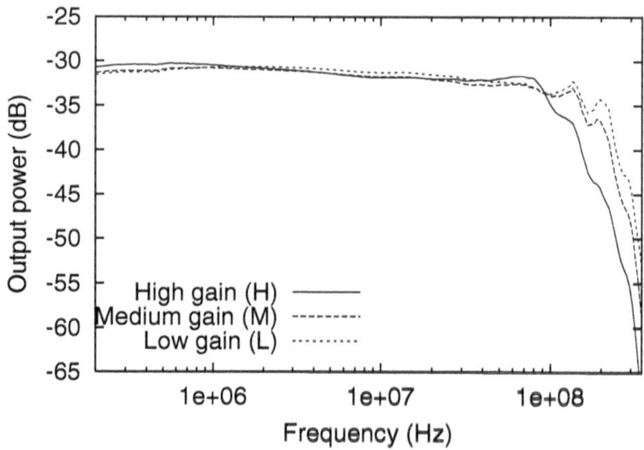

Fig. 6.71. Frequency responses of the fast channels A–D of a high-bandwidth BiCMOS OEIC for optical storage systems measured with three different optical input powers, i. e. three different gain factors [176]

The power consumption of the circuit in Fig. 6.69 is 7 mW at 5.0 V and with an antireflection coating a sensitivity of 10.5 mV/μW is achieved with the highest gain factor. The active die area of this circuit is 0.054 mm^2. An 8-channel OEIC with four of the above described fast amplifiers and four ten times more sensitive MOS amplifiers consumed a power of approximately 40 mW.

At the expense of a higher power consumption, an even higher speed of OEICs for optical storage systems is possible [176]. A high-bandwidth BiCMOS OEIC has been demonstrated, which implemented fast amplifiers with the same schematic shown in Fig. 6.69 but with larger bias currents than in [173]. These fast amplifiers exhibit −3dB bandwidths in excess of 90 MHz (Fig. 6.71).

An integrated double photodiode (Fig. 6.69) is connected to a transimpedance amplifier using an operational amplifier in order to obtain a low output offset voltage compared to a reference voltage of 2.5 V as is required for applications in optical storage systems. For a universal applicability, the gain is switchable by MOS elements between high (H, R3), medium (M, R4 || R3) and low (L, R5 || R4 || R3) with a ratio of approximately 1/3 each. Polysilicon-polysilicon capacitors are used for frequency compensation with C1, C2, and C3. Here, the bias current cancellation of the input transistors Q1 and Q2 reduces the systematic output offset of approximately 110 mV, which would result from the base current of Q1 across the resistor R3 $\approx$ 20 kΩ, to below 11 mV. According to simulations, the low-frequency open-loop gain of the operational amplifier is 27 dB and its transit frequency is 870 MHz in the case of a load of 1 kΩ and 10 pF. An OEIC was packaged, mounted together with these load elements on a printed circuit board, and the frequency responses were measured with a probe head having an input capacitance of 1.7 pF. The slight decrease in the frequency responses (Fig. 6.71) between about 5 MHz and 80 MHz is due to parasitic capacitors in the amplifier and is not due to slow diffusion of photogenerated carriers in the DPD. The measured bandwidths exceed a value of 92 MHz. This value is much larger than the bandwidth of 7.3 MHz of the circuit for 8× speed CD-ROMs fabricated in 0.8 µm CMOS technology with off-chip photodiodes [308].

Figure 6.72 shows the schematic of the four sensitive channels E–H with a ten times larger sensitivity for tracking control in the optical storage system. A double photodiode with approximately twice the size of the DPDs in the channels A–D is implemented in the channels E–H. The N-channel MOSFET source followers M1 and M2 are added in front of the bipolar difference amplifier Q1 and Q2 in order to avoid input currents and the resulting output offset voltages across the feedback resistors of about 200 kΩ. A high sensitivity of 100 mV/µW in combination with a low offset voltage can be realized in such a way. The PMOS load elements M3 and M4 are implemented for Q1 and Q2 in the difference amplifier in order to achieve a larger open-loop gain than with resistor load elements. The compensation is split between C1 and C2, C1 and C3, as well as C1 and C4. According to circuit simulations, the low-frequency open-loop gain of the circuit shown in Fig. 6.72 is 36 dB with a transit frequency of 130 MHz.

The frequency responses of the channels E–H are shown in Fig. 6.73. The values for the −3dB bandwidths are listed in Table 6.3 together with other results.

Each amplifier in the channels A–H covers an active die area of about 0.079 mm^2, and the total die area of the high-bandwidth BiCMOS OEIC amounts to 3.25 mm^2. The power consumption of the high-bandwidth OS-BiCMOS-OEIC is less than 75 mW at 5.0 V. Table 6.3 summarizes further

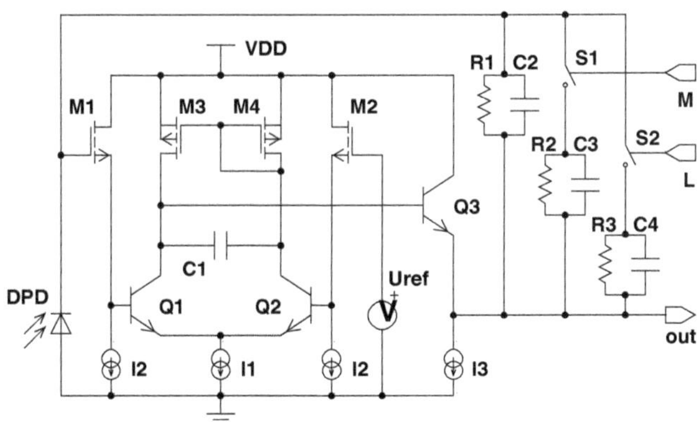

Fig. 6.72. Schematic of the sensitive channels E–H in a high-bandwidth OS-BiCMOS-OEIC [176]

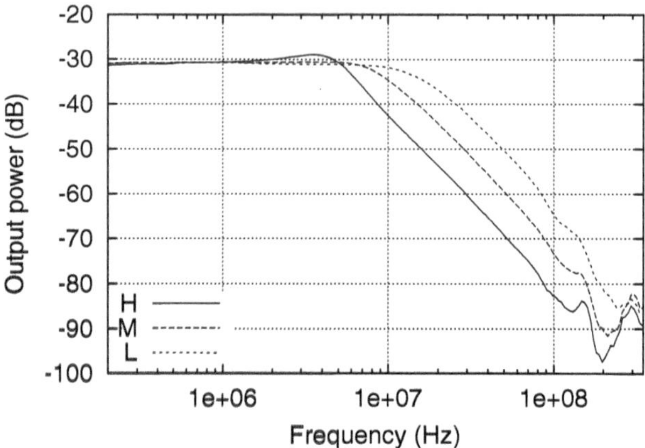

Fig. 6.73. Frequency responses of the sensitive channels E–H of a high-bandwidth BiCMOS OEIC for optical storage systems measured with three different optical input powers, i. e. three different gain factors

Table 6.3. Measured results of the high-bandwidth OS-BiCMOS-OEIC

	H	M	L
$f_{-3\,\mathrm{dB}}$ (MHz) A–D	92.0	94.9	95.1
$f_{-3\,\mathrm{dB}}$ (MHz) E–H	5.2	8.5	14.6
Sensitivity (mV/μW) A–D	8.8	2.9	0.9
Sensitivity (mV/μW) E–H	88.1	29.3	9.1
U_{Offset} (mV) A–D	< 10.8	< 9.5	< 9.0
U_{Offset} (mV) E–H	< 7.4	< 6.4	< 6.4
Noise (dBm) @10 MHz with 30 kHz RBW A–D	−81.5	−85.0	−85.2
Noise (dBm) E–H	−66.0	−67.5	−73.5

technical data of the fast and the sensitive channels of the high-bandwidth OS-BiCMOS-OEIC.

Figure 6.74 shows the microphotograph of the OS-BiCMOS-OEIC with bandwidths in excess of 90 MHz, which was realized in full custom design. The results demonstrate that it is possible to avoid the slow carrier diffusion problem by exploiting double photodiodes in standard BiCMOS technology.

A BiCMOS optoelectronic integrated circuit (OEIC) for optical storage systems containing photodiodes and transimpedance amplifiers with a reduced offset voltage is introduced next. A reference transimpedance amplifier and a low-offset operational amplifier are implemented for offset reduction. This technique can be called replica offset compensation.

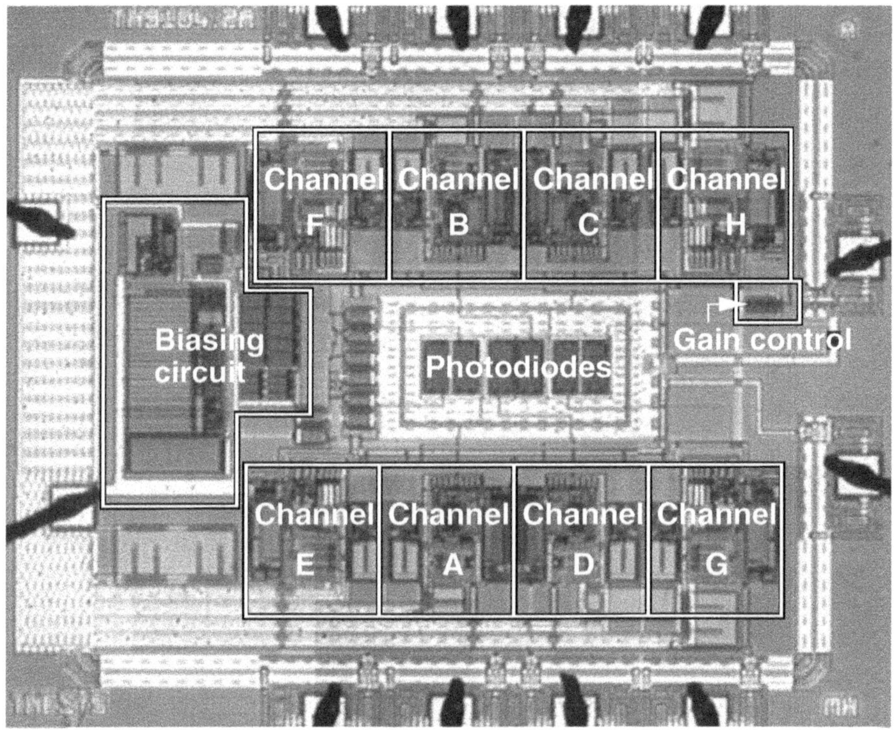

Fig. 6.74. Micrograph of a high-bandwidth BiCMOS DVD OEIC [176]

Bipolar or BiCMOS operational amplifiers are necessary in order to obtain a high -3dB frequency. Bipolar transistors in the input stage of such an amplifier, however, result in an input current in contrast to CMOS amplifiers. This input current flows across the feedback resistor and causes an output offset voltage in an operational transimpedance amplifier. A bias current concellation technique has been applied to reduce the output offset voltage from 100 mV to 9 mV in an OEIC with a sensitivity of 10.5 mV/μW at 638 nm [173] and to below 11 mV as shown in the last example above. For a sensitivity of 25 mV/μW the feedback resistor, however, has to be increased from about 24 kΩ to 60 kΩ leading to an increased output offset voltage when assuming the same input current of the operational amplifier. A circuit topology with a better performance implementing transistor transimpedance amplifiers instead of operational transimpedance amplifiers has, therefore, been investigated and will be shown here.

The innovative topology of an OEIC with an enhanced sensitivity for optical storage systems is shown in Fig. 6.75. This topology is called replica offset compensation. A current-to-voltage converter is connected to the integrated photodiode. The BOP1 (bias operating point) input allows to correct the operating point of this transimpedance amplifier in order to cancel

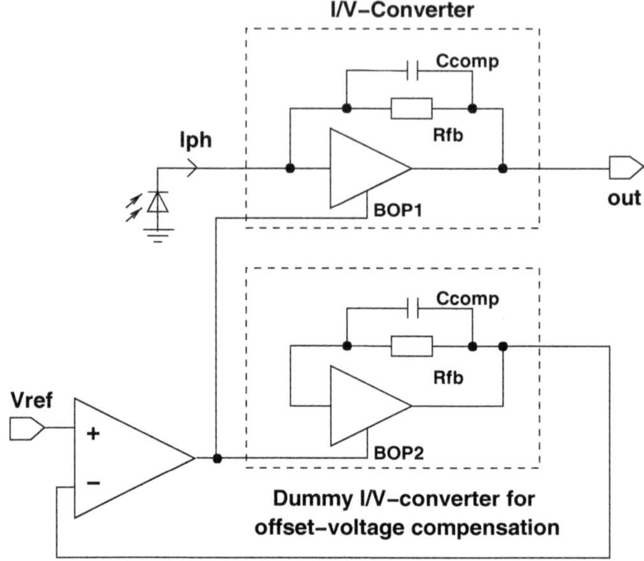

Fig. 6.75. Basic topology for output offset voltage reduction [181]

the output offset voltage. For this purpose an identical dummy current-to-voltage converter with an open input and a low-offset operational amplifier are integrated together with the transimpedance amplifier connected to the photodiode. The operational amplifier measures the output offset voltage of the dummy I/V converter compared to $V_{\mathrm{ref}} = 2.1\,\mathrm{V}$ and changes its operating point via the BOP2 input in order to minimize the output offset voltage. The BOP1 input of the actual transimpedance amplifier is also connected to the output of the operational amplifier. In such a way, the output offset voltage of the transimpedance amplifier can be reduced provided that a good matching between the two I/V converters and a low offset voltage of the operational amplifier can be guaranteed. The first requirement can be dealt with a careful layout reducing the distance between corresponding devices in the two I/V converters. The second requirement can be fulfilled, because the operational amplifier can be designed for a very slow response, i. e. transistors with a large emitter area or large gate lengths and widths for a good matching can be used. The replica offset compensation technique guarantes that the output voltage of the OEIC channels becomes independent of process tolerances — especially fast and slow cases — as well as independent of operating temperature.

An OEIC containing four fast channels (for data extraction) and four sensitive channels (for tracking control in the optical storage system) with larger Rfb values was fabricated in a 0.8 µm BiCMOS technology. The N-well/P-substrate photodiode (Fig. 2.53) with a light sensitive area of $2500\,\mu\mathrm{m}^2$ was implemented in the fast channels. The same photodiodes with an area of

9100 μm^2 were used in the sensitive channels. The schematic diagram of the fast and sensitive channels of the BiCMOS OEIC is shown in Fig. 6.76. The transistor transimpedance amplifier is formed by the NPN transistor Q1 in common emitter configuration, the PMOS load element M1, the emitter follower Q3, and the feedback resistor Rfb1. The NPN transistor Q3 provides a low output impedance of this input stage. The capacitor Cfb1 is used for compensation. The emitter follower Q11 decouples the feedback path from the output and provides a low overall output impedance. Q11 also leads to an output voltage close to the reference voltage. The dummy I/V converter is formed by Q2, M2, Q4, Rfb2, Cfb2, and Q12. Q5–Q10, R5–R10 and M3/M4 are used for biasing. The circuit diagram in Fig. 6.76 corresponds to the interlaced layout for good matching of the two I/V converters.

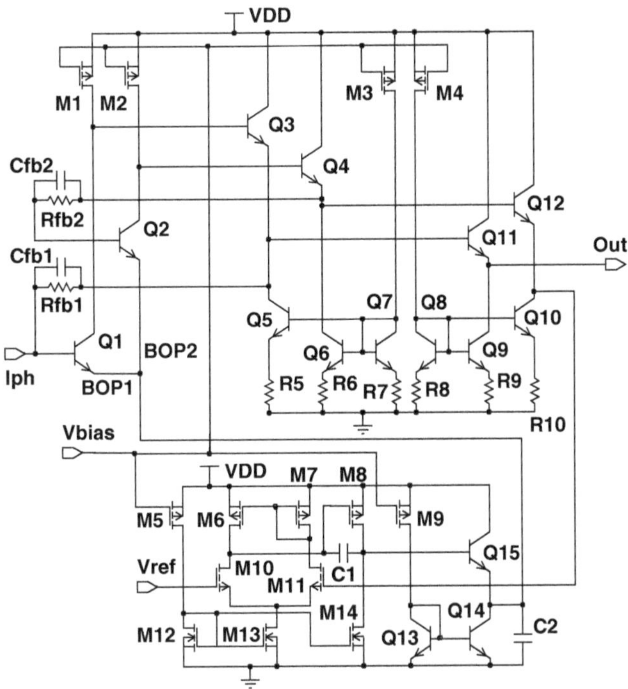

Fig. 6.76. Schematic diagram of one channel of the BiCMOS OEIC [181]

The BiCMOS operational amplifier is based on a CMOS Miller type topology with the NPN emitter follower Q15 at the output in order to supply a voltage source with a low output resistance for the emitter inputs BOP of the two I/V converters. The simulated low-frequency differential gain of the operational amplifier is 77 dB and its transit frequency is 7.3 MHz. The chipfoto of an eight-channel DVD OEIC implementing the replica offset compensation is shown in Fig. 6.77.

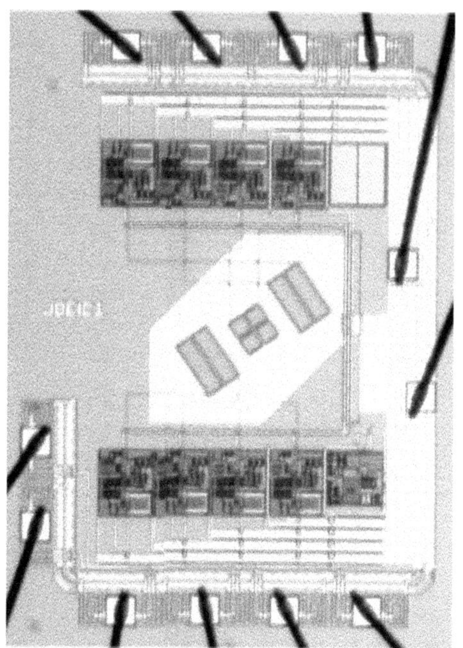

Fig. 6.77. Microphotograph of the BiCMOS OEIC implementing the replica offset compensation

For characterization, laser light with a wavelength $\lambda = 638$ nm was coupled into the photodiodes via a single-mode optical fiber adjusted on a wafer prober. A sensitivity for the fast channels of 25 and of 51 mV/μW for the sensitive channels at 638 nm was found [181] by determining the incident light power with a calibrated photodiode. A network analyzer HP8751A was used for laser modulation and for frequency response measurements. A −3 dB frequency of 58.5 MHz was measured for a fast channel A–D with a load of 10 kΩ and 10 pF (Fig. 6.78). The sensitive channels E–H showed a −3 dB frequency of 26.7 MHz with the same load [181]. All measured output offset voltages for photodiodes and OEICs in the dark were lower than 7.5 mV [181]. This value is considerably lower than the maximum simulated offset voltage of a reference amplifier without offset compensation of more than 72 mV for worst case transistor parameters at room temperature [181].

It should be mentioned that the BiCMOS operational amplifier and the emitter BOP inputs reduce the bandwidth of the I/V converters. This is due to the output resistance of this operational amplifier. The output resistance of the BiCMOS operational amplifier, however, is obviously rather low, since the bandwidth of 58 MHz is not considerably lower than that of a reference I/V converter of 82 MHz where the emitter of Q1 was connected to V_{ref} directly [181].

The bandwidth of the N-well/P-substrate photodiode may seem to be rather high compared to [64] where a bandwidth of 1.6 MHz was reported for a similar photodiode for $\lambda = 780$ nm. The high bandwidth determined here,

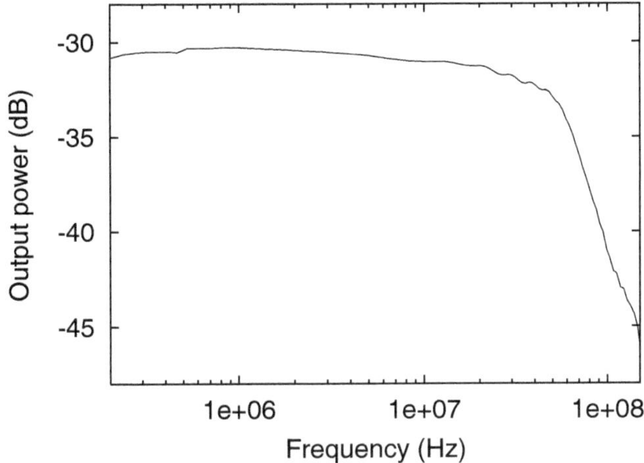

Fig. 6.78. Frequency response of a fast channel of the BiCMOS OEIC implementing replica offset compensation [181]

however, can be explained by the shorter wavelength and by an electric field, i. e. by a drift zone for photogenerated carriers, due to the doping gradient of the N-well similarly to that in the N-well of a double photodiode [67]. The external quantum efficiency of the N-well/P-substrate photodiode in the standard process for $\lambda = 638$ nm and a reverse bias of 2.5 V was 78 % corresponding to a responsivity R = 0.40 A/W. This value could be increased to 95 % (R = 0.49 A/W) by an antireflection coating layer.

Noise measurements at 10 MHz with a resolution bandwidth of 30 kHz revealed values of -76 dBm and -65 dBm for the fast and sensitive channels, respectively. The power consumption determined by simulation was 36 mW for a fast channel at a supply voltage of 5.0 V. The active die area of each channel is approximately 0.042 mm^2 and the total chip area is 3.1 mm^2.

The bandwidth-sensitivity product has been increased by factors of 1.8 and 3.2 compared to the circuits described in [176] and [173], respectively. Additionally, the new offset compensation technique allowed a reduction of the output offset voltage from 11 mV [176] to below 7.5 mV.

A newer BiCMOS OEIC with enhanced sensitivity for advanced optical storage systems was presented [245]. The OEIC in an industrial 0.8 µm BiCMOS process showed a sensitivity of 43 mV/µW in combination with a -3 dB-bandwidth of 60 MHz.

A schematic overview of the high-sensitivity DVD-OEIC is shown in Fig. 6.79. It contains six channels, each consisting of an N^+/N-well/P-substrate photodiode (Fig. 2.53) and a transimpedance amplifier.

The four photodiodes A–D for the fast channels cover an area of 75 × 75 μm^2 each, and the two larger photodiodes for the sensitive channels have an area of 190 × 250 μm^2. To reduce the chip area and the power consump-

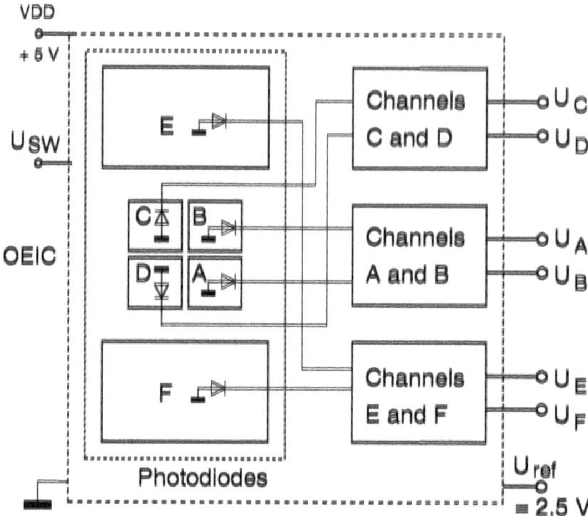

Fig. 6.79. Structure of the 6-channel OEIC for DVD [245]

tion, two channels are combined with a dummy amplifier for output voltage regulation, each.

Figure 6.80 shows the circuit diagram of the amplifier. The circuit mainly consists of two transimpedance preamplifiers for current/voltage conversion, a free running transimpedance amplifier without a photodiode and an operational amplifier for output voltage compensation. Two current sources provide the currents needed to adjust the operating points of the transimpedance amplifiers and the operational amplifier. The upper and the lower transimpedance preamplifiers convert the photocurrents of the N^+/N-well/P-substrate photodiodes A and B to corresponding voltages whereas the preamplifier in the middle is needed in order to refer the output voltage of the transimpedance amplifiers to the reference voltage U_{ref}. Without the dummy amplifier in the middle, the base current of the NPN transistor Q1 would cause an output offset voltage of about 0.3 V referred to U_{ref} when no light incidents into the photodiodes.

This offset voltage is reduced as follows: When the output voltage of the dummy amplifier, which is applied to the non-inverting input of the operational amplifier, is lower than the reference voltage U_{ref}, the output voltage U_{reg} has a low value. Since M_1 is a P-channel-MOSFET (Fig. 6.81), its gate-source voltage increases, and at the same time the voltage at the base of Q_2 increases. The consequence is that the output voltage of the transimpedance amplifiers becomes higher. Steady state is achieved when the voltages at the inverting and at the non-inverting inputs of the operational amplifier are equal. If good matching between the dummy amplifier and the other transimpedance amplifiers can be obtained and the operational amplifier has a low offset voltage, the offset voltage of the active channels can be minimised.

A detailed circuit diagram of the transimpedance amplifier for the fast channels can be seen in Fig. 6.81. The gain is switchable between high and

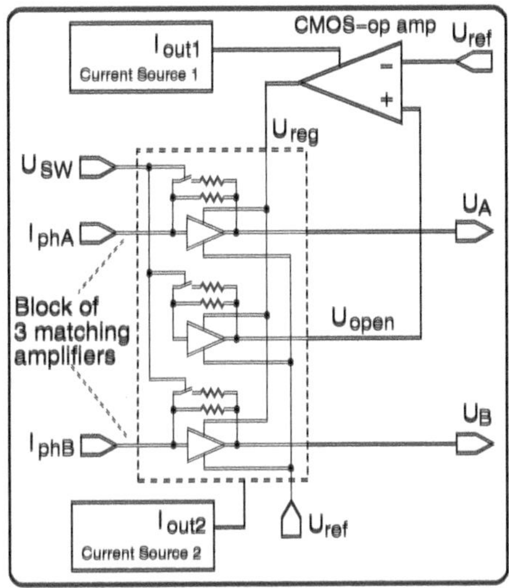

Fig. 6.80. Block diagram of two channels of the high-sensitivity DVD-OEIC [245]

low in order to detect two different photocurrents for read and write access. Again only NPN transistors are used in the signal path in order to obtain a high bandwidth. For the input stage the active load M_1 is applied.

A relation between the photocurrent and the output voltage can easily be found if the base currents are neglected. The voltage at the base of Q_1 is $U_{\text{ref}} + U_{\text{BE1}}$. A voltage of $I_{\text{ph}} \times R_{\text{fb}}$ is added, and the voltage U_{BE3} at the output emitter follower is subtracted. If we assume that the base-emitter voltages of Q_1 and Q_3 are equal, we obtain the relation

$$U_{\text{out}} = U_{\text{ref}} + I_{\text{ph}}\, R_{\text{fb}}. \tag{6.58}$$

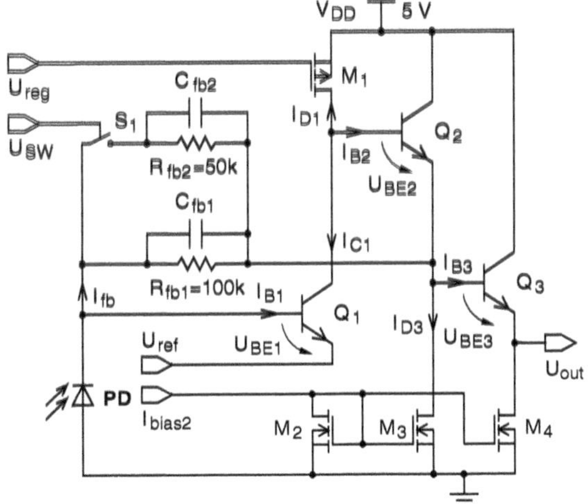

Fig. 6.81. Detailed circuit diagram of the transimpedance amplifier [245]

In a more detailed analysis, the base currents of the bipolar transistors have to be taken into account. This reveals a more complex DC transfer function which depends on numerous parameters of the transistors. Many of these parameters are subject to tolerances in the BiCMOS process making it difficult to design low-offset amplifiers without any control measures. Temperature changes within the operating temperature range also change the output voltage for constant or zero light levels.

Varying the drain current of PMOS load transistor M_1 has an impact on the DC transfer function and thus on the output voltage. Therefore the replica offset compensation scheme is applied to compensate the output voltage of the transimpedance amplifier with respect to the output voltage at dark photo current.

The main difference between the fast and the sensitive transimpedance amplifiers is the value of the resistors in the feedback path. In the fast channels feedback resistors of $R_{\mathrm{fb1}} = 100\,\mathrm{k\Omega}$ and $R_{\mathrm{fb2}} = 50\,\mathrm{k\Omega}$ are implemented. At high gain the sensitivity S^{high} of the amplifier can be calculated as

$$S^{\mathrm{high}} = \frac{I_{\mathrm{ph}}\,R_{\mathrm{fb1}}}{P_{\mathrm{opt}}}\,\eta_{\mathrm{tia}} = \frac{R_{\mathrm{fb1}}\,q\,\eta\,\lambda}{h\,c_0}\,\eta_{\mathrm{tia}} \approx 48\,\frac{\mathrm{mV}}{\mu\mathrm{W}}, \tag{6.59}$$

and at low gain a value of

$$S^{\mathrm{low}} = \frac{R_{\mathrm{fb1}}||R_{\mathrm{fb2}}\,q\,\eta\,\lambda}{h\,c_0}\,\eta_{\mathrm{tia}} \approx 15.9\,\mathrm{mV}/\mu\mathrm{W} \tag{6.60}$$

is obtained, where $\eta_{\mathrm{tia}} = r_{\mathrm{DS,M1}}\beta/(r_{\mathrm{DS,M1}}\beta + R_{\mathrm{fb}})$ [85] is the conversion efficiency of the transimpedance amplifier with a value of 98 % for $R_{\mathrm{fb1}} = 100\,\mathrm{k\Omega}$. In the case of the sensitive channels R_{fb1} is $150\,\mathrm{k\Omega}$ and R_{fb2} is $70\,\mathrm{k\Omega}$. This leads to sensitivities of $S^{\mathrm{high}} = 72\,\mathrm{mV}/\mu\mathrm{W}$ and $S^{\mathrm{low}} = 22.9\,\mathrm{mV}/\mu\mathrm{W}$.

Using equations (5.27) and (5.28) comparatively low values for the feedback capacitances of $C_{\mathrm{fb1}} = 15\,\mathrm{fF}$ and $C_{\mathrm{fb2}} = 10\,\mathrm{fF}$ were determined. A numerical circuit simulation considering the parasitic capacitances of the feedback resistors revealed that without these compensation capacitors a gain peaking in the frequency response curve would be visible [245]. For the sensitive channels lower values of the feedback capacitances could have been implemented according to equations (5.27) and (5.28). Instead, for additional compensation i. e. a larger phase margin even higher values of $C_{\mathrm{fb1}} = C_{\mathrm{fb2}} = 25\,\mathrm{fF}$ were chosen. The bandwidth is with 20 MHz still much higher than required for these channels.

The performance of the circuit was characterized on wafer level by the following experimental setup: red laser light with a wavelength of 638.3 nm was coupled into the photodiode via a single-mode optical fiber adjusted on a wafer prober. To measure the bandwidth, a network analyser was chosen, which modulated the laser driver and evaluated the electrical output signal of the OEIC picked up by a picoprobe. The picoprobe was characterised by an input impedance of $10\,\mathrm{M\Omega}$, an input capacitance of 100 fF and a -3 dB-bandwidth of 3 GHz.

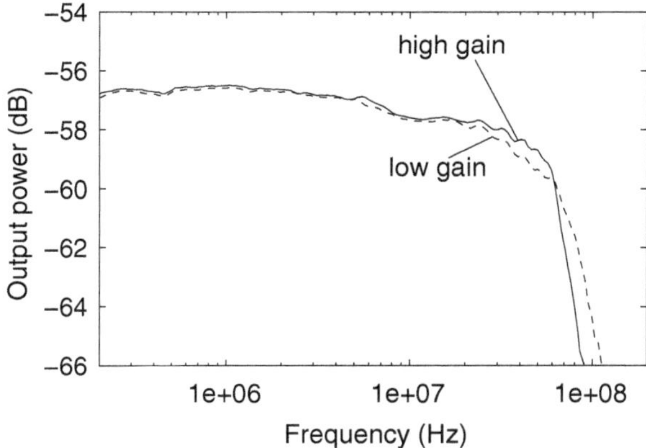

Fig. 6.82. Frequency response of fast channel [245]

For the fast channels of the OEIC a bandwidth of $f_{-3\,\mathrm{dB}}^{\mathrm{high}} = 60\,\mathrm{MHz}$ at high gain and $f_{-3\,\mathrm{dB}}^{\mathrm{low}} = 61.6\,\mathrm{MHz}$ at low gain were measured (Fig. 6.82), which were in good agreement with the simulated bandwidths [245].

The measured sensitivity of the fast and the sensitive channels at high gain was $43\,\mathrm{mV}/\mu\mathrm{W}$ and $60\,\mathrm{mV}/\mu\mathrm{W}$, respectively. This means that the fast channels are by a factor of 1.7 more sensitive than the OEIC described in [181] whereby the bandwidth is the same. The circuits can thus be applied in DVD systems with a lower laser power. The measured values for the sensitivity are lower than the calculated values which was lead back to process deviations for the polysilicon resistors. The measured results are summarised in Table 6.4.

Figure 6.83 shows a microphotograph of the six-channel OEIC with outside dimensions of $1333 \times 1595\,\mu\mathrm{m}^2$. In the upper row the pads for the supply and the reference voltage can be seen. The three pads T1, T2 and M on the left side are used for switching off selective channels for test purposes before packaging. The SW-pad in the lower row is needed for gain switching. All other pads are used for the output signals of the six transimpedance ampli-

Table 6.4. Measured results of high-sensitivity OEIC for $\lambda = 638.3\,\mathrm{nm}$). The noise values refer to a frequency of 10 MHz and a resolution bandwidth of 30 kHz [245]

	Offset (mV)	Sensitivity (mV/μW)	Noise (dBm)
A, Low Gain	17.8	15.3	−84.6
A, High Gain	8.4	43.3	−84.3
F, Low Gain	−12.7	20.7	−82.6
F, High Gain	8.2	60.4	−80.5

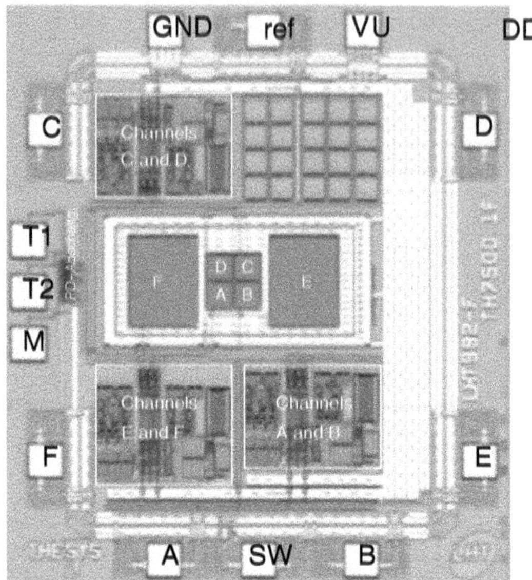

Fig. 6.83. Microphotograph of the high-sensitivity OEIC

fiers, which are located in groups of two each on the chip. The photodiodes are surrounded by a metal shield which prevents penetration of laser light into the silicon substrate outside the light sensitive areas of the photodiodes.

Summarizing and concluding, a high-sensitivity BiCMOS OEIC with bandwidths of more than 60 MHz has been realized. The bandwidth-sensitivity product has been increased by a factor of 1.7 compared to [181]. With a dummy channel for two read channels in the replica offset compensation scheme, the chip area was reduced from 3.1 to 2.1 mm^2 by a factor of almost 1.5. The N^+/N-well/P-substrate photodiode in conjunction with a BiCMOS amplifier is a good choice for high-speed, high-sensitivity and low-cost applications such as CD-ROM or DVD.

When a higher speed and a high sensitivity are required in addition to a low output offset voltage for the OS-OEICs, a two-stage optical receiver may be necessary. The circuit principle of such a two-stage amplifier [311] is shown in Fig. 6.84. The circuit consists of a transimpedance amplifier for the photocurrent and a reference I/U converter for offset compensation plus an operational amplifier in subtractor configuration. The subtractor can be used simultaneously as a voltage amplifier with the amplification factor RS2/RS1 enabling a high overall sensitivity of the OEIC. It must be mentioned, however, that offset voltages due to mismatch of the two transimpedance input amplifiers and due to mismatch in the operational amplifier are also amplified.

The circuit diagram of the complete circuit is shown in Fig. 6.85. In order to achieve a high bandwidth, only NPN transistors are used in the signal paths of the preamplifiers. Q1 is used in common-emitter configuration, Q3

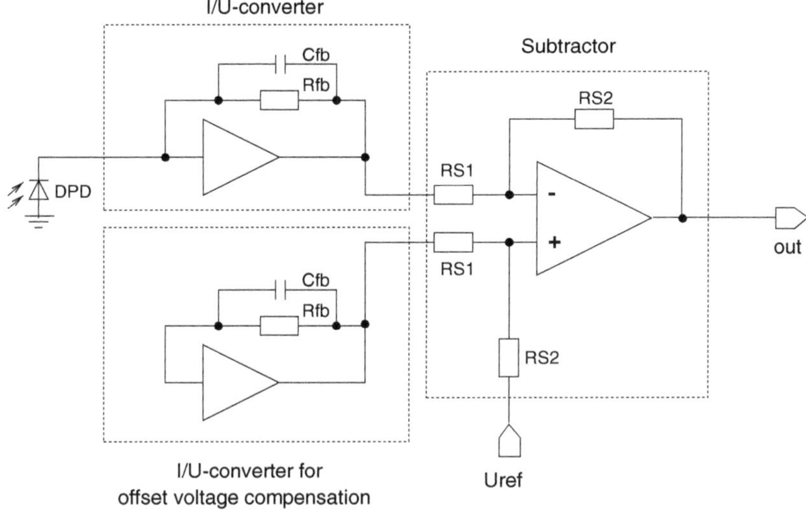

Fig. 6.84. Block diagram of a two-stage optical receiver for one fast channel of a high-speed OS-BiCMOS-OEIC

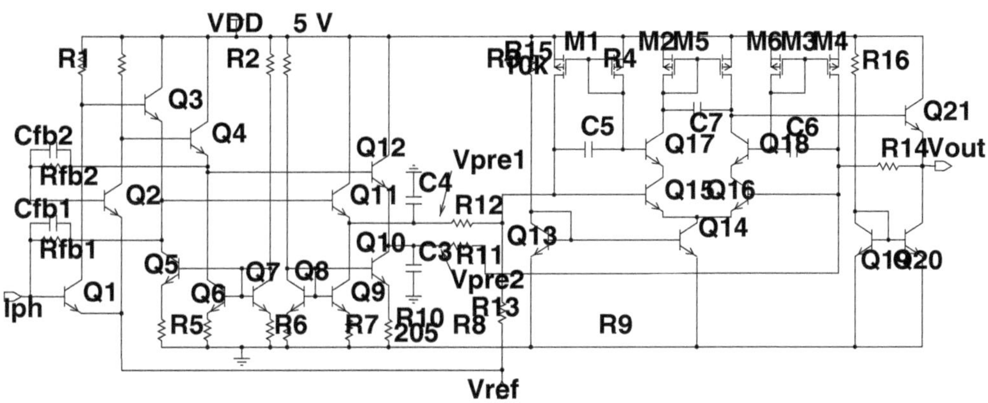

Fig. 6.85. Schematic of high-speed BiCMOS-OEIC for the fast channels A–D in an OS-BiCMOS-OEIC [177]

is used as emitter follower, and the feedback resistor R_{fb1} together with Q1 and Q3 represent a low input impedance for the photocurrent of the double photodiode (DPD). Thereby, the effect of the DPD capacitance is minimized. The reference voltage V_{ref} is chosen as the emitter potential of Q1 in order to increase the reverse voltage of the DPD to $V_{\mathrm{BE,Q1}} + V_{\mathrm{ref}}$. The values for R_1, R_{fb1}, and C_{fb1} were 3 kΩ, 27 kΩ, and 25 fF, respectively. A second emitter follower (Q11) is implemented for level shifting and decoupling of output and feedback path. The second preamplifier consists of transistors Q2, Q4, and Q12 as well as the current mirror with Q6 and Q7 and the feedback resistor R_{fb2} plus the compensation capacitor C_{fb2}. At the outputs of the preamplifiers, C3 and C4 are added as further compensation capacitors.

The large signal DC transfer functions of the preamplifiers are given by

$$V_{\mathrm{pre1}} = V_{\mathrm{ref}} + \eta_{\mathrm{tia}} I_{\mathrm{ph}} R_{\mathrm{fb1}} + I_{\mathrm{B,Q1}} R_{\mathrm{fb1}}, \tag{6.61}$$

and

$$V_{\mathrm{pre2}} = V_{\mathrm{ref}} + I_{\mathrm{B,Q2}} R_{\mathrm{fb2}}, \tag{6.62}$$

when we assume $U_{\mathrm{BE,Q1}} = U_{\mathrm{BE,Q11}}$ and $U_{\mathrm{BE,Q2}} = U_{\mathrm{BE,Q12}}$. The efficiency factor η_{tia} of the preamplifier is given by $\eta_{\mathrm{tia}} = R_1\beta/(R_{\mathrm{fb1}}+R_1\beta)$. For $\beta = 100$ and for the resistor values given above the efficiency factor η_{tia} of the transimpedance preamplifier is equal to 0.917. The base current of Q1 (and Q2) causes a voltage of about 0.1 V across R_{fb1} (and R_{fb2}). In order to achieve a low output offset voltage compared to V_{ref}, therefore, the second preamplifier without a photodiode and the subtractor operational amplifier are necessary. Perfect matching of the two preamplifiers ($I_{\mathrm{B,Q1}} = I_{\mathrm{B,Q2}}$, $R_{\mathrm{fb1}} = R_{\mathrm{fb2}}$, $R_1 = R_2$, $\beta_1 = \beta_2$, $U_{\mathrm{BE,Q3}} = U_{\mathrm{BE,Q4}}$, $U_{\mathrm{BE,Q11}} = U_{\mathrm{BE,Q12}}$) is, however, necessary in order to obtain $V_{\mathrm{pre1}} = V_{\mathrm{pre2}}$ for a dark photodiode. This perfect matching of the two preamplifiers requires a careful layout to achieve a low output offset voltage.

The preamplifiers are connected to the subtractor operational amplifier via R5 and R6. The bias current cancellation introduced in Fig. 6.69 is applied to reduce the input currents of the operational amplifier necessary for a low output offset voltage. A PMOS current mirror load with M5 and M6 is used here in order to obtain a higher open loop gain of the operational amplifier. For $R_{11} = R_{12}$ and $R_{13} = R_{14}$, an analysis of the subtractor amplifier yields the transfer function

$$V_{\mathrm{out}} = V_{\mathrm{ref}} + \frac{R_{13} A_{\mathrm{d}}(s)}{(R_{12} + R_{13}) + R_{12} A_{\mathrm{d}}(s)} (V_{\mathrm{pre1}} - V_{\mathrm{pre2}}) \tag{6.63}$$

and

$$V_{\mathrm{out}} \approx V_{\mathrm{ref}} + \frac{R_{13}}{R_{12}} \eta_{\mathrm{tia}} I_{\mathrm{ph}} R_{\mathrm{fb1}}, \tag{6.64}$$

when we assume a large open loop voltage gain $A_{\mathrm{d}}(s)$ of the operational amplifier. A −3dB frequency of 189 MHz was determined by numerical prelayout simulation for the complete amplifier with a load of $R_{\mathrm{L}} = 1\,\mathrm{k\Omega}$ and $C_{\mathrm{L}} = 10\,\mathrm{pF}$. The complete two-stage amplifier was designed for a sensitivity of 10 mV/μW and an offset voltage of less than 10 mV for $R_{11} = R_{12} = R_{13} = R_{14}$.

The OEIC with the DPD and the two-stage amplifier was fabricated in a 0.8 μm BiCMOS technology. The measured frequency response of this two-stage optical receiver is shown in Fig. 6.86. A −3dB frequency of 147.7 MHz is determined from this frequency response. The power consumption of the two-stage optical receiver is 35 mW at a supply voltage of 5 V. The active die area of the two-stage optical receiver is $340 \times 140\,\mu\mathrm{m}^2$.

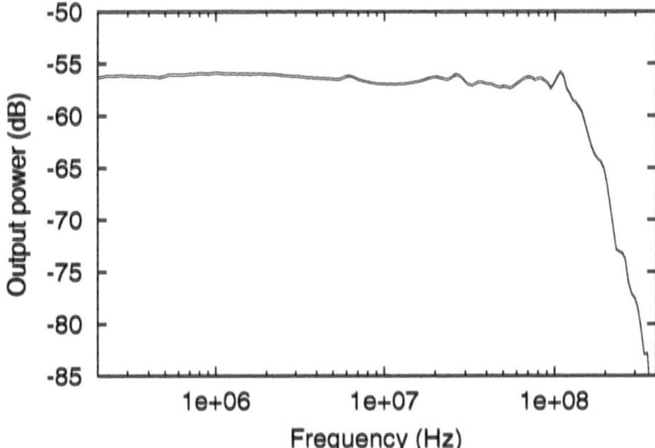

Fig. 6.86. Frequency response of a two-stage optical BiCMOS receiver for the fast channels A–D of a high-speed BiCMOS OEIC for optical storage systems [177]

An even faster OEIC for optical storage systems has been presented recently. A bandwidth of 250 MHz has been achieved with a four-stage current amplifier in a standard 0.6 μm BiCMOS technology with fast vertical PNP transistors [182]. This OEIC can be used for channel data rates up to 600 Mb/s corresponding to 22 X DVD with 12,000 RPM. The photodiode shown in Fig. 2.54 was optimized to a low capacitance of 0.25 pF. Nevertheless, the high bandwidth is an astonishing result when we consider that no bias currents were superimposed to the photocurrents to reduce the offset of the amplifiers. Lowest power consumption, therefore, also could be achieved.

Figure 6.87 shows the topology of the current preamplifier used in the 8-channel OEIC. Since the gain-bandwidth product of the BiCMOS process is 400 GHz and a gain over 1,000 was needed, a four-stage amplifier with a bandwidth of at least 400 MHz for each stage was necessary [182]. The current

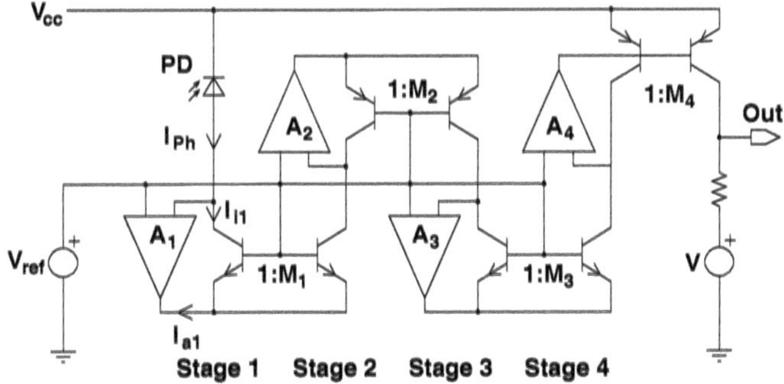

Fig. 6.87. Circuit topology of one segment in the fast OEIC for optical storage systems [182]

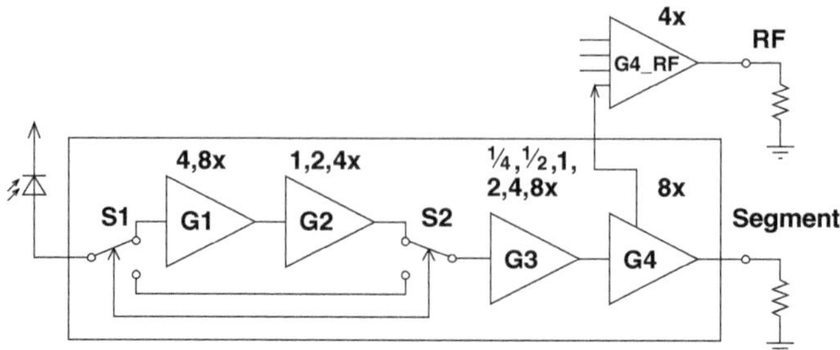

Fig. 6.88. Gain partitioning of one segment in the fast OEIC for optical storage systems [182]

amplifiers are no typical current mirrors. Instead the base-emitter voltage is the same of both transistors in each amplifier stage of which the second transistor actually is a parallel circuit of up to four transistors and the emitter currents are supplied by V/I amplifiers A1–A4. With MOS switches between and in the stages, the gain was switchable between 2 and 2048 for different laser powers during read and write as well as for different media according to the scheme illustrated in Fig. 6.88.

A sum amplifier (RF amplifier) for the channels A–D was also implemented [182]. The bandwidth of this RF amplifier only weakly depended on the gain. The gain of the satellite amplifiers in channels E–H was always four times higher than shown in Fig. 6.88.

It has to be mentioned that a trick was necessary to achieve the high bandwidth for low photocurrents, since photocurrents of 100 nA to 1 μA cause the transit frequencies of the NPN transistors T1 and T2 to drop to tens of kHz to 1 MHz [182]. The large attenuation resulting at 400 MHz had to be eliminated. This was done by bypass capacitors which were not included in the circuit diagram shown in [182], however. The V/I amplifiers A1 to A4 used NMOS input transistors to avoid base input currents, their DC-current offset, and their base-current shot-noise.

Gain peaking around 200 MHz was caused by crosstalk between the output drivers and the first two stages [182]. It was stated that this problem can be solved in a redesign by separated front-end and back-end supplies. The output stage delivered a current of 8 mA into a 150 Ω load with a one-sigma offset value of 117 μV [182].

6.4.10 Continuous-Mode Fiber Receivers

Optical multimode fibers for optical data transmission usually possess a core diameter of 50 or 62.5 μm. The core diameter of single-mode fibers for wavelengths shorter than 1.1 μm actually is less than 10 μm. Small-area photodiodes therefore can be used in order to realize a low capacitance at the input of the receiver circuits. The bondpad capacitances of wire-bonded receivers are much larger than the capacitance of PIN photodiodes. Monolithically integrated optical fiber receivers should be the first choice as a consequence. In the following, a bipolar OEIC as well as several NMOS OEICs, BiCMOS OEICs, and many CMOS OEICs for application as continuous-mode fiber receivers will be described.

Bipolar SiGe Receiver

The superior speed of SiGe HBTs compared to Si bipolar transistors has already been mentioned and the structure of a monolithic SiGe-Si PIN-HBT receiver was described in Chap. 4. Here, the bipolar transimpedance amplifier circuit of this receiver (see Fig. 6.89) will be discussed.

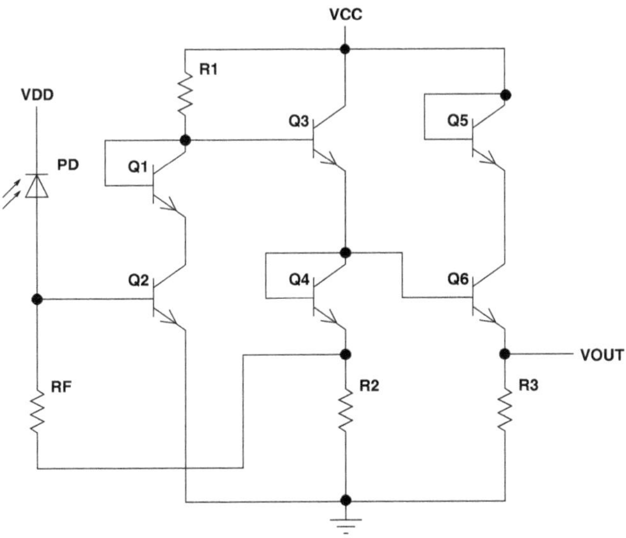

Fig. 6.89. Circuit diagram of a bipolar photoreceiver [231]

The receiver consists of a PIN photodiode, a common emitter gain stage, two emitter follower buffers, and a resistive feedback loop. NiCr thin-film resistors were used in the monolithic SiGe HBT receiver [231]. The transistors Q1, Q4, and Q5 are used as level shifting diodes. Q1 and Q5 reduce U_{CE} of Q2 and Q6, respectively, because the breakdown voltages of high-speed transistors are quite low.

The two voltage sources VDD and VCC were necessary to optimize the PIN transient behavior and the operating point of the amplifier. The value of the feedback resistor R_{F} determines the bandwidth, gain, and noise characteristics of the photoreceiver. The value of R_{F} is usually chosen based on a trade-off between these three parameters. In [231], a value of $640\,\Omega$ was chosen for R_F resulting in a transimpedance gain of 52.2 dBΩ. The bandwidth of 1.6 GHz was obtained for the transimpedance amplifier with a f_T of 25 GHz for the HBTs with an emitter area of 5×5 μm. The optical bandwidth of 460 MHz of the PIN-HBT receiver was measured for VDD = 9 V and VCC = 6 V. The bandwidth of the receiver was limited by the photodiode and the trade-off mentioned above might be improved with respect to an increased gain, i. e. a larger sensitivity. An input noise spectral density of 8.2 pA/$\sqrt{\text{Hz}}$ up to 1 GHz caused by shot noise from the base current and thermal noise from the feedback resistor was given. With these values, the photoreceiver sensitivities of −24.3 and −22.8 dBm were estimated for 0.5 and 1 Gb/s, respectively, for a bit error rate (BER) of 10^{-9} and $\lambda = 850$ nm.

NMOS Receivers

The next example is an NMOS OEIC. A lateral PIN photodiode was integrated in a 1.0 μm NMOS technology using a nominally undoped substrate, which was actually P-type with $N_A = 6 \times 10^{12}$ cm^{-3} [72,73]. This lateral PIN photodiode is described in Sect. 2.2.5.

An NMOS transimpedance preamplifier (Fig. 6.90) implementing a depletion transistor at the input was integrated together with a lateral PIN photodiode (see Fig. 2.17). The second and third stages were formed by source followers. The source follower stage M3/M4 with M4 as a constant current source is used to establish the correct operating point across the feedback loop with MF as an active resistor. The source follower stage M5/M6 at the output was sized to match a 50 Ω load impedance. Open-eye operation for bit-rates of up to 40 Mb/s with a photocurrent of 3 μA for $\lambda = 870$ nm and for a transimpedance of 3 kΩ was reported [72].

The authors of [72] later improved their photoreceiver [313]. An N-type Si substrate with a resistivity of 1000–3000 Ωcm was taken and an interdigitated lateral PIN structure with a finger width of 2 μm and a finger spacing of 10 μm instead of the ring structure (see Fig. 2.17) was implemented. The total area of the photodiode was 50 × 50 μm^2. A dark current of 1.3 pA at 5 V and of 63 nA at 30 V was found. The quantum efficiency was increased to 84 and 74 % at 800 and 870 nm, respectively, due to a SiO_2 antireflection coating with a thickness of 150 nm. A bit-rate of 500 Mb/s was achieved from the interdigitated PIN photodiode at 30 V, however, its frequency response showed a diffusion tail below 100 MHz [313].

The preamplifier has also been modified. Figure 6.91 shows the improved three-stage preamplifier. The feedback via M3 is only across the first stage with the common-source amplifier M1 and the depletion load M2. The second stage with the enhancement mode MOSFETs M4 and M5 further amplifies the signal and is used as a buffer to drive the depletion source follower M7 with an output impedance of 50 Ω. At the optimum feedback, $V_F = 1.25$ V, the transimpedance was 6.5 kΩ and a 45 μA dynamic range of the photocurrent was obtained. The bandwidth of the photoreceiver was 130 MHz for 870 nm

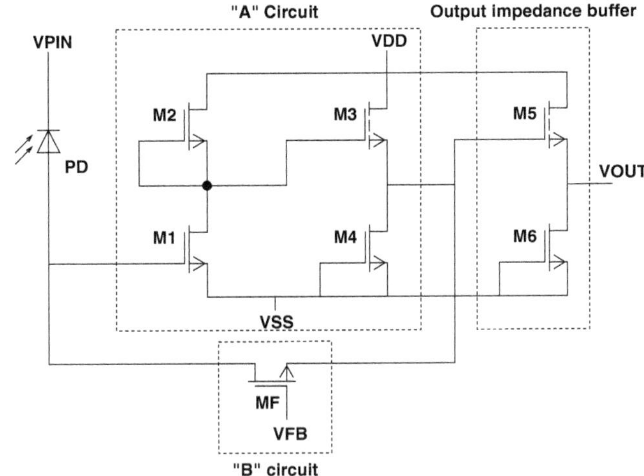

Fig. 6.90. Schematic of an NMOS fiber receiver OEIC [72]

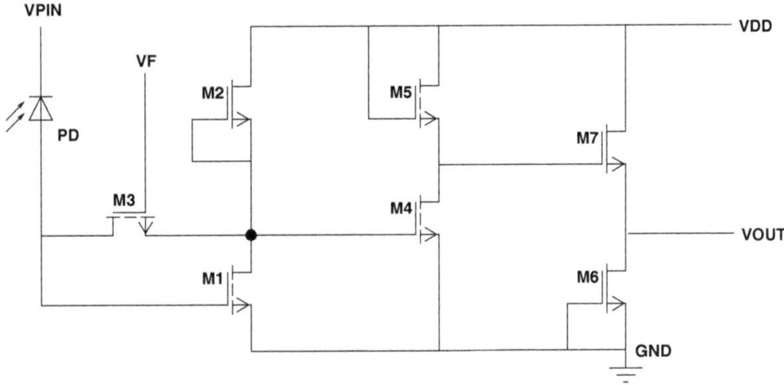

Fig. 6.91. Circuit diagram of an improved NMOS fiber receiver OEIC [313]

light, when biased with VDD = 8 V, $V_{\mathrm{F}} = 1.25$ V, and $V_{\mathrm{PIN}} = 30$ V. Open-eye operation under these conditions was demonstrated up to 300 Mb/s. The sensitivity of the photoreceiver was −33 dBm at 155 Mb/s and −25.5 dBm at 300 Mb/s at a bit error rate (BER) of 10^{-9} for a pseudo-random bit sequence (PRBS) of $2^{23} - 1$ under the same conditions. At a bias of VDD = 8 V, the power dissipation was 44 mW, with 2 mW from the first two stages. A redesigned circuit [314] achieved sensitivities of −22.8, −15, and −9.3 dBm at bit rates of 622, 900, and 1000 Mb/s, respectively. This redesigned preamplifier had a bandwidth of 500 MHz and dissipated only 10.8 mW at a power supply voltage of 1.8 V. $V_{\mathrm{PIN}} = 30$ V, however, was still necessary for a −3dB frequency of 150 MHz and a −6dB frequency of about 750 MHz for the lateral PIN photodiode.

Newer results were presented for photoreceivers with the same circuit topology as in Fig. 6.91 on high resistivity and on SOI substrates [315]. With the lateral PIN photodiode on high-resistivity silicon a sensitivity of -23.2 dBm at a data rate of 1 Gb/s and a BER of 10^{-9} was achieved with a reverse bias of 30 V at the photodiode and a VDD of 3.5 V. The same sensitivity at a data rate of 622 Mb/s was reported with a photodiode reverse bias of 10 V.

With a lateral PIN photodiode in a 3.0 µm thick SOI layer of N-type and orientation (100) with a resistivity of 50–80 Ωcm on a buried oxide with a thickness of 3 µm and an NMOS transimpedance amplifier with the circuit topology of Fig. 6.91 a sensitivity of -15.3 dBm at 622 Mb/s and at a BER of 10^{-9} was obtained for VDD = 3 V and $V_{PIN} = 3$ V with $\lambda = 850$ nm [315]. With higher supply voltages of VDD = 5 V and $V_{PIN} = 5$ V a sensitivity of -12.0 dBm at a data rate of 1.5 Gb/s was found. A sensitivity of -12.2 dBm finally was reported with VDD = 5 V and $V_{PIN} = 20$ V for a data rate of 2.0 Gb/s. These impressive results are due to the optimized SOI layer thickness allowing the supression of the slow drift in the depth between the N^+- and P^+-fingers, which is present in the lateral PIN photodiode on high-resistivity bulk silicon.

It can be concluded that the approach of [313–315] results in a rather good performance of the photoreceiver at the cost, however, of a rather large supply voltage of up to 30 V for the lateral PIN photodiode. This voltage is usually not present in modern electronic systems and advanced microelectronic circuits, which operate at 5 V, 3.3 V, 2.5 V, 1.8 V or even lower voltages.

BiCMOS Receivers

Results of BiCMOS-OEICs, consisting of a PIN photodiode and an amplifier, which exploited only MOSFETs and no bipolar transistors, have been published [171, 172]. The BiCMOS technology, therefore, was chosen merely for the integration of the PIN photodiode. A standard BiCMOS technology with a minimum effective channel length of 0.45 µm was used without any modifications [171, 172]. This effective channel length corresponds to a drawn or nominal channel length of about 0.6 µm. The buried N^+ collector in Fig. 2.44 was used for the cathode of the PIN photodiode, the P^+-source/drain island served for the anode, and the intrinsic zone of the PIN photodiode was formed by the N well (see Sect. 2.3).

The circuit diagram of the CMOS preamplifier used in the BiCMOS OEIC is shown in Fig. 6.92. The input stage of this amplifier with the NMOS transistors N1, N2, and N3 is a single-ended transimpedance amplifier featuring DC input coupling. The feedback resistor R3 had a value of 1.4 kΩ. However, with the additional voltage gain of the following circuit elements, the effective transimpedance of the amplifier was about 3.0 kΩ. The circuit with the transistors N5 and N6 produces a reference voltage for the gate of P5. This voltage is close to the midpoint of the voltage swing at the gate of P3

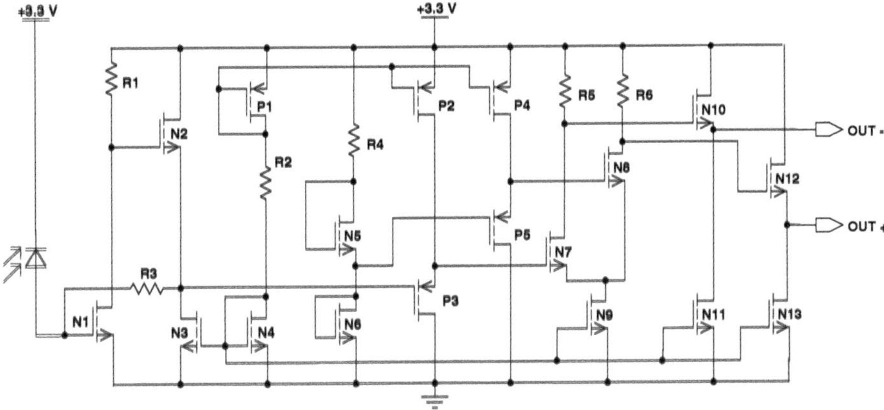

Fig. 6.92. Circuit diagram of an OEIC in BiCMOS technology with a buried N^+ collector as cathode of a PIN photodiode [171]

and can be considered as a kind of decision threshold. P3 and P5 are source followers, which are biased by the current sources P2 and P4, respectively. N4–N6 and P1–P5 perform a single-ended to differential conversion. N7 and N8 form a differential amplifier. N10–N13 are source-follower drivers with an output impedance of 50 Ω.

The power dissipation of the core amplifier circuit was 30 mW from a 3.3 V supply, with an additional 57 mW in the output source followers. The −3dB bandwidth of the receiver was 300 MHz. The OEIC reached a bit rate of 531 Mb/s with a bit error rate of 10^{-9} and a sensitivity of −14.8 dBm for $\lambda = 850$ nm.

In [172] a laser with a wavelength of 670 nm was used for the characterization of the same OEIC as in [171]. The data rate was increased to 622 Mb/s for this wavelength. This bit rate was limited by the capacitance of the photodiode and the feedback resistor of 1.4 kΩ in the amplifier transimpedance input stage.

Another BiCMOS fiber receiver OEIC has been described containing a double photodiode (Fig. 2.47) and a bipolar preamplifier (Fig. 6.93). This OEIC has been fabricated in a 0.8 μm BiCMOS technology.

The transistor Q1 is used in common-emitter configuration. Together with the emitter follower Q2 and the feedback resistor R_{fb}, Q1 forms a transimpedance input stage. The emitter follower Q4 is used for level shifting and for decoupling the feedback loop from the output. A reference voltage V_{REF} of 2.5 V has been applied to the emitter of Q1 resulting in a reverse bias of about 3.3 V for the double photodiode lying between the input and ground.

The transient response of a double photodiode with an area of 530 μm^2 shown in Fig. 2.51 has been measured with this amplifier. A bandwidth of 367 MHz has been determined for the OEIC. Wide-open eye patterns with a turn-on delay of 0.2 ns at a bit rate of 531 Mb/s for $\lambda = 638$ nm (Fig. 6.94)

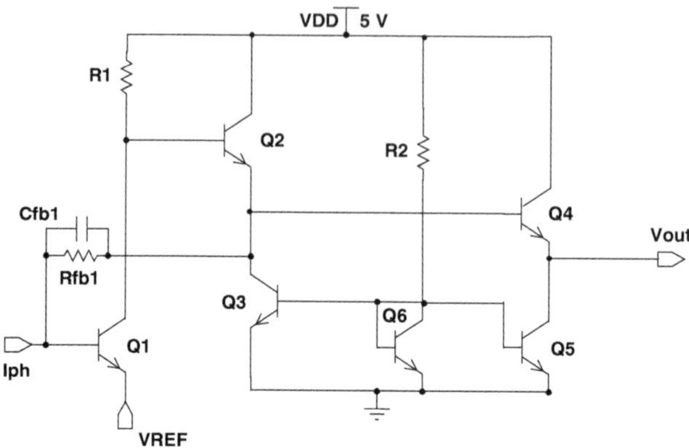

Fig. 6.93. Schematic of fiber receiver OEIC fabricated in BiCMOS technology [316]

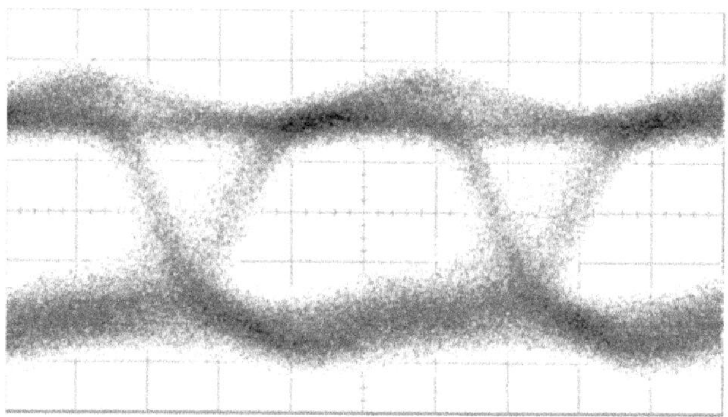

Fig. 6.94. Eye diagram of a BiCMOS fiber receiver OEIC recorded at 531 Mb/s with a PRBS word length of $2^{23}-1$ (time scale: 400 ps/div.; amplitude: 100 mV/div.) [316]

were obtained with this BiCMOS OEIC containing a double photodiode with an area of 530 μm^2 in a standard technology without any modifications [316].

A low-noise monolithically integrated 1.5 Gb/s optical receiver in 0.6 μm BiCMOS technology with a PIN photodiode similar to that described in [35] was presented recently [317]. It can be used in optical interconnects and in short-range data transmission via optical fibers in local area networks and for fiber-in-the-home. Optical interconnects may offer a solution for the increased bandwidth demands in communication within electronic systems e. g. between electronic boards via optical backplanes or via free space. High-bandwidth and high-sensitivity optical receivers are especially necessary when many star couplers or many beam splitters are implemented in the optical transfer path. The same demands hold for free space optical interconnects when the emitted optical power is spread over a large area to reach many receivers. Optical free-space data transmission between different computers, printers, scanners and so on within larger rooms also requires sensitive receivers. Low-cost, high-reliability, and high-performance photo receivers are needed to support the growth of optical interconnects.

A possibility to speed up integrated photodiodes is to use a second higher supply voltage for the photodiode to increase the electric field , i. e. the drift velocity. A second supply voltage of 24 V was also necessary for lateral PIN photodiodes on SOI [315, 318]. Due to the thickness of the SOI layer of 2 µm the photodiode's quantum efficiency, however, was small at 850 nm. In [188]), the sensitivities of −26.2 dBm, −20.3 dBm, and −12 dBm at data rates of 622 Mb/s, 1 Gb/s, and 2 Gb/s, respectively, for the receiver in [318] were given for a pseudo-random-bit-sequence (PRBS) of 2^7-1. Below an OEIC in 0.6 µm BiCMOS technology [317] will be described, which achieves a sensitivity of −23.9 dBm at a single 5 V supply and −24.3 dBm at VCC = 5 V and VPD = 17 V both at a data rate of 1 Gb/s and a PRBS of $2^{31}-1$.

In [317] a vertical PIN photodiode suggested in [35] with a large I-layer thickness was implemented leading to a quantum efficiency of 74 % at a wavelength of 670 nm and 50 % at 850 nm. This photodiode uses a P-type substrate. The cathode is formed by an N^+ region on which a low-doped N^- epitaxial layer is grown. A deep P-type region is implemented to reconstruct the P-type substrate doping concentration in the circuit regions of the OEIC [35]. This photodiode has the advantage that a reverse bias higher than the circuit supply voltage can be applied for a high drift speed.

The schematic of the amplifier implemented in the BiCMOS OEIC is shown in Fig. 6.95. The OEIC consists of a PIN photodiode followed by a transimpedance amplifier and two post amplifier stages. The PIN photodiode had a responsivity of 0.40 A/W at 670 nm and a capacitance of about 50 fF already at a reverse bias of 3 V. The diameter of this diode was 50 µm. To maintain low noise the first amplifier stage is a transimpedance amplifier with a feedback resistor of $R_1 = 3.2\,\mathrm{k}\Omega$. Further amplification is obtained with two amplifier stages to enhance the signal amplitude and drive the bondpad and the load. This two-stage post amplifier had an overall gain of 20 dB and every stage had the same structure as the input stage to obtain low offset. In this application low offset was only necessary to ensure that the bias points of the amplifiers have the optimal values and that the output voltage range is not significantly limited by process variations.

An NMOS transistor (M1) instead of a NPN transistor was implemented at the input of the amplifier in order to achieve a larger output voltage range (about 1.3 V) compared to less than 0.8 V in case of an NPN transistor. The MOS transistors M1 to M6 had the shortest possible gate length (0.6 µm) for highest bandwidth and maximum conductance. T1 to T6 were bipolar transistors which have a transit frequency of about 25 GHz. Emitter followers were used instead of source followers to increase the bandwidth due to the lower output impedance. A second reason was, that gain reduction could be

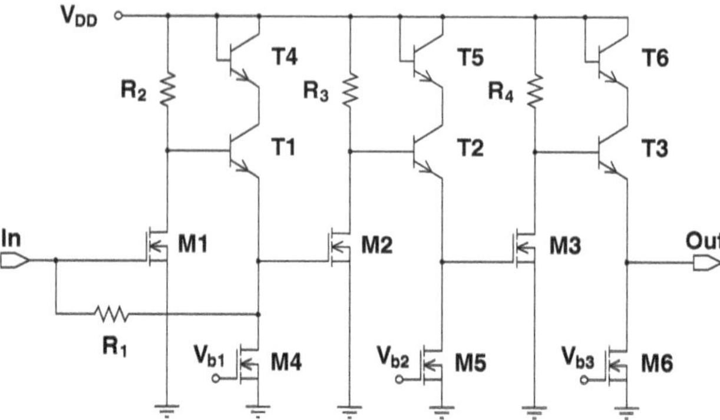

Fig. 6.95. Schematic of low-noise fiber receiver OEIC fabricated in 0.6 μm BiCMOS technology [317]

avoided which would have resulted due to the back-gate effect of the available NMOS transistors without a complimentary buried layer below the P well. T1,T3 and T5 were used as diodes to reduce the collector-emitter voltage of the emitter followers (T2,T4 and T6) to prevent breakdown.

The areas of the MOS- and bipolar transistors were exponentially expanded with progressive stages to obtain the same relative load for each stage and finally a low impedance output stage. This was done to have the possibility to drive low-impedance loads during measurement. This option, however, lead to a high supply current. For applications where the signal is processed internally on chip the output stage can be reduced in size and current consumption.

Figure 6.96 shows the resulting chip which occupies an area of approximately 870 μm × 410 μm. The photodiode together with its metal shield against light incidence outside of its light-sensitive area with a diameter of 50 μm has a size of 140 μm × 140 μm. The light shield can be made smaller or omitted entirely. The active area of the amplifier, however, is only 80 μm × 350 μm allowing easily a pitch of 125 μm or 150 μm in one dimension for massively parallel optical interconnects. The output pad is small compared to the supply pads to maintain a low capacitive load for the data signal.

For measurements, the OEIC was bonded on a printed circuit board (PCB) to minimize the stray inductance of the connection from the supply blocking capacitors to the chip [317]. The evaluation of the behavior of this OEIC was done under two load conditions. First the bandwidth and transient response were measured with a 3-GHz picoprobe on the output pad. The picoprobe loaded the OEIC output with 100 fF and 10 MΩ. The same measurements were done with an AC-coupled resistor divider connected directly to the 50 Ω oscilloscope input. Therefore the output pad was connected with a bond wire to a PCB, where an external 220 Ω resistor and a 100 nF

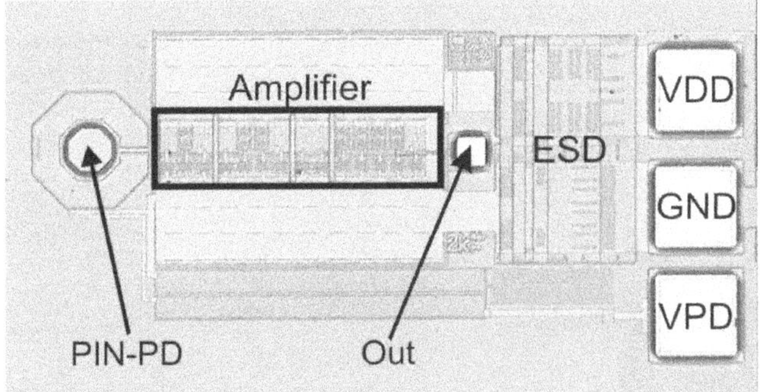

Fig. 6.96. Chipfoto of fiber low-noise receiver OEIC fabricated in 0.6 µm BiCMOS technology [317]

capacitor were mounted. The output signal was lead to the oscilloscope via a 50 Ω micro-strip line on the print, which resulted in a 220 Ω-50 Ω voltage divider. This was done to overcome the problem of the noise from the pico-probe which degrades the receiver sensitivity significantly. The lower cutoff frequency of this capacitively coupled voltage divider was 6 kHz. The 220 Ω resistor together with the characteristic impedance of our measurement system (50 Ω) resulted in an overall load of 270 Ω at the OEIC output. The bandwidth and rise time was slightly degraded due to this load, but the noise was only limited to thermal noise of the 220/50 voltage divider, which was much lower than the picoprobe noise [317]. The results of the frequency responses with VPD = 17 V can be seen in Fig. 6.97. In Fig. 6.98 the transient behavior is shown for the two cases.

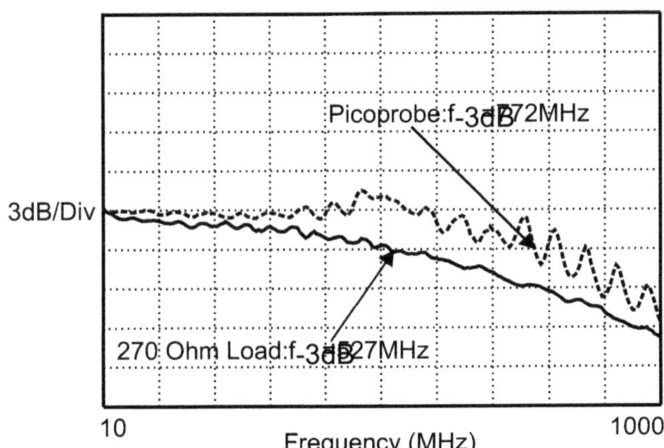

Fig. 6.97. Frequency response of low-noise fiber receiver OEIC fabricated in 0.6 µm BiCMOS technology under different load conditions with VPD = 17 V [317]

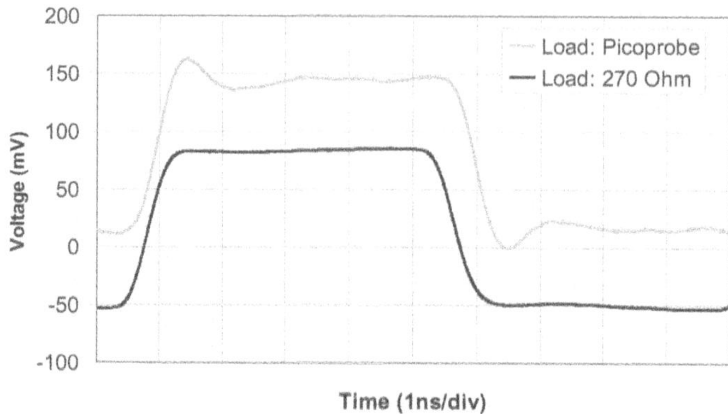

Fig. 6.98. Transient response of low-noise receiver OEIC fabricated in 0.6 μm BiCMOS technology for different loads with VPD = 17 V [317]

Table 6.5 shows the bandwidth dependence on photodiode bias voltage. With the supply voltage of the photodiode increased to 17 V, the bandwidth extends to 772 MHz. Table 6.6 lists the measured rise and fall times in de-

Table 6.5. Bandwidth of the OEIC in dependence on the photodiode bias voltage [317]

V_{PD} (V)	Load: 270 Ω (MHz)	Load: picoprobe (MHz)
5	442	658
7.5	487	723
10	489	765
12	518	769
15	519	770
17	527	772

Table 6.6. Rise and fall times of the OEIC in dependence on the photodiode bias voltage [317]

V_{PD}	Load: 270 Ω		Load: picoprobe	
	t_r	t_f	t_r	t_f
(V)	(ps)	(ps)	(ps)	(ps)
5	669	760	531	547
7.5	602	679	515	533
10	589	661	473	493
12	580	653	463	481
15	575	648	465	484
17	571	648	462	482

pendence on the photodiode reverse voltage. With 17 V, rise and fall times of 0.46 ns and 0.48 ns, respectively, were achieved [317].

The OEIC consumed 31 mA of supply current at a supply voltage of 5 V which results in a power consumption of 155 mW. The DC-sensitivity of this OEIC was 13.7mV/μW at 670 nm which leads together with the responsivity of the photodiode to an effective transimpedance of 34.1 kΩ. Together with a bandwidth of 772 MHz, a 26 THzΩ effective transimpedance-bandwidth product was obtained. This OEIC had a separate supply pad (VPD) for the cathode of the photodiode. The receiver, therefore, could be used either in a single 5 V environment at a reduced data rate of 1.25 Gb/s or the full data rate of 1.5 Gb/s could be achieved with 17 V at the cathode of the photodiode. All measurements were performed with a 670nm laser diode modulated by a 2.5 Gb/s ECL pattern generator for the bit error measurements or via a bias-T by a HP8753E network analyzer for the frequency response measurements.

The sensitivity measurements for a BER of 10^{-9} were performed with a pseudo random bit sequence (PRBS) with a length of $2^{31}-1$. The eye diagram was observed with a Tektronix communication system analyzer CSA8000. Due to the noisy picoprobe, these measurements were taken with the 220/50 Ω resistor divider at the output. The results for different data rates and photodiode voltages are shown in Table 6.7. The result for a data rate of 1.5 Gb/s with a dual supply of 5 V and 17 V was a sensitivity value of −22.1 dBm. For a single supply of 5 V and a data rate of 1.25 Gb/s at a sensitivity of −22.7 dBm was obtained. At a data rate of 1.0 Gb/s, the sensitivity at a single 5 V supply was −23.9 dBm which is slightly better than the value of −23.3 dBm reported in [315] with two supply voltages (VPD = 30 V and VDD = 3.9 V, see Table 6.13). Table 6.8 compares the results achieved in [317] with that reported in [188, 318, 319]. In [318] a PRBS of 2^7-1 was used and a penalty of 1.5 dB was mentioned for a PRBS of $2^{23}-1$ leading to a sensitivity of −18.7 dBm at 1 Gb/s compared to a value of −24.3 dBm for a PRBS of $2^{31}-1$ of the 0.6 μm PIN BiCMOS OEIC reported in [317]. For a photodiode bias of 5 V, a penalty of 2–4 dB was reported in [318] leading to a sensitivity of about −18 dBm, which compares to a value of −23.9 dBm of the 0.6 μm PIN BiCMOS OEIC reported in [317].

Table 6.7. Summary of the sensitivity of the low-noise BiCMOS OEIC for a BER of 10^{-9} and a PRBS of $2^{31}-1$ [317]

Data rate (Mb/s)	Sensitivity at $V_{PD} = 5$ V (dBm)	Sensitivity at $V_{PD} = 17$ V [317] (dBm)
622	−24.5	−24.5
1000	−23.9	−24.3
1250	−22.7	−24.1
1500	-	−22.1

Table 6.8. Comparison of the sensitivity of the OEIC described in [188,318] with the low-noise BiCMOS OEIC in [317] for a data rate of 1.0 Gb/s and VDD = 5 V

	Sensitivity (dBm)	PRBS	V_{PD} (V)
1 µm NMOS on SOI [188,318]	−20.2	2^7-1	20
1 µm NMOS on SOI [188,318]	−18.7	$2^{23}-1$	20
1 µm NMOS on SOI [188,318]	≈ −18	2^7-1	5
0.6 µm BiCMOS [317]	−24.3	$2^{31}-1$	17
0.6 µm BiCMOS [317]	−23.9	$2^{31}-1$	5

The measured incident optical powers reported in [317] were averaged values of a laser diode with an extinction ratio of 6.5. The results, therefore, can be corrected to values that are 1.3 dB lower, if a source with infinite extinction ratio is used. Further improved sensitivity at the lower bit rates can be achieved when filters limiting the bandwidth according to the desired data rate are applied between the output of the OEIC and the signal analyzer [317]. The results in Table 6.7 were measured without a filter. Figure 6.99 shows the eye diagrams at the data rates of 1.0 Gb/s and 1.5 Gb/s at a bit error ratio (BER) of 10^{-9} with a photodiode bias of 17 V.

Figure 6.100 shows the dependence of the BER on the average optical input power for various data rates. Due to intersymbol interference (ISI), the sensitivity at 1.5 Gb/s is reduced by 2 dB. At 1.25 Gb/s a BER below 10^{-12} was verified. At 1.0 Gb/s a BER of 10^{-13} was achieved at an averaged incident optical power of −23.1 dBm [317].

It was shown that low noise and high data rates can be achieved with silicon PIN BiCMOS OEICs, which make them useful for application in optical interconnections. The low active area of photodiode and amplifier allow the application in massively parallel optical interconnects. The sensitivity, bandwidth and current consumption can be further optimized [317].

CMOS Receivers

Another field of application for OEICs in addition to fiber receivers is optical interconnect technology, because electrical interconnects on a board or system level are becoming a problem due to steadily increasing clock frequencies. In particular, signal reflections on long interconnects on large boards and the cross-talk on high-density electronic boards with conductor widths of 0.1 mm and conductor spacings of 0.1 mm are critical. Optical interconnects, e.g.

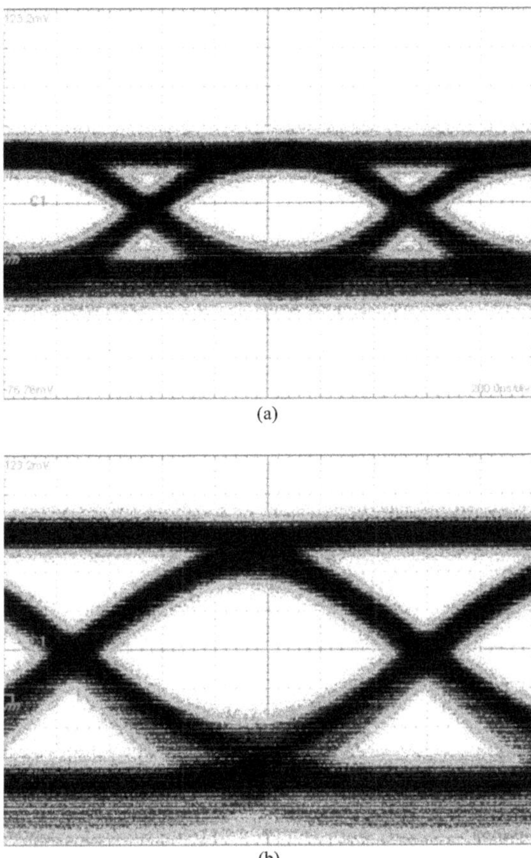

Fig. 6.99. Eye diagrams of low-noise receiver OEIC fabricated in 0.6 μm BiCMOS technology with VPD = 17 V, $\lambda = 670$ nm, $2^{31}-1$ PRBS at a data rate of **(a)** 1 Gb/s with $P_{\text{in}} = -24.3$ dBm (H: 200 ps/div, V: 20 mV/div) **(b)** 1.5 Gb/s with $P_{\text{in}} = -22.1$ dBm (H: 100 ps/div, V: 20 mV/div) [317]

via waveguide-in-board or fiber-in-board and optical backplanes, avoid the problems of electrical interconnects.

For the application in optical interconnect technology, a CMOS preamplifier OEIC was developed choosing the innovative monolithic integration of vertical PIN photodiodes in a twin-well CMOS-process (Fig. 2.19) [76], which uses epitaxial wafers. The integration of PIN photodiodes in such a CMOS technology requires much less additional process complexity than the published approaches to standard-buried-collector (SBC) based bipolar OEICs [35, 305]. Three additional masks were necessary for the PIN-bipolar integration and for the avoidance of the Kirk effect [35]. Only one photodiode protection mask is added for the PIN-CMOS integration in order to

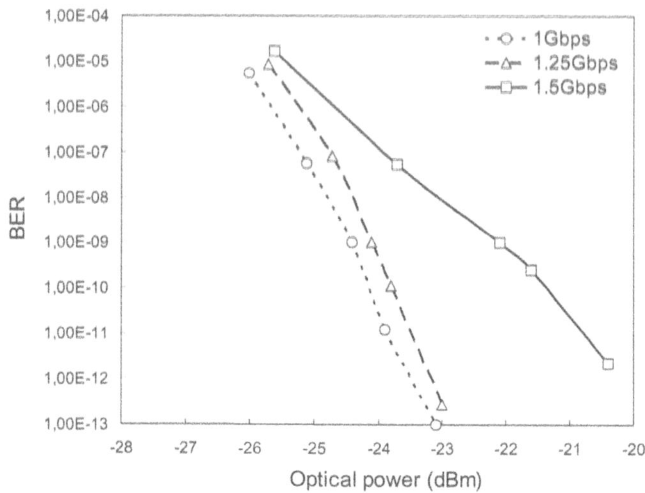

Fig. 6.100. BER of low-noise receiver OEIC fabricated in 0.6 μm BiCMOS technology with VPD = 17 V in dependence on the average optical input power at $\lambda = 670$ nm, $2^{31}-1$ PRBS, VPD = 17 V [317]

block out an originally unmasked threshold implantation from the photodiode area. A reduction of the standard doping concentration C_e of approximately $1 \times 10^{15}\,\mathrm{cm}^{-3}$ in the epitaxial layer was necessary in order to obtain fast integrated PIN photodiodes. In contrast to a reduction of the current gain and of the transit frequency of bipolar transistors in bipolar OEICs due to the Kirk effect, the electrical performance of the N- and P-channel MOSFETs is not degraded when the doping level in the epitaxial layer is reduced, because these MOSFETs are placed in wells [76]. Reach-through and electrostatic discharge (ESD) aspects in the CMOS OEICs can be dealt with using appropriate design measures.

In contrast to the OEIC of [313], here, only a single power supply of 5 V is needed. The circuit of the preamplifier OEIC is shown in Fig. 6.101.

The amplifier consists of three stages. The input stage with M1–M3 is a transimpedance configuration. The source followers M2, M5, and M8 as well as the current sources M3, M6, and M9 are used for level shifting. Without these transistors, V_{GS} of transistor M1 would be larger and a lower voltage across the PIN photodiode would result and would increase the rise and fall times of its photocurrent. The threshold implant in Fig. 2.18g was omitted in order to reduce the threshold voltage of transistors M2, M5, and M8 intentionally to about 0.4 V by using the photodiode protection mask in order to obtain lower V_{GS} values and to realize the optimum level shifting in such a way. Due to the feedback across the 2.2 kΩ transimpedance resistor R_F and to the identical dimensions of the transistors in the different stages, a good

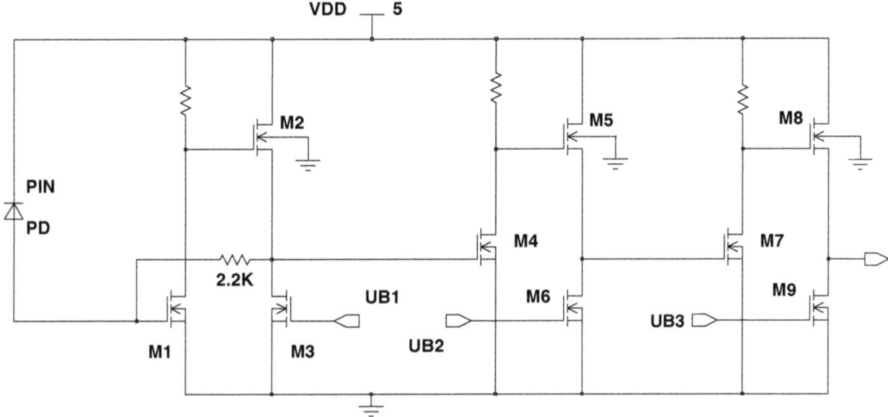

Fig. 6.101. Circuit diagram of a CMOS preamplifier OEIC [77]

independence from process deviations within the relatively large specified process tolerances of the used digital CMOS process is obtained [77].

The small-signal transimpedance of the input stage can be calculated from the small-signal equivalent circuit (Fig. 6.102a).

For lower frequencies the small-signal equivalent circuit simplifies because the inner transistor capacitances can be considered as open circuit (Fig. 6.102a). This leads to the node equations [320]:

$$i_{\text{pd}} - \frac{v_1}{r_1} - \frac{v_1 - v_{\text{out1}}}{R_{\text{F}}} = 0 \quad \text{(Node 1)} \tag{6.65}$$

$$g_{m1} v_1 + \frac{v_2}{r_2} = 0 \quad \text{(Node 2)} \tag{6.66}$$

$$g_{\text{m2}} v_2 - \frac{v_{\text{out1}}}{r_3} + \frac{v_1 - v_{\text{out1}}}{R_{\text{F}}} = 0 \quad \text{(Node 3)}, \tag{6.67}$$

where

$$r_1 = R_{\text{D}}$$

$$r_2 = \frac{1}{g_{\text{ds1}} + \frac{1}{R1}}$$

$$r_3 = \frac{1}{g_{\text{m2}} + g_{\text{mbs2}} + g_{\text{ds2}} + g_{\text{ds3}}} .$$

C_1 includes the capacitances between input node 1 and ground; C_2 between node 1 and 2. C_3 are all capacitances between node 2 and ground, C_4 between node 2 and 3, and C_5 between node 3 and ground. There are the following equations:

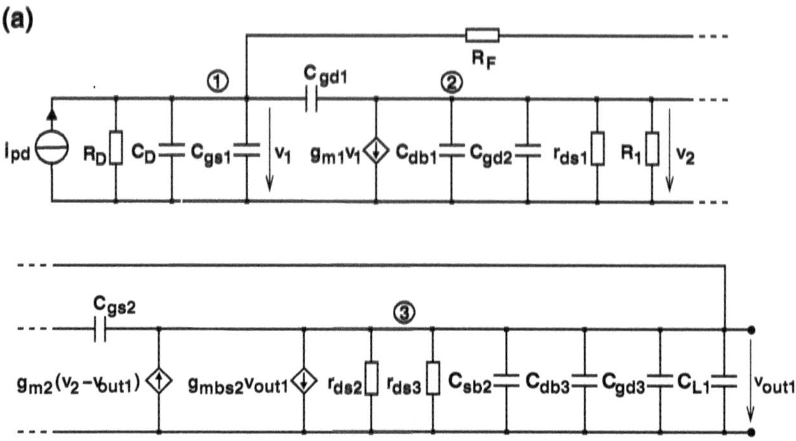

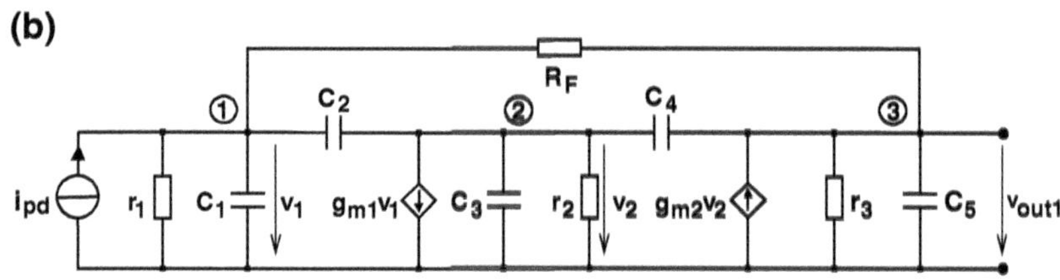

Fig. 6.102. Small-signal equivalent circuit of the first stage in the amplifier shown in Fig. 6.101 **(a)** and its simplified small-signal equivalent circuit **(b)** [320]

$$
\begin{aligned}
C_1 &= C_{\mathrm{D}} + C_{\mathrm{gs1}} \\
C_2 &= C_{\mathrm{gd1}} \\
C_3 &= C_{\mathrm{db1}} + C_{\mathrm{gd2}} \\
C_4 &= C_{\mathrm{gs2}} \\
C_5 &= C_{\mathrm{sb2}} + C_{\mathrm{db3}} + C_{\mathrm{gd3}} + C_{\mathrm{L1}} \quad .
\end{aligned}
$$

C_5 does not explicitely include C_{gs3} which is in series to C_{gd3} with a value of $\approx C_{\mathrm{gd3}}$. These capacitances are considered in the load capacitance C_{L1}, which also considers the input capacitance C_{in2} of the second amplifier stage.

Solving the node equations leads to the transimpedance $v_{\mathrm{out1}}/i_{\mathrm{pd}}$ of the input stage. With the resistance R_{D} of the photodiode much bigger than R_{F}, it follows:

$$
\frac{v_{\mathrm{out1}}}{i_{\mathrm{pd}}} = R_{\mathrm{F}} \frac{\left(\frac{1}{R\mathrm{F}} - A\right)}{\left(\frac{1}{r3} + A\right)} \quad , \tag{6.68}
$$

with

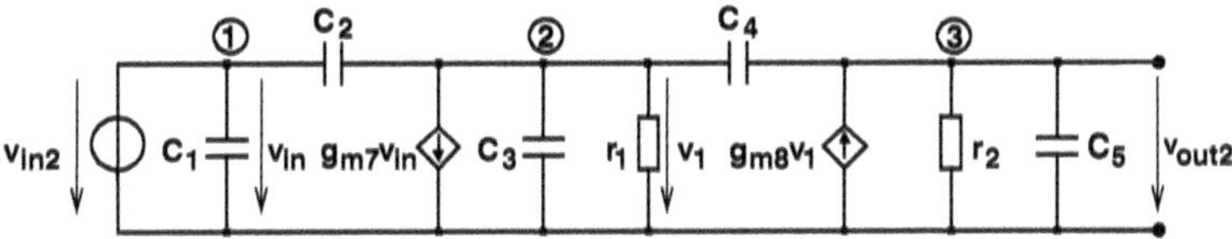

Fig. 6.103. Small-signal equivalent circuit of the second stage in the amplifier shown in Fig. 6.101 [320]

$$A = g_{m1} g_{m2} r_2 \quad . \tag{6.69}$$

If $A \gg \frac{1}{RF}$ and $A \gg \frac{1}{r3}$, the transimpedance of the input stage is just $-R_F$.

The small-signal gain of the second and third stage is identical considering an unloaded output of the third stage and can be extracted from Fig. 6.103 to

$$A_{v2} = \frac{v_{out2}}{v_{in2}} = -\frac{g_{m7} g_{m8}}{\left(g_{ds7} + \frac{1}{R2}\right)\left(g_{m8} + g_{mbs8} + g_{ds8} + g_{ds10}\right)} \quad . \tag{6.70}$$

Assuming an ideal source follower, i. e. the well of the transistor is connected to its source: $g_{mbs8} = 0$, and assuming low output conductances g_{ds}, the expression simplifies to

$$\frac{v_{out2}}{v_{in2}} \approx -\frac{g_{m7} g_{m8}}{\frac{1}{R2} g_{m8}} = -g_{m7} R_2 \quad . \tag{6.71}$$

However, it is necessary to take the bulk-source voltage of the source follower M2, M5, and M8 into account, such (6.70) has to be used. With a single power supply of 5 V and parameter values from a numerical circuit simulation, the transimpedance of the input stage can be calculated with (6.68) to $-1.616\,\mathrm{k}\Omega$, and the gain of the second and third stage with (6.70) to -3.61 [320]. Thus the overall transimpedance of the amplifier results to $-21.06\,\mathrm{k}\Omega$.

The transfer function $v_{out1}(s)/i_{pd}(s)$ of the input stage of the amplifier shown in Fig. 6.101 can be written with the three node equations including the capacitances according to Fig. 6.102(a+b) [320]:

$$v_1 \left[\frac{1}{r_1} + \frac{1}{R_F} + s(C_1 + C_2)\right] - v_2(sC_2) - v_{out1}\left(\frac{1}{R_F}\right) = i_{pd} \tag{6.72}$$

$$v_1 \left(g_{m1} - sC_2\right) + v_2 \left[\frac{1}{r_2} + s(C_2 + C_3 + C_4)\right] - v_{out1}(sC_4) = 0 \tag{6.73}$$

$$v_1 \left(\frac{1}{R_F}\right) + v_2(g_{m2} + sC_4) - v_{out1}\left[s(C_4 + C_5) + \frac{1}{r_3} + \frac{1}{R_F}\right] = 0 \tag{6.74}$$

(6.75)

or as a matrix equation:

$$\begin{pmatrix} a_{11} & a_{12} & a_{13} \\ a_{21} & a_{22} & a_{23} \\ a_{31} & a_{32} & a_{33} \end{pmatrix} \begin{pmatrix} v_1 \\ v_2 \\ v_{out1} \end{pmatrix} = \begin{pmatrix} i_{pd} \\ 0 \\ 0 \end{pmatrix} \quad . \tag{6.76}$$

The solution of this equation system with $a_{13} = a_{31}$ is

$$\frac{v_{\text{out1}}(s)}{i_{\text{pd}}(s)} = \frac{a_{21}(-a_{12}a_{13} + a_{32}a_{11}) + a_{13}(a_{12}a_{21} - a_{22}a_{11})}{(a_{13}a_{21} - a_{11}a_{23})(-a_{12}a_{13} + a_{32}a_{11}) - (a_{12}a_{21} - a_{22}a_{11})(-a_{13}^2 + a_{33}a_{11})} \quad . \tag{6.77}$$

This can also be written in the following form

$$\frac{v_{\text{out1}}(s)}{i_{\text{pd}}(s)} = \frac{Z(s)}{N(s)} = \frac{a + bs + cs^2 + ds^3}{e + fs + gs^2 + hs^3 + is^4} \quad . \tag{6.78}$$

Here, it can be seen, that the input stage has already four poles and three zeros. The disadvantage of algebraic solving of the node equations is the high complexity with just small circuits. In addition, the solution may be correct, but hard to "understand". For a better understanding there are two main possibilities to estimate the cutoff frequency of the circuit. (i) If there is a dominant pole, the cutoff frequency is determined mainly by this pole. This is true when a zero in the denominator of the transfer function is much smaller than the others in absolute values. In addition the zeros of the numerator have to be at higher frequencies such that the cutoff frequency is not influenced. The dominant pole is formed by the quotient of the first two coefficients of the denominator polynomial. (ii) With no dominant pole the cutoff frequency is determined by several poles and zeros. A simple estimation of the cutoff frequency can be done with the so-called *Open-Circuit Time-Constant* method [321]. The cutoff frequency of a circuit is then

$$f_{-3\text{dB}} \approx \frac{1}{2\pi \sum_{i=1}^{m} R_{i0} C_i} \quad . \tag{6.79}$$

The resistances R_{i0} have to be determined at the nodes of the respective capacitances C_i while all other capacitances are removed. After determination of all terms $R_{i0}C_i$ it is obvious which node is dominant. For the input stage of the amplifier (Fig. 6.101, Fig. 6.102) this shows that the cutoff frequency is dominated by the time constant $R_{10}C_1$. It is [320]

$$R_{10} = \frac{R_{\text{F}} + r_3}{1 + A r_3} \quad , \tag{6.80}$$

with A from equation (6.69). For a high cutoff frequency, R_{10} should be small, and for a high transimpedance (6.68) has to be considered. Because of the coupled parameters, a high cutoff frequency and a high transimpedance cannot be chosen freely from each other. Therefore the shown design is a compromise between these two goals.

The cutoff frequency of the first stage of the amplifier in Fig. 6.101 after (6.79) is 1034 MHz. The result obtained by numerical circuit simulation for the cutoff frequency of the first stage is 1080 MHz. Such a high cutoff frequency is possible because of the very low capacitance C_D of the integrated PIN photodiode (about 26 fF). A discrete setup of photodiode and amplifier would add more capacitances at the input node of the amplifier and therefore reduce the cutoff frequency of the circuit.

Simulated results of the complete amplifier of Fig. 6.101 with all three stages are: the cutoff frequency is 419 MHz and its overall transimpedance is $-20.7\,\mathrm{k\Omega}$ [320]. The lower cutoff frequency is caused by the capacitive load of 0.3 pF. This low capacitive load is justified with an internal postamplification of the output voltage. There is a good agreement between the simulated and analytically calculated transimpedance of $-20.7\,\mathrm{k\Omega}$ and $-21.1\,\mathrm{k\Omega}$, respectively.

The CMOS receiver OEIC was fabricated in a 1.0 μm industrial CMOS process [78]. The microphotograph of the CMOS preamplifier OEIC can be seen in Fig. 6.104.

The photodiode, having a light sensitive area of $2700\,\mu\mathrm{m}^2$, together with its metal shield around covers an area of approximately $150 \times 150\,\mu\mathrm{m}^2$. The preamplifier occupies an active area of less than $130 \times 200\,\mu\mathrm{m}^2$. The sensitivity of the PIN CMOS preamplifier OEIC was 4.7 mV/μW without ARC and its power consumption was 19 mW at 5.0 V.

The oscilloscope extracted a rise time $t_r^{\mathrm{osc,disp}} = 15.5\,\mathrm{ns}$ and a fall time $t_f^{\mathrm{osc,disp}} = 17.6\,\mathrm{ns}$ for the OEIC with the doping concentration of $1 \times 10^{15}\,\mathrm{cm}^{-3}$ in the epitaxial layer for a laser wavelength of 638.3 nm. These large values for t_r and t_f are due to the slow carrier diffusion in the standard epitaxial layer of the photodiode. The corresponding values for the concentration of $2 \times 10^{13}\,\mathrm{cm}^{-3}$ in the epitaxial layer, where the depletion region spreads through the whole epitaxial layer and carrier diffusion in the photodiode is eliminated, were 1.05 ns and 1.26 ns, respectively [77].

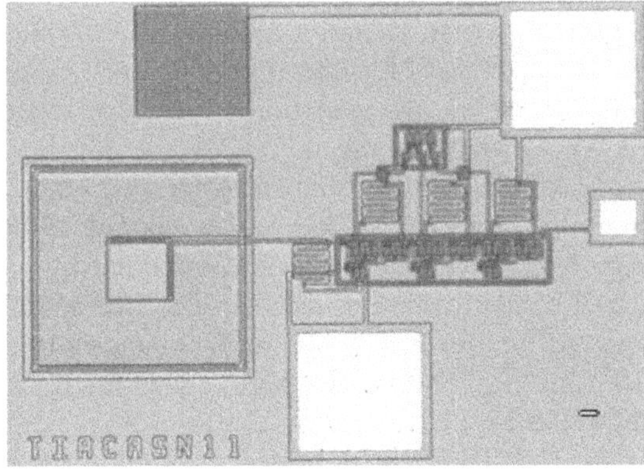

Fig. 6.104. Microphotograph of a CMOS receiver OEIC [77]

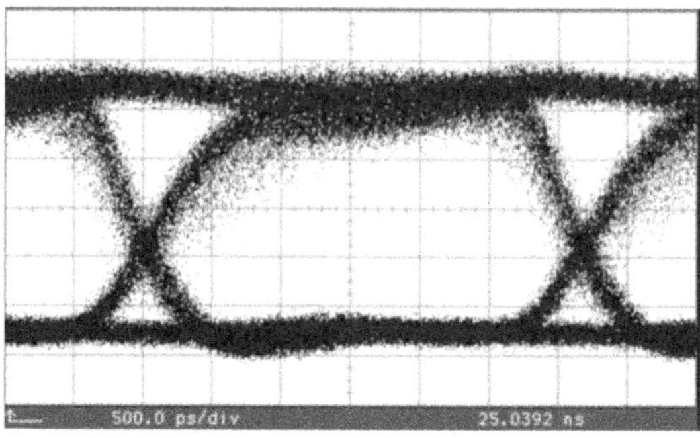

Fig. 6.105. Measured eye diagram of a CMOS preamplifier OEIC for the application in a fiber and interconnect receiver (time: 500 ps/div, amplitude: 0.2 V/div)

Figure 6.105 shows the eye diagram of the CMOS preamplifier OEIC in Figs. 6.101 and 6.104. The eye diagram was measured on the wafer level with a picoprobe (input capacitance: 0.1 pF) at the output of the OEIC, with an HP54750/51 digital sampling oscilloscope, and with an ECL bit pattern generator, which modulated a red semiconductor laser with a wavelength of 638.3 nm. A pseudo-random bit sequence (PRBS) of $2^{23}-1$ in a non-return-to-zero (NRZ) bit rate of 320 Mb/s was used. The doping concentration in the epitaxial layer of the wafer which was used for the fabrication of the OEIC was $2 \times 10^{13}\,\mathrm{cm}^{-3}$.

Simulations showed that shorter rise and fall times can be expected for an improved preamplifier and bit rates in excess of 600 Mb/s seem feasible for a wavelength of 638 nm. For wavelengths of 780 and 850 nm bit rates of 500 Mb/s were estimated by simulations [77]. Meanwhile, the OEIC has been redesigned and characterized by eye diagram measurements at a bit rate of 622 Mb/s with a wavelength of 638 nm [80]. The redesigned OEIC containing a PIN photodiode with an area of 2700 $\mu\mathrm{m}^2$ and the amplifier of Fig. 6.101 in 1.0 μm CMOS technology with an active area of $230 \times 150\ \mu\mathrm{m}^2$ showed a measured sensitivity of 5.7 mV/μW for 638 nm and 7.1 mV/μW for 850 nm [320]. With the responsivity of the pin photodiode without antireflection coating (ARC) of 0.24 A/W for 638 nm and 0.30 A/W for 850 nm an overall transimpedance of 23.75 kΩ results for the amplifier in Fig. 6.101 compared to a simulated transimpedance of 20.7 kΩ and a calculated transimpedance of 21.1 kΩ. The higher measured value can be explained by process tolerances. For a wavelength of 638 nm, rise and fall times of 1.08 ns and 0.94 ns, respectively, were measured for the PIN OEIC with the amplifier in Fig. 6.101 and with $C_e \approx 2 \times 10^{13}\,\mathrm{cm}^{-3}$. The eye diagram of this OEIC for a data rate of 622 Mbit/s is shown in Fig. 6.106.

A sensitivity of -14.9 dBm was obtained for an non-return-to-zero (NRZ) data rate of 622 Mb/s by measurement of the bit error rate in dependence on the optical input power for a BER of 10^{-9} at 638 nm (Fig. 6.107).

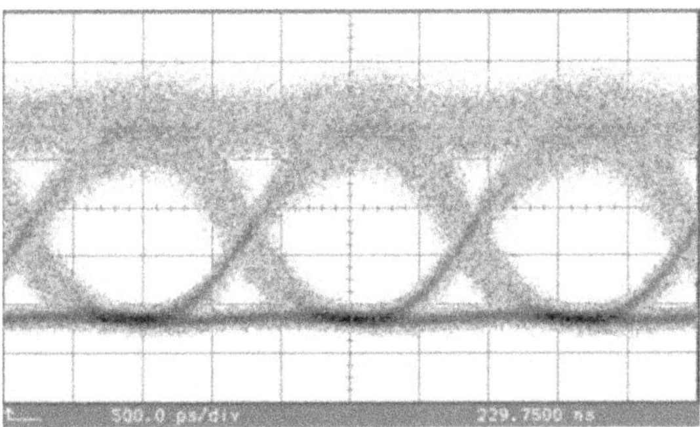

Fig. 6.106. Eye diagram of the redesigned PIN OEIC with the amplifier shown in Fig. 6.101 (622 Mbit/s, 500 ps/div, 200 mV/div, $\lambda = 638$ nm, $C_e = 2 \times 10^{13}$ cm^{-3}) [320]

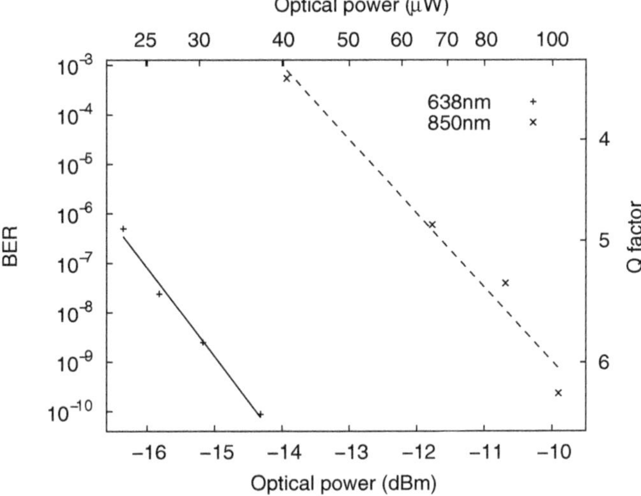

Fig. 6.107. Bit error rate of the PIN OEIC with the amplifier from Fig. 6.101 in dependence on the optical input power with VDD = 5 V and $C_e = 2 \times 10^{13}$ cm^{-3} at 622 Mb/s [320]

For 850 nm, the sensitivity was -10.0 dBm for the same data rate and BER. For the redesigned PIN OEIC after Fig. 6.101, a power consumption of 23 mW was observed for a supply voltage of 5 V [320].

An sophisticated OEIC in standard CMOS technology using the SML detector described in Sect. 2.2.4 was reported [70]. Nine receiver channels for optocoupler applications were implemented. In each channel, the photocurrent of the immediate detector (I in Fig. 6.108) and the diffusion current of the deferred detector (D in Fig. 6.108) are first amplified and then an OTA (operational transconductance amplifier) is used to subtract the diffusion signal from the immediate detector signal.

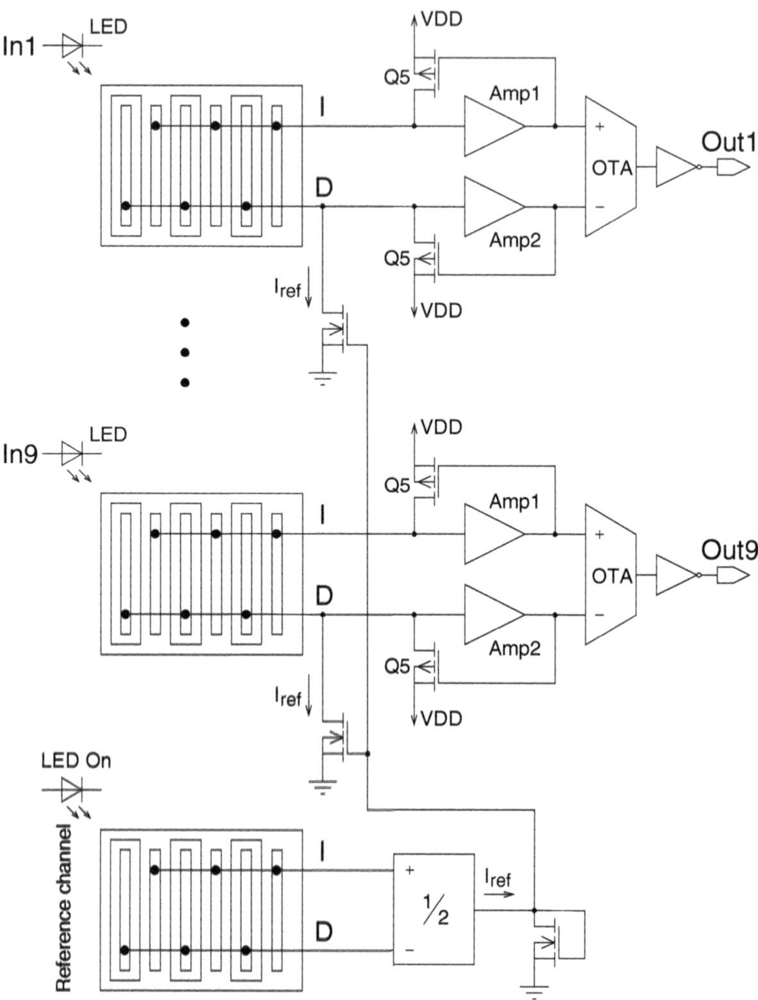

Fig. 6.108. Schematic overview of the SML detector OEIC [70]

A reference channel is needed since data transmitted by optocouplers are not coded, and hence not dc balanced. The average of the signal, therefore, cannot be used as a threshold level. A reference channel has to be used to determine the threshold level I_{ref} (see Fig. 6.108). The LED in this channel is always on and the reference receiver takes half of the difference between the photocurrent of the immediate detector and the diffusion current of the deferred detector. The current I_{ref} is mirrored into the active channels 1–9 as a threshold value for these channels. When the difference current of immediate and deferred detector ($I - D$) exceeds this threshold, the OTA produces a positive output signal. If no light incidents in the SML detectors of channels 1–9, the threshold is not reached and the output signal of the OTA remains negative [70].

The circuit diagram of the two identical amplifiers of each channel is shown in Fig. 6.109. This amplifier has active transistor feedback (Q5) and therefore acts as a transimpedance amplifier [70]. Transistor Q5 connected as a one-stage amplifier in common-source configuration was degenerated with a channel length of 6.3 µm [70]. A transimpedance amplifier is obtained since Q5 (with the constant current load Q9) transforms a voltage into a current (g_{m5} corresponds to $1/R_F$).

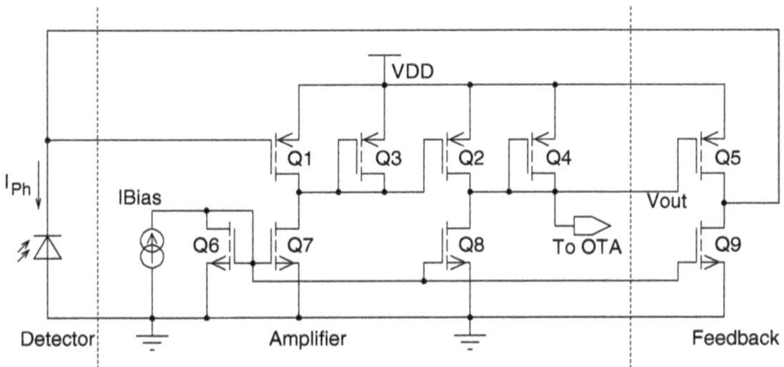

Fig. 6.109. Circuit diagram of the transimpedance amplifier used in the SML OEIC [70]

For a large open-loop gain A_{OL} the transimpedance

$$\frac{V_{out}}{I_{Ph}} = \frac{1/(-g_{m5})}{1 - i(f2\pi C_{det})/(A_{OL}g_{m5})} \tag{6.81}$$

is obtained [70] with V_{out} and I_{Ph} defined in Fig. 6.109. If $-g_{m5}$ is replaced by $1/R_F$, the behavior of a conventional TIA is obtained. MOS diode loads Q3 and Q4 are used to achieve stability (Fig. 6.109).

The circuit was implemented in a 0.6 µm standard CMOS technology. The SML detectors had an area of $50 \times 50\,\mu m^2$. The receivers including the output driver fitted easily into an area of $50 \times 230\,\mu m^2$ each [70]. At 860 nm, sensitivities of −23 dBm for 200 Mb/s and −18 dBm for 250 Mb/s, respectively, were reported for a BER of 10^{-9}. The power consumption of each channel was about 4 mW [70]. A data rate of 700 Mb/s was achieved in a 0.25 µm CMOS technology with a sensitivity of −18.7 dBm for a BER of 10^{-9} at 860 nm. The sensitivity for a BER of 10^{-12} was about −18.0 dBm. The power consumption of a channel in this chip was 2.5 mW [322], which is a remarkable low value.

A fast fully integrated single-beam optical bulk CMOS receiver (Fig. 6.110) was realized in the Lucent 0.35 µm production process [69]. This receiver contained the photodetector shown in Fig. 2.15. The amplifier of this receiver

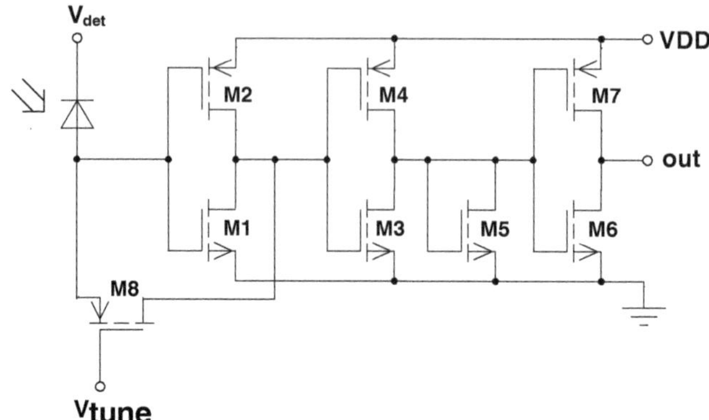

Fig. 6.110. CMOS inverter receiver OEIC [69]

is formed by three inverter stages. The first stage with M1 and M2 is an inverter-based transimpedance stage with the P-channel MOSFET M8 as the feedback element. The gate voltage of M8 could be adjusted for optimum performance (V_{tune}) at a given optical power and bit rate. This first stage converts the photocurrent, flowing from the anode of the photodiode through the active resistor M8, into a voltage. The second stage with the inverter M3, M4 amplifies this voltage. The N-channel MOSFET load M5 reduces the gain and increases the bandwidth. With M5, the technology dependence of the amplifier gain is reduced and the switching threshold of this stage is stabilized [323]. The third stage with M6 and M7 supplies a high gain. This stage acts as an asynchronous decision circuit resulting in a fully digital logic output level. For a correct and optimum performance of this DC-coupled three-stage preamplifier circuit, the dimensions of the transistors M5 and M8 have to be chosen carefully [69].

The supply voltage of the preamplifier was varied between 1.8 and 3.3 V. The best sensitivity was obtained at a supply bias of VDD = 2.2 V. A bit error rate of 10^{-9} for a bit rate of 1 Gb/s was obtained with an average optical input power of −6.3 dBm and with a detector bias $V_{\mathrm{det}} = 10$ V. This low sensitivity results from the low responsivity of the photodiode of less than 0.04 A/W. Another disadvantage of this OEIC is the high detector bias of 10 V.

For an OEIC fabricated in a 0.25 µm fully-depleted CMOS SOI process technology, a single 2 V supply was sufficient [192]. The photodetector of this OEIC was a lateral PIN photodiode in a 50 nm thick SOI layer (see Fig. 3.5). The authors ascribed the high photodiode responsivity of 0.4 A/W to an avalanche effect [192]. The amplifier circuit of this OEIC is shown in Fig. 6.111.

The preamplifier circuit is based on a transimpedance amplifier with a feedback resistor R_{fb} of 5 kΩ. The N-channel MOSFET N1 is used in a common-source circuit. The PMOS transistor P1 forms the active load for

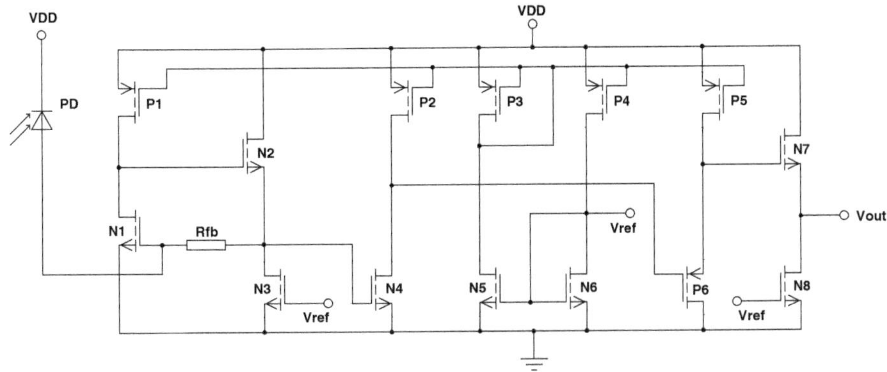

Fig. 6.111. CMOS receiver OEIC on SOI [192]

N1. N2 is used as source follower in order to obtain a low output impedance of the transimpedance input stage. N3 is a constant-current source. N4 amplifies the output voltage of the transimpedance input stage. P6 is used as a source follower for shifting the signal level towards VDD and reducing the output impedance. This shifting towards VDD is necessary to allow the implementation of the second source follower N7 for further reducing the output impedance. P5 and N8 are constant-current sources. P3, P4, N5, and N6 supply the reference voltage $V_{\rm ref}$ for biasing the amplifier. The constant-current sources P1, P2, and P5 are set by P3. These four transistors form current mirrors.

Using a 850 nm wavelength, a bandwidth of 1 GHz was measured for a supply voltage of 2 V for an average optical input power of −13 dBm (50 µW) [192]. The OEIC occupied an area of $650 \times 400\,\mu\text{m}^2$.

An OEIC for inter-chip communication using silicon-on-sapphire (SOS) CMOS technology was introduced [194, 195]. Similar to SOI technology, SOS technology reduces parasitic bulk capacitances. It, therefore, increases the bandwidth. Further advantages are that SOS reduces cross-talk, power and substrate noise. Especially instabilities due to bulk substrate feedack are avoided [194]. A receiver circuit fabricated in an SOS 0.5 µm CMOS process operating at 1 Gb/s while consuming only 5 mW of power at a 3.3 V supply with an effective transimpedance of 50 kΩ was presented [194]. This power consumption is low compared to 26 mW of a 1.5 Gb/s receiver [324] and to over 100 mW of other designs [310, 325].

The circuit diagram of the SOS receiver is shown in Fig. 6.112. The first stage has a transimpedance of 9 kΩ and was designed for Gb operation with 0.5 pF input capacitance generated by a MSM or PIN photodiode. The photodiode is represented by a reverse biased diode in parallel with its capacitance $C_{\rm D}$ in Fig. 6.112. There are five high-bandwidth gain stages following the transimpedance input stage. A differential-to-single ended conversion is done in a sixth stage and a source follower output buffer (not capable of driv-

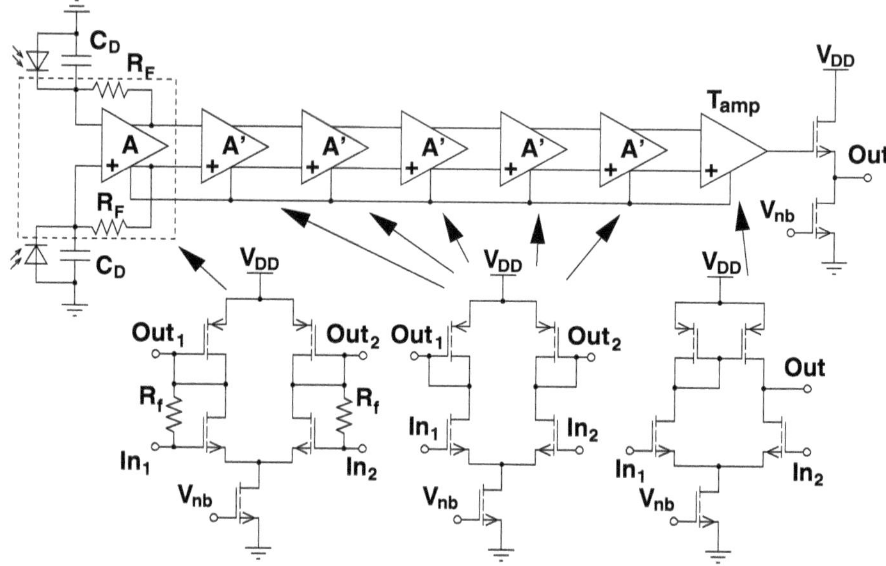

Fig. 6.112. CMOS receiver OEIC on SOS [194]

ing a 50 Ω load rail-to-rail) forms the output [194]. 10 μA of photocurrent at each receiver input or 20 μA of photocurrent at one input caused a rail-to-rail output swing for a 3.3 V supply [194]. PMOS load elements are used in a MOS diode configuration to avoid biasing circuitry. The receiver, therefore, required only one bias (V_{nb}). The transimpedance amplifier had a phase margin of 45 % for limited gain peaking [194].

The transistors in the gain stages had smaller widths than that in the transimpedance stage. The chain of gain stages, therefore, could be made fast while consuming little power since the input transistors of each stage comprise the load of the previous stage. Large input transistors in the transimpedance amplifier improve the gain of the input stage and the sensitivity of the receiver due to reducing electronic noise. The bandwidth of the transimpedance amplifier is dominated by the input node capacitance, feedback resistance, and the gain of the differential amplifier.

The Peregrine Ultra-thin SOS process provided N and P channel devices with three channel doping level, i. e. with three different threshold voltages [194]. Zero-threshold MOSFETs with no channel doping with the same output currents as doped devices with larger widths could therefore be exploited with smaller widths. Thus, gate capacitances could be minimized. The P-channel load devices were zero-threshold devices in order to limit capacitive loading on the outputs of the amplifier stages. The NMOS input transistors, however, were regular threshold devices with $V_{\mathrm{Th}} = 0.7\,\mathrm{V}$ since they exhibit a higher transconductance at equal currents than the zero-threshold devices [194]. Higher g_m input transistors result in a higher gain-bandwidth product.

The differential topology was chosen to make it well suited to optical interconnect applications. The differential design is robust to process variations, which may vary device parameters, to power supply noise, and to ambient light conditions. Due to its low power consumption the SOS receiver is ideal for use among a large array of VCSEL inputs switched about a laser threshold.

The SOS receiver was operated at 650 Mb/s with a differential optical input and at 900 Mb/s with a single optical input [194]. The experiments were performed with a 0.5 pF MSM photodiode bonded to each input. The SOS receiver area was $45 \times 70\,\mu\mathrm{m}^2$ [194]. No sensitivity data were reported. The quality of eye diagrams for 100 µW optical input power was not very good due to jitter [194].

Based on the circuit shown in Fig. 6.101, an improved CMOS photoreceiver (Fig. 6.113), which contains a vertical PIN photodiode and which combines a data rate of 622 Mb/s with a photodiode quantum efficiency of 94 %, has been developed. The innovative monolithic integration of vertical PIN photodiodes in a twin-well CMOS process (Fig. 2.19) [76] which uses epitaxial wafers, has been demonstrated to combine both high speed and large quantum efficiency of the photodiode [77]. In contrast to the OEICs of [69, 313], there, only a single power supply of 3.3 V was needed. The circuit of the high-bandwidth preamplifier with an integrated PIN photodiode is shown in Fig. 6.113. It is a typical high-frequency amplifier. Only N-channel MOSFETs are used in order to obtain a high bandwidth. The input stage with the transistors M1–M4 is a transimpedance configuration, which converts the

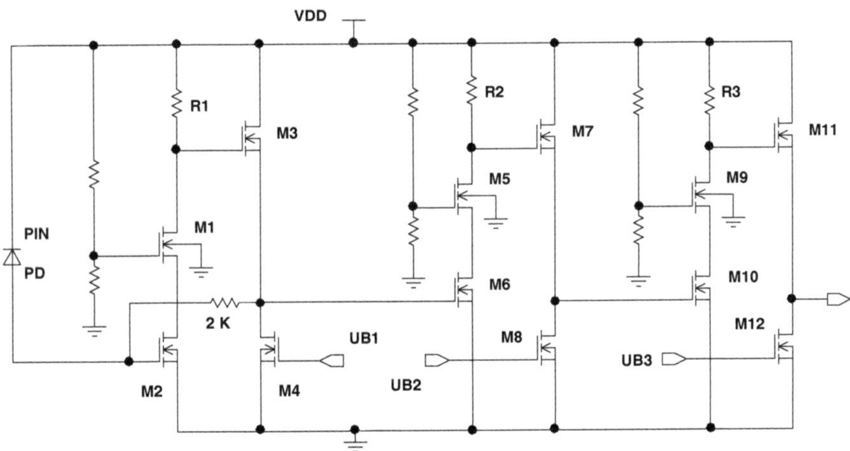

Fig. 6.113. Circuit diagram of a CMOS preamplifier OEIC with a possible data rate of 622 Mb/s for application in a fiber and interconnect receiver [77]

photocurrent change in the integrated PIN photodiode to a voltage change. The cascode transistors M1, M5, and M9 reduce the Miller effect and increase the bandwidth correspondingly. The source followers M3, M7, and M11 as well as the current sources M4, M8, and M12 are used for level shifting. The threshold voltage of transistors M3, M7, and M11 has been reduced intentionally to about 0.4 V by the photodiode protection mask in order to obtain lower V_{GS} values. Polysilicon resistors were employed as load elements, since depletion transistors were not available in the digital CMOS process. Due to the feedback across the 2 kΩ resistor and to the identical dimensions of the transistors in the different stages, a good independence from process deviations within the relatively large specified process tolerances of the digital CMOS process used has been obtained. Three identical biasing circuits (UB1=UB2=UB3) are used instead of one in order to minimize parasitic coupling between the stages.

The following calculations and estimations can be made for the circuit schematic of Fig. 6.113: The small-signal transimpedance of the first stage is calculated in the same way as described for the amplifier shown in Fig. 6.101. In Fig. 6.114 the complete and the simplified small-signal equivalent circuit of the first stage is shown.

Caused by the cascode transistor M1, another node is present compared to the circuit in Fig. 6.103. The small-signal output resistance r_{ds1c} between node 2 and 3 has been neglected in Fig. 6.114b. It is [320]

$$r_1 = R_D$$
$$r_2 = \frac{1}{g_{ds1} + g_{m1c} + g_{mbs1c}}$$
$$r_3 = \frac{1}{g_{m2} + g_{ds2} + g_{ds3}} \quad .$$

The same expression as in equation (6.68) follows for the transimpedance. But now it is

$$A = g_{m1} g'_{m1c} g_{m2} r_2 R_3 \quad . \tag{6.82}$$

with $g'_{m1c} = g_{m1c} + g_{mbs1c}$. Compared to the amplifier in Fig. 6.101, the transimpedance of the first stage is now a little bit lower caused by the smaller feedback resistance R_F (2.0 kΩ instead of 2.2 kΩ). This was chosen to achieve a higher cutoff frequency. With $V_{DD} = 5\,V$ and the parameter values taken from a numerical circuit simulation and used in eq. (6.68), the transimpedance of the input stage is −1.585 kΩ. The small-signal amplification of the second stage is [320]

$$A_{v2} = \frac{v_{out2}}{v_{in2}} = -\frac{g_{m4} g_{m5} (g_{m4c} + g_{mbs4c}) R_3}{(g_{m5} + g_{ds5} + g_{ds6})(g_{m4c} + g_{mbs4c} + g_{ds1})} \quad . \tag{6.83}$$

Here a value of −4.64 can be calculated. With the second and third stage designed identically, the overall transimpedance of the amplifier in Fig. 6.113

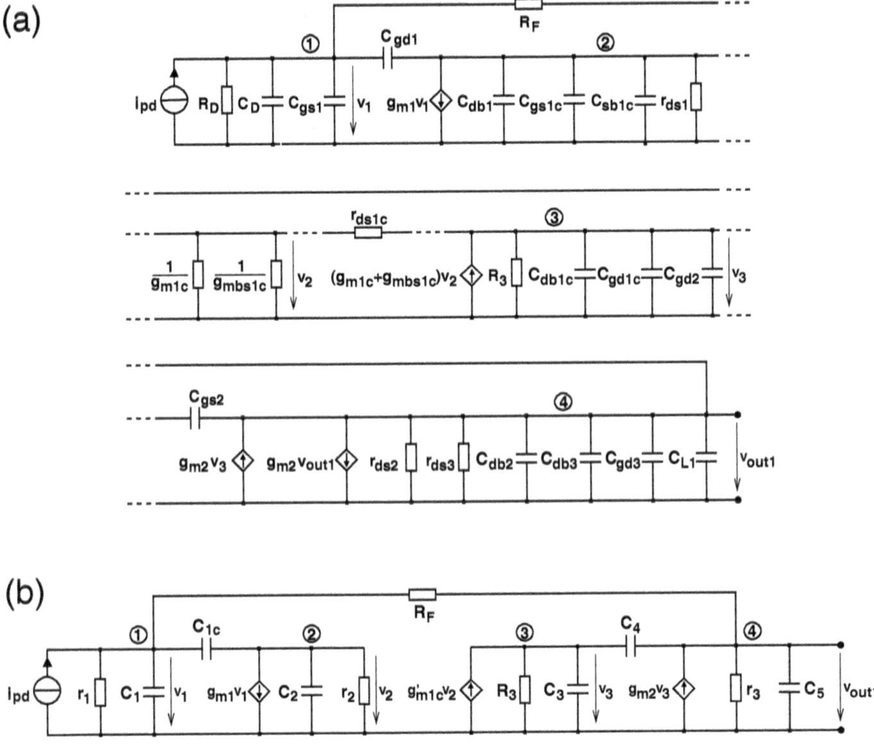

Fig. 6.114. Small-signal equivalent circuit of the first stage in the amplifier of Fig. 6.113 **(a)** and its simplified small-signal equivalent circuit **(b)** [320]

results to $-34.1\,\mathrm{k}\Omega$. The higher amplification in the second and third stage is responsible for a 50 % increase of the overall transimpedance compared to the amplifier shown in Fig. 6.101.

Estimating the frequency response of the amplifier shown in Fig. 6.113, it is necessary to consider the capacitances of the amplifier. For the first stage they are (Fig. 6.114):

$$\begin{aligned}
C_1 &= C_\mathrm{D} + C_\mathrm{gs1} \\
C_\mathrm{1c} &= C_\mathrm{gd1} \\
C_2 &= C_\mathrm{db1} + C_\mathrm{gs1c} + C_\mathrm{sb1c} \\
C_3 &= C_\mathrm{db1c} + C_\mathrm{gd1c} + C_\mathrm{gd2} \\
C_4 &= C_\mathrm{gs2} \\
C_5 &= C_\mathrm{db2} + C_\mathrm{db3} + C_\mathrm{gd3} + C_\mathrm{L1} \quad .
\end{aligned}$$

The four node equations to calculate the transfer function are [320]:

$$v_1\left[\frac{1}{r_1}+\frac{1}{R_\mathrm{F}}+s(C_1+C_{1\mathrm{c}})\right]-v_2(sC_{1\mathrm{c}})-v_{\mathrm{out1}}\left(\frac{1}{R_\mathrm{F}}\right)=i_{\mathrm{pd}} \quad (6.84)$$

$$v_1\left(g_{\mathrm{m1}}-sC_{1\mathrm{c}}\right)+v_2\left[\frac{1}{r_2}+s(C_{1\mathrm{c}}+C_2)\right]=0 \quad (6.85)$$

$$-v_2 g'_{\mathrm{m1c}}+v_3\left[\frac{1}{R_3}+s(C_3+C_4)\right]-v_{\mathrm{out1}}(sC_4)=0 \quad (6.86)$$

$$-v_1\frac{1}{R_\mathrm{F}}-v_3(g_{\mathrm{m2}}+sC_4)+v_{\mathrm{out1}}\left[\frac{1}{R_\mathrm{F}}+\frac{1}{r_3}+s(C_4+C_5)\right]=0\,. \quad (6.87)$$

The first stage of the amplifier shown in Fig. 6.113 has five poles and four zeroes. The transfer function is therefore even more complex as the one for the amplifier of Fig. 6.101. As with the amplifier of Fig. 6.101 the Open-Circuit Time-Constant method [321] was used. The time constant related with node 1 was again the one with the biggest influence on the cutoff frequency [320]. The resistance R_{10} results in the same expression as in equ. (6.80) with A from equ. (6.82). Now the difference to the amplifier shown in Fig. 6.101 is obvious. Although the capacitance C_1 is the same, a higher value for A could be achieved through the insertion of the cascode transistor M1. This leads to a smaller time constant $R_{10}C_1$. Additionally the feedback resistance R_F was lowered from 2.2 kΩ in the amplifier of Fig. 6.101 to 2 kΩ in the amplifier of Fig. 6.113. This results in a cutoff frequency of 1.27 GHz for the input stage of the amplifier shown in Fig. 6.113 [320]. A simulation for the complete three-stage topology of the amplifier in Fig. 6.113 showed an overall cutoff frequency of 922 MHz with a capacitive load of 0.3 pF.

The sensitivity of the PIN preamplifier OEIC was 4.7 mV/μW for a wavelength $\lambda = 638$ nm increasing to 9.0 mV/μW with ARC, which corresponds to an overall transimpedance of 18.4 kΩ. Its power consumption was 44 mW at 5.0 V reducing to 17 mW at 3.3 V [320].

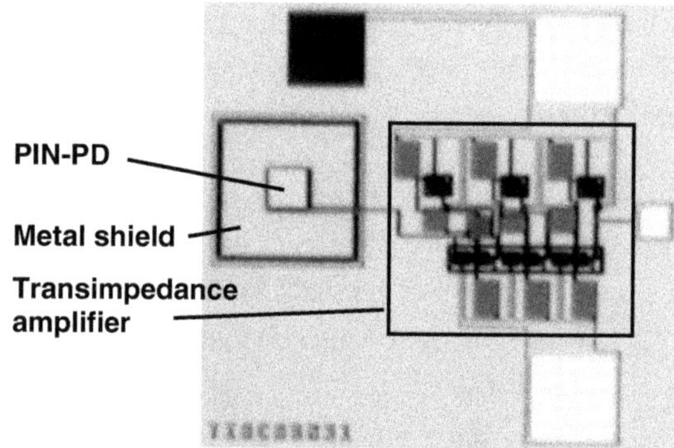

Fig. 6.115. Microphotograph of a high-speed CMOS receiver OEIC [326]

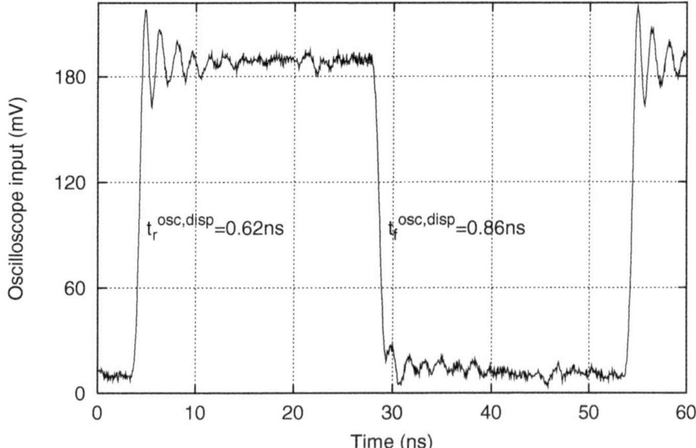

Fig. 6.116. Waveform at the output of the CMOS preamplifier OEIC with a possible data rate of 622 Mb/s for application in a fiber and interconnect receiver [85]

The photodiode, together with its metal shield around, covers an area of approximately $150 \times 150\,\mu\text{m}^2$ (see Fig. 6.115). The preamplifier occupies an active area of less than $190 \times 200\,\mu\text{m}^2$.

For $\lambda = 638$ nm, values of 0.62 ns and 0.86 ns for the rise and fall times, respectively, at the output of the preamplifier (Fig. 6.116) were obtained for a concentration of $2 \times 10^{13}\,\text{cm}^{-3}$ in the epitaxial layer, where the depletion region spreads through the whole epitaxial layer already with a supply voltage of 3.3 V [77].

The correction of the $t_{\text{r/f}}$ values for the laser and picoprobe rise and fall times results in $t_{\text{r}}^{\text{OEIC}} = 0.53$ ns and $t_{\text{f}}^{\text{OEIC}} = 0.69$ ns. These values indicate that CMOS OEICs with a reduced doping concentration in the epitaxial layer having an appropriate output buffer can be used as receivers for optical data transmission via fibers or for optical interconnects on a board level up to a bit rate BR of 622 Mb/s in the non-return-to-zero (NRZ) mode, verified by a measured eye diagram with a pseudo-random bit sequence (PRBS) of $2^{23} - 1$ (Fig. 6.117).

An redesigned OEIC implementing the PIN photodiode with an area of $2700\,\mu\text{m}^2$ and the amplifier of Fig. 6.113 in 1.0 μm CMOS technology with an active area of $190 \times 200\,\mu\text{m}^2$ showed a measured sensitivity of 10.4 mV/μW for 638 nm and 10.9 mV/μW for 850 nm. With the responsivity 0.24 A/W of the photodiode without ARC at 638 nm, we obtain an overall transimpedance of 43.2 kΩ for the amplifier of Fig. 6.113 compared to a calculated transimpedance of 34.1 kΩ [320]. The higher measured value can be explained by process tolerances of the resistors' sheet resistance and of the transistor para-

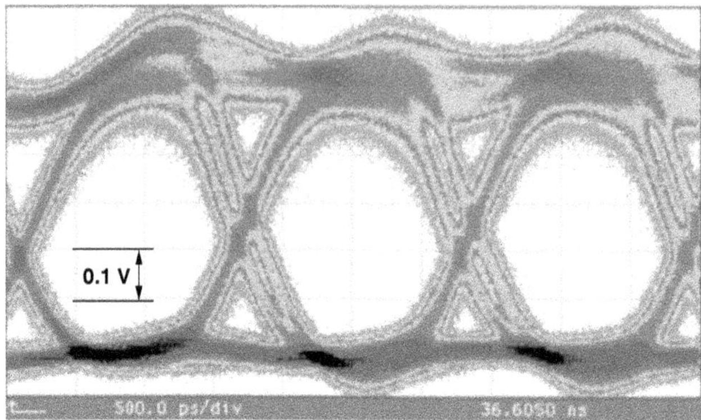

Fig. 6.117. Measured eye diagram of a CMOS preamplifier OEIC with a possible data rate of 622 Mb/s for application in a fiber and interconnect receiver (time: 500 ps/div, amplitude: 0.1 V/div) [77]

meters. The transient response of this PIN OEIC with VDD = 2.5 V is shown in Fig. 6.118.

The measured rise and fall times of the PIN OEIC with the amplifier shown in Fig. 6.113 and with a supply voltage of 2.5 V are 0.89 ns and 1.14 ns, respectively.

A sensitivity of -14.0 dBm for a data rate of 622 Mb/s at a wavelength of 638 nm for a BER of 10^{-9} was determined by measurements for the PIN OEIC with the amplifier shown in Fig. 6.113 working with a supply voltage of 3.3 V (Fig. 6.119). For 850 nm, the sensitivity was -12.3 dBm for the same

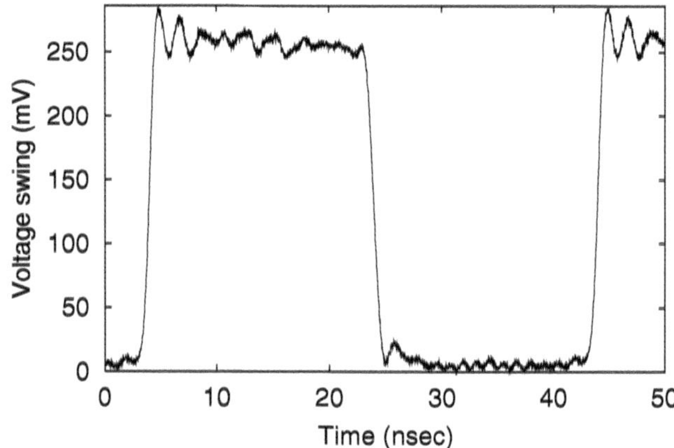

Fig. 6.118. Transient response of the PIN OEIC with the amplifier of Fig. 6.113 at VDD = 2.5 V ($\lambda = 638$ nm, $C_e = 2 \times 10^{13}$ cm^{-3}) [320]

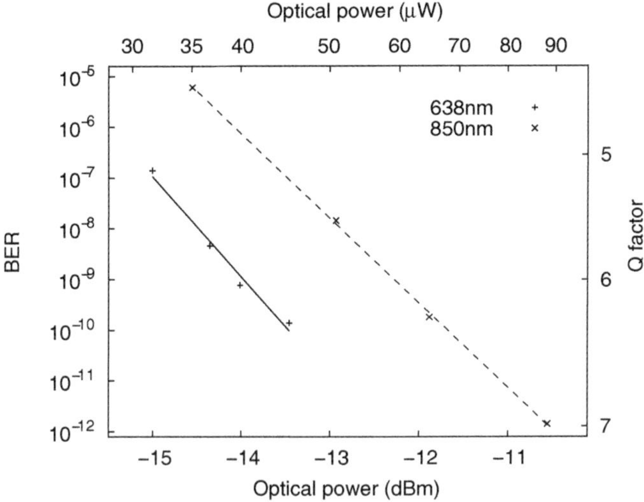

Fig. 6.119. Bit error rate of the PIN OEIC with the amplifier of Fig. 6.113 at VDD = 3.3 V in dependence on the optical input power with $C_e = 2 \times 10^{13}\,\mathrm{cm}^{-3}$ at 622 Mb/s [320]

supply voltage, data rate, and BER. The power consumption of the redesigned amplifier (Fig. 6.113) in 1.0 μm technology was 42 mW for a supply voltage of 5 V reducing to 15 mW for 3.3 V and 8 mW for 2.5 V. The sensitivity, however, increased to -10.3 dBm for a data rate of 622 Mb/s, for 638 nm, and for the supply voltage of 2.5 V.

OEICs in a 1.0 μm technology implementing the amplifier according to Fig. 6.113, however, with the double photodiode, i. e. $C_e \approx 1 \times 10^{15}\,\mathrm{cm}^{-3}$, also were investigated [320]. A sensitivity of about -8.4 dBm for a BER of 10^{-9} was found for a data rate of 622 Mb/s at 638 nm with a supply voltage of 5 V [320]. For 850 nm, the time response of the DPD was too slow for data rates of 622 Mb/s and 531 Mb/s. Even at 311 Mb/s, a BER of 10^{-9} could not be achieved for a maximum optical power of 180 μW at 850 nm [320].

OEICs in a 0.8 μm technology implementing the double photodiode and an amplifier according to Fig. 6.120 also were fabricated and characterized. The measured rise and fall times of these OEICs with the double photodiode and the standard doping concentration $C_e \approx 1 \times 10^{15}\,\mathrm{cm}^{-3}$ in the epitaxial layer were 1.21 ns and 1.14 ns, respectively, for 638 nm and for a supply voltage of 5 V. A data rate of 622 Mb/s at 638 nm was possible with this OEIC using a standard doping concentration $C_e \approx 1 \times 10^{15}\,\mathrm{cm}^{-3}$ in the epitaxial layer. With a reduced doping concentration $C_e \approx 2 \times 10^{13}\,\mathrm{cm}^{-3}$ in the epitaxial layer, i. e. with the double-pin photodiode a BER of 10^{-9} could not be achieved for 850 nm at a data rate of 622 Mb/s due to limited laser power. At a data rate of 531 Mb/s, however, a Q factor of 6.85 corresponding to a BER of 3.77×10^{-12} for an optical input power of about 180 μW was determined for 850 nm [320].

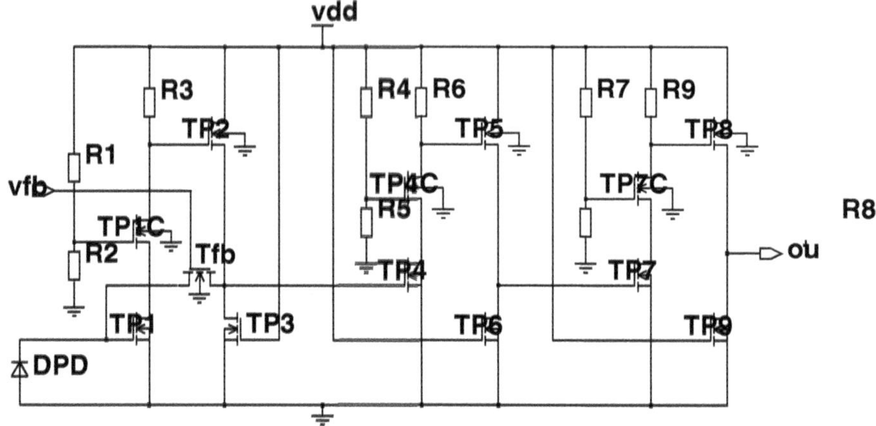

Fig. 6.120. Circuit schematic of the transimpedance amplifier with cascode and double photodiode in 0.8 μm-BiCMOS (p-substrate)

For the application in optical interconnects, OEICs implementing the double or double-pin photodiode were processed in a modular 0.8 μm BiCMOS technology, however only devices also present in the corresponding 0.8 μm CMOS technology were implemented [327]. The circuit diagram of the interconnect receiver OEICs is shown in Fig. 6.121.

Only N-channel MOSFETs with minimum channel lengths (L = 0.8 μm) were used in order to achieve a high data rate. The NMOS transistors M1–M5 operate in common source configuration (Fig. 6.121). The input stage with transistor M1 in transimpedance configuration converts changes in the photocurrent of the photodiode to voltage changes. The feedback resistor in the input stage is formed by the NMOS transistor Mfb. Two further stages with two transistors each are used as voltage amplifiers. The active feedback

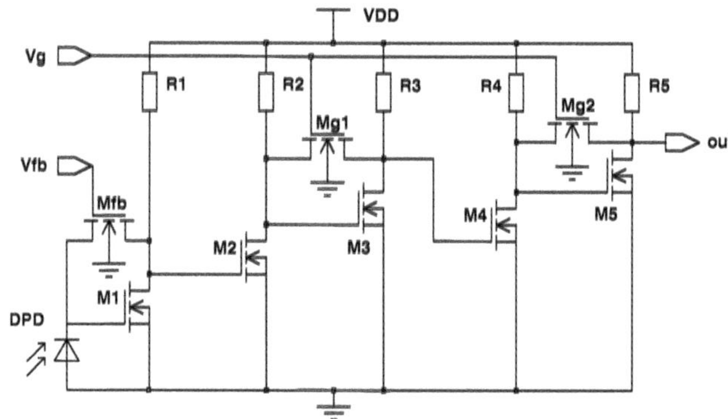

Fig. 6.121. Circuit diagram of interconnect receiver OEIC [327]

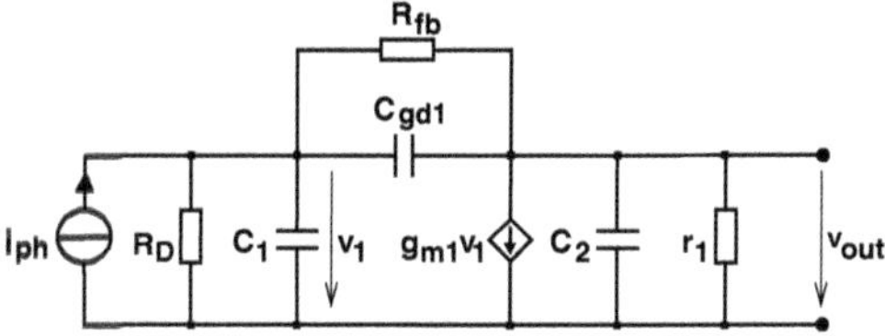

Fig. 6.122. Small-signal equivalent circuit of the first amplifier stage [327]

resistor in each of the two stages (Mg1, Mg2) limits the gain and therefore enhances the bandwidth.

In the following, expressions for the transimpedance and for the poles of the circuit will be presented. For that purpose the circuit is divided into three stages. The first stage is an inverter with NMOS transistor M1 and with resistor load R1. The feedback device Mfb with the value Rfb is connected between drain and gate of M1. The small-signal transimpedance can be calculated for the small-signal equivalent circuit shown in Fig. 6.122.

For a large value of the space-charge resistance R_D, which is justified for the integrated photodiodes with low leakage currents (for small-sized integrated silicon photodidoes: $R_\mathrm{D} > 10^{10}\ \Omega$), the result for the low-frequency small-signal transimpedance of the first stage is [327, 328]:

$$\frac{v_\mathrm{out1}}{i_\mathrm{pd}} = \frac{1 - g_\mathrm{m1} R_\mathrm{fb}}{\frac{1}{r1} + g_\mathrm{m1}} \quad , \tag{6.88}$$

with

$$r_1 = \frac{1}{g_\mathrm{ds1} + \frac{1}{R1}} \quad .$$

For $g_\mathrm{m1} R_\mathrm{fb} \gg 1$ and $g_\mathrm{m1} \gg \frac{1}{r1}$, the transimpedance corresponds to the negative value of the feedback resistor R_fb. The value of the feedback resistor R_fb determines the ideal transimpedance of the input stage in a first approximation.

A calculation of the dominant pole, which can be assumed according to simulations, leads to [327, 328]:

$$\omega_\mathrm{p1} = \frac{1 + g_\mathrm{m1} r_1}{R_\mathrm{fb}\left[C_1(1 + \frac{r1}{Rfb}) + C_\mathrm{gd1}(1 + g_\mathrm{m1} r_1) + \frac{C2\,1}{Rfb}\right]} \quad , \tag{6.89}$$

for the first stage with

$$C_1 = C_\mathrm{D} + C_\mathrm{gs1} + C_\mathrm{sb,Mfb} + C_\mathrm{gs,Mfb} \quad \text{and}$$
$$C_2 = C_\mathrm{db1} + C_\mathrm{L1} + C_\mathrm{db,Mfb} + C_\mathrm{gd,Mfb} \quad ,$$

where C_D is the space-charge capacitance of the integrated photodiode and C_L1 is the input capacitance of the second amplifier stage. From this expression for the dominant pole it can be seen easily that a larger value of R_fb for a larger transimpedance results in a lower bandwidth as one expects.

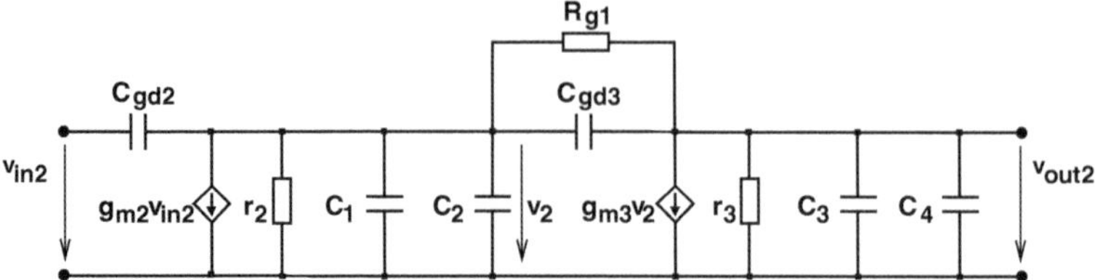

Fig. 6.123. Small-signal equivalent circuit of the second amplifier stage [327]

The second and third amplifier stages are constructed identically. They consist of two NMOS-resistor inverters each with a resistive feedback from the output of the second inverter to the input of the first inverter. The feedback resistors M_{g1} and M_{g2} with the values R_{g1} and R_{g2}, respectively, determine the voltage gain as well as the bandwidth of the second and third stage. Larger values of R_{g1} and R_{g2} result in a larger gain but also in a lower bandwidth. The choice of R_{g1} and R_{g2} by setting V_{g}, therefore, is a trade-off between a high bandwidth and a high transimpedance of the interconnect receiver OEIC. The low-frequency small-signal voltage gain A_{v2} of the second stage can be derived from the small-signal equivalent circuit shown in Fig. 6.123.

The following result is obtained [328]:

$$A_{\mathrm{v2}} = \frac{v_{\mathrm{out2}}}{v_{\mathrm{in2}}} = \frac{g_{\mathrm{m2}}(R_{\mathrm{g1}} g_{\mathrm{m3}} - 1)}{g_{\mathrm{m3}} + \frac{1}{r_2} + \frac{1}{r_3} + \frac{R_{\mathrm{g1}}}{r_2 r_3}} \approx g_{\mathrm{m2}} \left(R_{\mathrm{g1}} - \frac{1}{g_{\mathrm{m3}}} \right) \tag{6.90}$$

with

$$r_2 = \frac{1}{g_{\mathrm{ds2}} + \frac{1}{R_2}} \quad \text{and} \quad r_3 = \frac{1}{g_{\mathrm{ds3}} + \frac{1}{R_3}} \quad . \tag{6.91}$$

Here, R_{g1} and g_{m2} are the major factors influencing the voltage gain. The transconductance g_{m3} of transistor M3, however, also should be made large in order to increase the voltage gain. The voltage gain of the third stage can be calculated correspondingly. The total or effective transimpedance of the receiver OEIC is the product of the transimpedance of the first stage and the voltage gains of the second and third stage. A value of 65 kΩ for the total transimpedance is estimated using Eqs. (6.88) and (6.90). The corresponding value obtained by numerical circuit simulation was 71 kΩ [328].

The small-signal equivalent circuit of the second stage also can be used to determine its bandwidth under the assumption that a dominating pole exists [327, 328]:

$$\omega_{\mathrm{p2}} = \frac{g_{\mathrm{m3}} + \frac{1}{r_2} + \frac{1}{r_3} + \frac{R_{\mathrm{g1}}}{r_2 r_3}}{\left(1 + \frac{R_{\mathrm{g1}}}{r_3}\right)(C_{\mathrm{gd2}} + C_{\mathrm{gd3}} + C_1 + C_2) + \left(1 + \frac{R_{\mathrm{g1}}}{r_2}\right)(C_{\mathrm{gd3}} + C_3 + C_4) - 2C_{\mathrm{gd3}}}\ , \tag{6.92}$$

with

$$C_1 = C_{db2} + C_{gs3} \quad ,$$
$$C_2 = C_{sb,Mg1} + C_{gs,Mg1} \quad ,$$
$$C_3 = C_{db,Mg1} + C_{gd,Mg1} \quad \text{and}$$
$$C_4 = C_{db3} + C_{L2} \quad ,$$

where C_{L2} is the input capacitance of the third stage. The pole of the third stage can be calculated accordingly. Considering all three stages, a bandwidth of about 350 MHz can be estimated [327, 328].

An OEIC with the CN-well double photodiode having a size of 2700 μm^2 and the circuit shown in Fig. 6.121 was fabricated in a 0.8 μm MOS technology. The active die area of the OEIC was $130 \times 70\,\mu m^2$ and the OEIC consumed a current of 13 mA from VDD = 5 V. The transient response of this DPD OEIC is presented in Fig. 6.124. A large voltage swing of 4 V is achieved for an optical input power of 200 μW. A large-signal sensitivity of 20 mV/μW results from these values for the output voltage swing and for the optical input power. The effective large-signal transimpedance of 44.4 kΩ for the amplifier can be obtained when we consider a realistic quantum efficiency of 90% for $\lambda = 638$ nm resulting in a photocurrent of the DPD of approximately 90 μA for an optical input power of 200 μW. This value of 44.4 kΩ is lower than the above estimated value of 65 kΩ for the over-all small-signal transimpedance. This difference can be explained by a non-linear behavior for large signals. The value of 44.4 kΩ for the large-signal over-all transimpedance, however, is a remarkably high value at the data rates achieved.

Let us note that the MOS amplifier topology is advantageous compared to bipolar amplifier topologies at high optical input power. High optical input powers can be handled without slowing down the maximum data rate since the MOS amplifier does not show saturation effects known from bipolar amplifiers. The MOS topology, therefore, allows a high dynamic range of the optical input power without a special limiter or automatic gain control circuit. The eye diagram of the interconnect DPD OEIC is shown in Fig. 6.125.

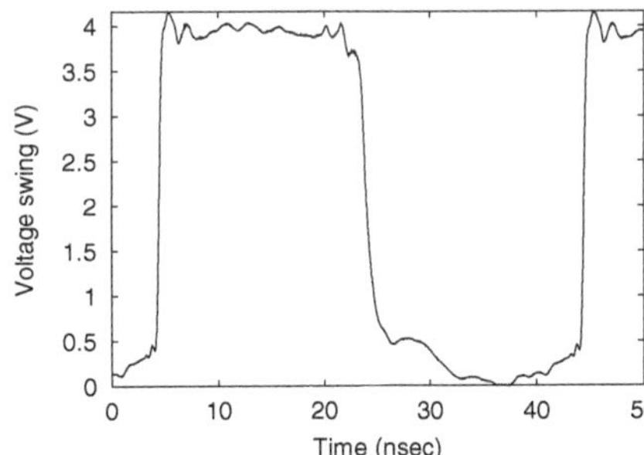

Fig. 6.124. Transient response of a DPD OEIC with $\lambda = 638$ nm, $C_e \approx 1 \times 10^{15}$ cm^{-3}, $V_{fb} = 4$ V, $V_g = 4$ V and $P_{opt} = 200\,\mu$W [327]

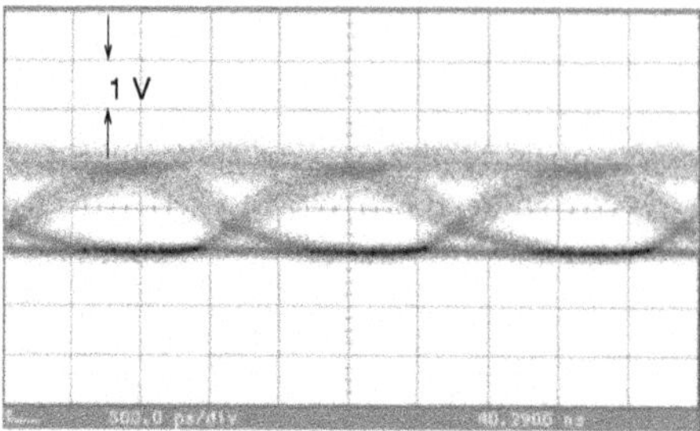

Fig. 6.125. Eye diagram of a DPD OEIC for a data rate of 622 Mbit/s with $\lambda = 638$ nm, $C_e \approx 1 \times 10^{15}$ cm^{-3}, $V_{fb} = 4$ V, $V_g = 4$ V and $P_{opt} = 200$ μW (500 ps/div, 1 V/div) [320]

The bit error rate of the DPD OEIC was determined for the NRZ data rates of 531 and 622 Mb/s in dependence on the optical input power. The results for the bit error rate (BER) are shown in Fig. 6.126. It can be seen from this diagram that the optical input power of -10.0 dBm is necessary for a bit error rate of 10^{-9} at the data rate of 622 Mb/s. At 531 Mb/s, the sensitivity of the DPD OEIC is -11.2 dBm for the same bit error rate. These values are very good results for an OEIC in a standard MOS technology with the standard doping concentration in the epitaxial layer of the order of 10^{15} cm^{-3}.

A PIN CMOS OEIC in 1.0 μm technology with improved properties has been presented [329] whereby the same PIN photodiode as in [77] has been used. For $\lambda = 850$ nm and for the standard epitaxial layer doping $C_e \approx 10^{15}$ cm^{-3} with $|V_{PD}| = 3.0$ V, rise and fall times of 11.5 ns and 11.7 ns, respectively, were measured for the CMOS-integrated PIN photodiode. These large values for t_r and t_f are due to the slow diffusion of photogenerated carriers in the standard epitaxial layer of the photodiode. The corresponding values for $C_e = 2 \times 10^{13}$ cm^{-3}, where the depletion region spreads through the whole epitaxial layer and carrier diffusion in the photodiode is eliminated, were 0.55 ns and 0.67 ns, respectively, for $\lambda = 850$ nm [329]. These values for the rise and fall times were not corrected for the rise and fall times of the 850 nm light source and of the picoprobe.

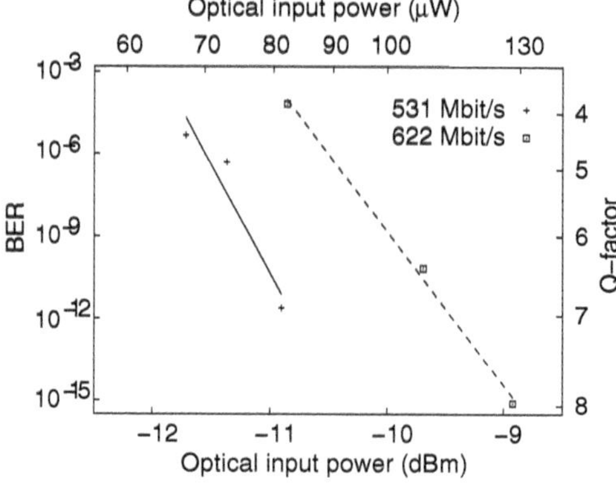

Fig. 6.126. Bit error rate (BER) of the DPD OEIC in dependence on the optical input power with $\lambda = 638$ nm and $C_e \approx 1 \times 10^{15}$ cm^{-3} ($V_{fb} = 4$ V, $V_g = 4$ V) [320]

The quantum efficiency for 638 nm was 47 % and 44 % for 850 nm without antireflection coating. A special antireflection coating (ARC) on the PIN photodiodes was optimized for $\lambda = 638.3$ nm resulting in a quantum efficiency of 94 % for this wavelength [77]. The responsivity at 850 nm measured for the PIN photodiodes with this ARC is 0.49 A/W corresponding to a quantum efficiency of 71.8 % [329]. A thicker antireflection coating optimized for 850 nm, however, will result in a larger quantum efficiency according to optical simulations. The quantum efficiency of the PIN photodiodes for $\lambda = 638.3$ nm would then be reduced of course [329].

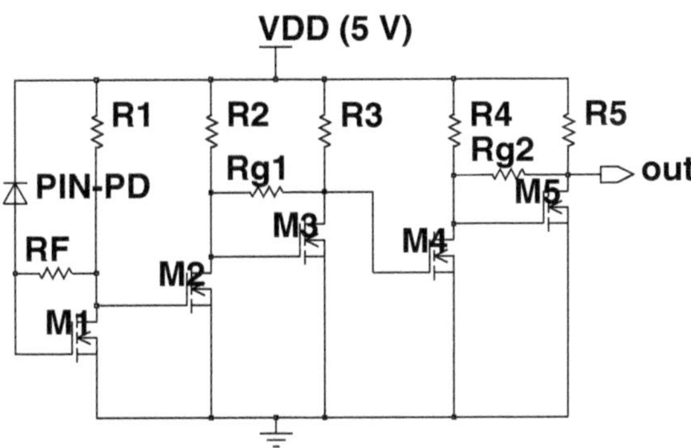

Fig. 6.127. Circuit diagram of a CMOS preamplifier OEIC with a possible data rate of 1 Gb/s for application in a fiber and interconnect receiver [329]

The high-speed preamplifier circuit with an integrated PIN photodiode is shown in Fig. 6.127. Only n-channel MOSFETs with minimum channel lengths are used in order to achieve a high bandwidth. The transistors M1–M5 operate in common-source configuration. The input stage with M1 in transimpedance configuration converts changes in the photocurrent of the photodiode to voltage changes. The feedback resistor RF in the input stage is formed by gate polysilicon. Two further stages with two transistors each are used as voltage amplifiers. The polysilicon feedback resistor in each of the two stages (Rg1, Rg2) limits the gain and therefore boosts the bandwidth. A bandwidth of 480 MHz was simulated for the OEIC. Due to the feedback resistors, a good independence from process deviations within the relatively large specified process tolerances of the used digital CMOS process is obtained. The sensitivity of the PIN preamplifier OEIC was 13.8 mV/μW without ARC, which corresponds to an overall transimpedance of 45.9 kΩ. Compared to [77], the sensitivity of the new OEIC has been increased by a factor of more than 2.5. Its power consumption was 22.5 mW at 5.0 V. The photodiode had an area of 2700 μm^2. The preamplifier occupied an active area of $245 \times 140\,\mu m^2$ due to the large area necessary for the resistors formed by low-resistivity (gate) polysilicon.

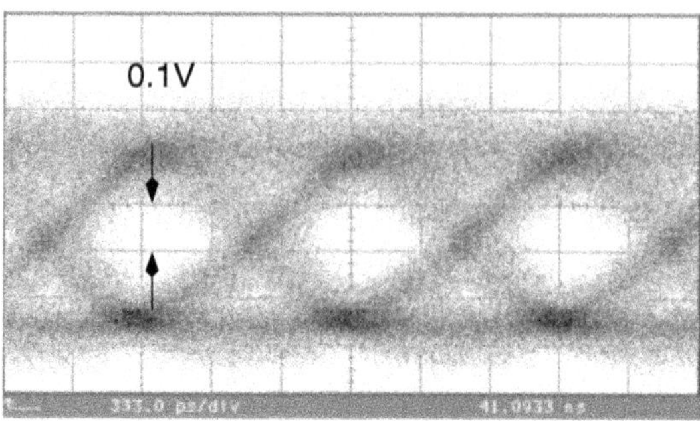

Fig. 6.128. Measured eye diagram of a 1.0 µm CMOS preamplifier OEIC for the application in a fiber and interconnect receiver (time: 333 ps/div, amplitude: 0.1 V/div) [329]

The rise and fall times measured at the OEIC output with the picoprobe were 0.78 ns and 0.76 ns, respectively, with a supply voltage of 5.0 V. The eye diagram measured at the OEIC output is shown in Fig. 6.128 for an NRZ data rate of 1 Gb/s with a PRBS of $2^{23}-1$. The higher data rate of this OEIC is due to less transistors in the preamplifier than in the typical high-frequency circuit with cascode transistors and source followers described in Ref. [77] and shown in Fig. 6.113. In Ref. [77], 12 NMOS transistors were used in the preamplifier, which therefore contained more parasitic capacitances. Here, the number of parasitic capacitances is considerably lower. In addition, the gain per amplifying transistor is lower here, allowing for a higher bandwidth of the new preamplifier.

The so-called Q-factor is defined as $(B_1 - B_0)/(\sigma_1 + \sigma_0)$ [330], where B_1 and B_0 are the mean values and σ_1 and σ_0 are the standard deviations of the output signal for a '1' and '0'. B_1, B_0, σ_1, and σ_0 can be determined from eye diagram measurements. The bit error rate (BER) then is available from BER $\approx \exp(-Q^2/2)/Q/\sqrt{2\pi}$ [330]. The BER of the OEIC has been determined with an HP54750/51 digital sampling oscilloscope for the NRZ data rates of 622 Mb/s and 1 Gb/s in dependence on the optical input power [329]. The results for the bit error rate are shown in Fig. 6.129. From this diagram for a $\lambda = 638$ nm being especially interesting for data transmission over plastic optical fibers and for optical interconnects, it can be seen that the optical input power of -15.4 dBm is necessary for a BER of 10^{-9} at a data rate of 1 Gb/s. At 622 Mb/s, the sensitivity of the OEIC for the same BER is -17.2 dBm. The sensitivity for a BER of 10^{-10} at a data rate of 1 Gb/s is -14.8 dBm. The sensitivity for a BER of 10^{-11} at a data rate of 622 Mb/s is -16.5 dBm. For $\lambda = 850$ nm, a sensitivity of -15.3 dBm for a BER of 10^{-9} at a data rate of 622 Mb/s has been verified. These values are very good results for an OEIC in a 1.0 µm CMOS technology.

The overall transimpedance of 46 kΩ and the data rate of 1 Gb/s result in a 46 TbitΩ/s transimpedance data rate product. With the simulated bandwidth

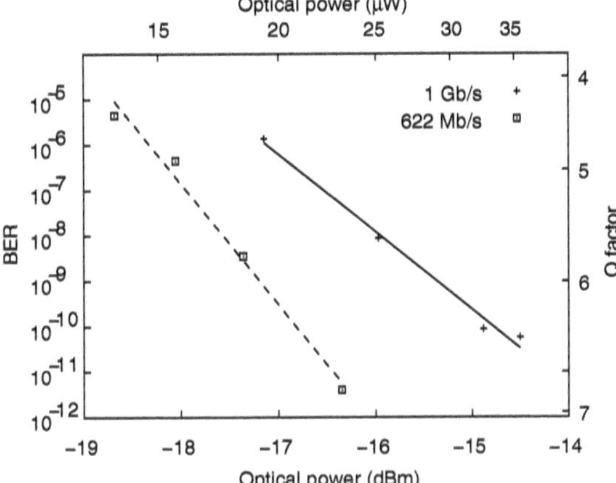

Fig. 6.129. Bit error rate in dependence on the optical input power of a 1.0 μm CMOS preamplifier OEIC for the application in a fiber and interconnect receiver [329]

of 480 MHz a 23 THzΩ effective transimpedance-bandwidth product results which exceeds the value of 18 THzΩ reported in [309].

Table 6.9 summarizes the sensitivity results for the PIN CMOS OEIC. The results of an OEIC with a similar circuit and a double photodiode in a 0.8 μm CMOS technology [327] are included for comparison. Active feedback resistors for flexible adjustment of the sensitivity were implemented in this DPD OEIC instead of polysilicon feedback resistors (see Fig. 6.121). The DPD OEIC fabricated in a standard MOS technology achieved a data rate of 622 Mb/s at a quite good sensitivity value of -10 dBm for 638 nm [327]. A sensitivity of -7.6 dBm for 850 nm at a data rate of 531 Mb/s was verified for the DPD OEIC with a reduced epitaxial doping concentration $C_e = 2 \times 10^{13}\,\text{cm}^{-3}$. It shall be mentioned that this sensitivity value for 850 nm can be improved

Table 6.9. Comparison of the sensitivity of a PIN OEIC and of a DPD OEIC for a BER of 10^{-9} (^{a}: $C_e \approx 1 \times 10^{15}\,\text{cm}^{-3}$, ^{b}: $C_e = 2 \times 10^{13}\,\text{cm}^{-3}$)

λ	PIN OEIC	DPD OEIC
	without ARC	with ARC
(nm)	(dBm @ Mb/s)	(dBm @ Mb/s)
638	-15.4 @ 1000 ^{b}	-10.0 @ 622 ^{a}
638	-17.2 @ 622 ^{b}	-11.2 @ 531 ^{a}
850	-15.3 @ 622 ^{b}	-7.6 @ 531 ^{b}

by 2–3 dB by optimizing the thickness of the antireflection coating for the wavelength of 850 nm. The sensitivity at 638 nm then would be reduced of course. It also has to noted that the PIN CMOS OEIC was optimized whereas the DPD OEIC has not been redesigned.

The sensitivity of the PIN CMOS OEIC for a data rate of 1 Gb/s and a BER of 10^{-9} with a value of -15.4 dBm for 638 nm exceeds the sensitivity of -6.3 dBm of a 0.35 μm OEIC for 850 nm [69] by a factor of 8. Compared to an OEIC with a lateral PIN photodiode [314] the sensitivity was increased by a factor of 4. It shall be mentioned that the implementation of an antireflection coating will improve the sensitivity of the OEICs by a value of 2–3 dBm. The minimum sensitivity of -17 dBm specified in the Gigabit Ethernet networking standard, therefore, should be achievable with PIN CMOS OEICs [329].

The data rate of the PIN CMOS OEICs has been limited by the amplifier in a 1.0 μm technology. With sub-micrometer PIN-CMOS-OEICs, however, data rates in excess of 1 Gb/s are possible.

An OEIC in a 130 nm unmodified CMOS process flow on SOI wafers with a 2 μm thick Si film was reported recently [188]. The resistivity of the Si film was larger than 10 Ωcm. Since the thickness of the SOI film was larger than the SOI film thickness of 100 nm used for the SOI-CMOS process, the transistors essentially had the performance of bulk devices. The P^+ and N^+ drain/source regions were used for the anode and cathode stripes of lateral PIN photodiodes in the 2 μm thick Si film. Due to the only 2 μm thick Si film, the external quantum efficieny was about 10 % for a wavelength of 850 nm [188]. The total area of the photodiodes was 2500 μm^2, whereby the electrode width and spacing was varied. The electrodes of the lateral PIN photodiode were salicided to reduce their series resistance. The integration process was performed to the first copper level [188].

The NMOS transistor had a saturation transconductance of 550 mS/mm for a width/length ratio of 1/0.175 and a threshold voltage of 0.330 V. The cutoff frequency f_T of this transistor was 60 GHz. The P-channel transistor had a cutoff frequency of 40 GHz [188]. These values make them attractive for Gb/s-electronics.

The circuit diagram of the CMOS transimpedance amplifier used in the SOI OEIC is shown in Fig. 6.130. A PMOS load M_{P1} is used for the NMOS transistor M_1 in the transimpedance input stage. The output voltage of this input stage is amplified by a second stage with M_2 and M_{P2}. A diode load (M_3) is used for this stage to limit the gain and boost the bandwidth. A source follower (M_4) with its current source M_{NB} is used at the output. A 1 GHz design and a 3 GHz design was made for the circuit shown in Fig. 6.130 [188]. The 1 GHz design achieved a sensitivity of -19 dBm at 1 Gb/s for a PRBS of $2^7 - 1$. The 3 GHz design achieved maximum sensitivities of -16.6, -15.4, -10.9, 0.9, and 2 dBm at 2, 3.125, 5, 6, and 8 Gb/s, respectively, for a BER of 10^{-9} with a wavelength of 850 nm. It was stated that the receiver characterisitics was measured with a circuit supply voltage VDD of 3.2 V [188].

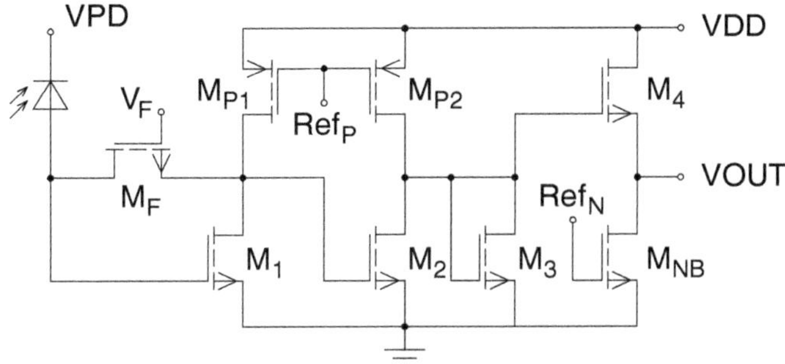

Fig. 6.130. Circuit diagram of a 130 nm CMOS OEIC [188]

The photodetector bias was 24 V for the mentioned performance [188]. It was reported that single supply 3 V operation was possible up to 3.125 Gb/s. A penalty of 0.5 dB was incurred for a PRBS of $2^{31} - 1$ [188]. The total dissipated power was 10 mW for the 1 GHz design and 35 mW for the 3 GHz design [188].

6.4.11 Receiver for Wireless Infrared Communication

Certain applications as for instance wireless infrared data communication require rejection of photocurrents generated by ambient light. Variable gain transimpedance amplifiers are also required for this application. In [331] a preamplifier with a 70 MHz bandwidth, a dynamic range of 77 dB, and a maximum transimpedance gain of 19 kΩ was presented, whereby the bandwidth was controlled within a factor of two over a 31-dB variation in gain. Free-space infrared (IR) wireless data links in laptop computers, computer peripherals, and digital cameras pushed new developments in optical receiver design. Such wireless communications emphasize system integration and low cost. Commercial digital CMOS consequently is preferred over higher-speed technologies such as GaAs or bipolar, especially since high speed is not required and is not possible since large-area photodiodes are required. Current Infrared Data Association (IrDA) standards support 4 Mb/s [332] and future standards for data rates of 100 Mb/s are being investigated [331]. A wide dynamic range is essential in order to accommodate variable link distances. In situations where the signal is weak, the photocurrent generated by ambient light can overwhelm the signal. IrDA standards specify a maximum ambient light level being more than 100 times larger than the minimum signal. The inherent dynamic range of fixed-transimpedance amplifiers is typically not sufficient for an IR wireless receiver.

A technique placing a variable signal attenuator before the preamplifier [333, 334] has been described. A bipolar differential pair as a current attenuator was used. Bipolar transistors are necessary in this technique because their exponential voltage-to-current characteristics ensures a well-regulated photodiode bias voltage across many orders of magnitude. Varying the preamplifier gain [335, 336] poses its own problems. Reported variable-gain transimpedance amplifiers based on traditional designs are difficult to stabilize [335, 336]. In [331] a two-stage differential transimpedance amplifier was presented whose stability depends only on the tracking of identical pairs of resistors. In contrast to the designs reported in [335] and [336], the bandwidth of this amplifier is well controlled over the entire gain range. A controlled bandwidth improves the sensitivity by rejecting out-of-band noise without additional filtering.

For rejecting ambient light at the preamplifier, there are two main alternatives. The first possibility is to place a high-pass resistance-capacitance (RC) network at the input of the preamplifier (see Fig. 6.131a, [334, 337, 338]). The high-frequency signal current i_s passes through while the dc part I_{dc} is blocked and shunted away by resistor R. This solutuion has two major drawbacks. Large on-chip resistors and capacitors are required to obtain a sufficiently low cut-off frequency and the photodiode bias varies with the ambient light intensity reducing the bandwidth for a high ambient light level.

The second possibility is to use an active feedback loop as shown in Fig. 6.131b. Peak level detection is effective [339, 340], but assumes that the average current is constant and requires a reset mechanism. Alternatively, average level detection can be applied [62]. The structure described in [331] represents another solution, where the feedback loop does not impose con-

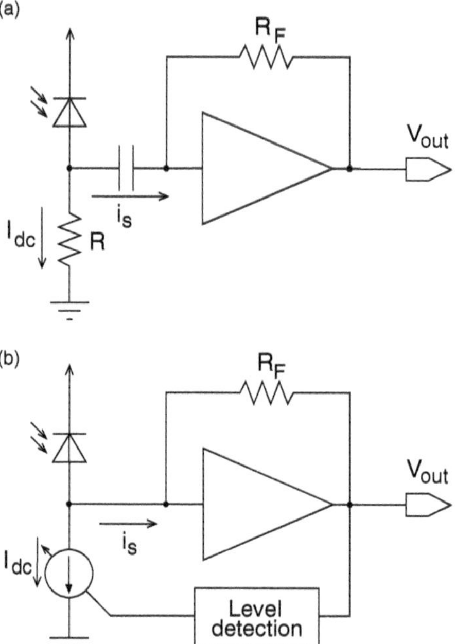

Fig. 6.131. Ambient photocurrent rejection technique with **(a)** a passive RC network and with **(b)** an active feedback loop [331]

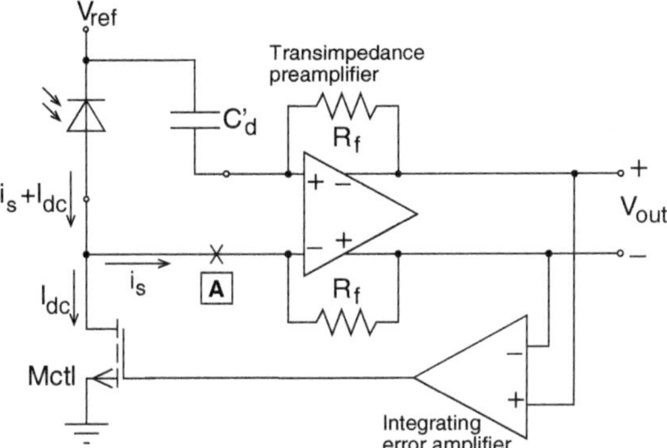

Fig. 6.132. Optical preamplifier with ambient photocurrent rejection [331]

straints on the design of the transimpedance amplifier. The ambient light-rejection circuit shown in Fig. 6.132 requires a low chip area and effectively regulates the photodiode bias voltage.

The transimpedance amplifier converts the current from the photodiode into a differential voltage. The rejection circuit formed by the error amplifier and transistor M_{ctl} opeartes as follows: The dc component of the photocurrent produces an offset in the average levels of the differential outputs. This offset is integrated over time by the error amplifier. Transistor M_{ctl} acts as a variable-current sink which eliminates the average photocurrent I_{dc} at steady state. The average photocurrent consists of the ambient photocurrent and the dc component of the signal. The differential signal path allows a high immunity to noise from the power supply and substrate. The photodiode, however, can only be connected to one terminal creating an unsymmetry at the input of the differential structure. Therefore, an additional capacitor C_{d}' is required at the other input of the transimpedance amplifier to match the photodiode capacitance and to rebalance the circuit [331]. Perfect matching ensures that noise injected at the bias voltage V_{ref} appears as a common-mode signal that is rejected effectively by the differential structure.

To study the characteristics of the ambient photocurrent rejection circuit, we can break the loop at point A in Fig. 6.132. The feedback loop is effective at low to midband frequencies where the loop gain is given by:

$$L(s) = A_{\mathrm{timp}} A_{\mathrm{err}}(s) g_{\mathrm{mctl}} \tag{6.93}$$

with A_{timp} being the passband gain of the transimpedance amplifier and g_{mctl} the transconductance of M_{ctl}. The gain of the error amplifier $A_{\mathrm{err}}(s)$ has a dominant-pole response

$$A_{\mathrm{err}}(s) = \frac{A_{\mathrm{dc}}}{1 + s/\omega_{\mathrm{p1}}} \tag{6.94}$$

with A_{dc} being the dc voltage gain, and ω_{p1} the dominant pole frequency. The closed-loop response of the preamplifier with ambient light rejection is

$$\frac{v_{out}}{i_s}(s) = \frac{A_{timp}}{1+L(s)} \approx A_{timp} \times \frac{s+\omega_{p1}}{s+A_{timp}A_{dc}\omega_{p1}g_{mctl}} \tag{6.95}$$

where the zero located at $\omega = \omega_{p1}$ is much lower in frequency than the pole at $\omega = A_{timp}A_{dc}\omega_{p1}g_{mctl}$. The preamplifier, therefore, exhibits a high-pass response with a cut-off frequency $\omega_{HP} = A_{timp}A_{dc}\omega_{p1}g_{mctl}$, a passband gain A_{timp}, and a dc gain $1/A_{dc}g_{mctl}$. g_{mctl} is proportional to $\sqrt{I_{dc}}$. Therefore, both the preamplifier's dc gain and high-pass cutoff frequency increases and the dc gain decreases with increasing I_{dc} [331].

An adequate phase margin is required for stability of the feedback loop. The error amplifier has to be compensated so that ω_{p1} is low enough for maximum values of A_{timp} and g_{mctl} [331] (see eq. 6.93). An upper limit to the transconductance g_{mctl} can be set by limiting the gate voltage of M_{ctl}. Accordingly a maximum current in M_{ctl} results, which must be larger than the current generated by the photodiode under the brightest ambient light conditions.

As mentioned above, variable-gain transimpedance amplifiers are difficult to stabilize. Figure 6.133a shows the traditional transimpedance amplifier with transistor M1 in common-source configuration and a source follower M2 to obtain a lower output impedance. Figure 6.133b shows the transimpedance amplifier proposed in [331]. The main difference is transistor M2 in common-source and in shunt-feedback configuration. This structure was proposed in [341,342] for fixed-gain preamplifiers. The three-stage amplifier in Fig. 6.133b consists of a transconductance first stage, a transimpedance second stage, and an output stage for small-signal inversion. The overall voltage gain of the amplifier is [331]

$$A = -\left(\frac{v_{out}}{i_0}\frac{i_0}{v_{in}}(-1)\right) \approx g_{m1}R_1 \quad . \tag{6.96}$$

The same result is obtained for the traditional design assuming unity gain for the output source follower.

The stability of these circuits can be investigated by breaking the feedback loops at points B and C. Figure 6.134 shows the behavior of the loop gain of both circuits.

The TIA unity-gain frequency is approximately

$$\omega_t \approx \frac{A}{R_F C_D} \approx \frac{g_{m1}R_1}{R_F C_D} \tag{6.97}$$

The stability of the amplifiers is determined by the relative position of ω_t to the non-dominant pole ω_{p2}. According to eq. 6.97, ω_t increases when R_f reduces. The non-dominant pole, however, is not significantly affected by R_F. Gain-peaking at high frequency occurs when ω_t comes too close to

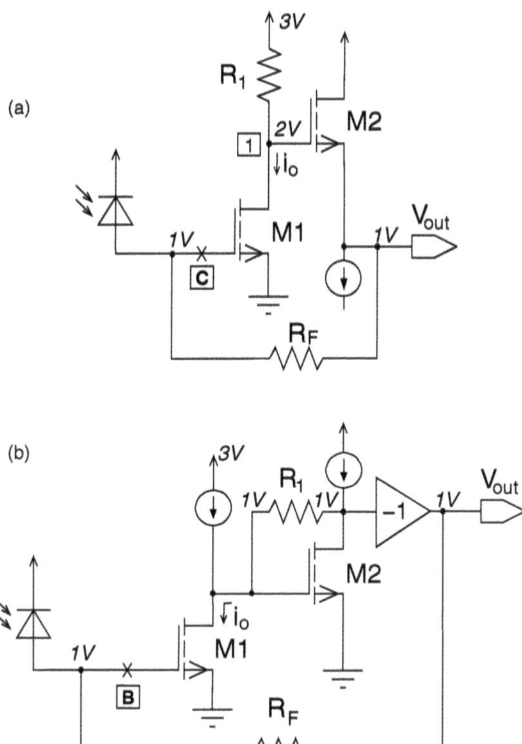

Fig. 6.133. Traditional **(a)** and in [331] proposed **(b)** transimpedance amplifier

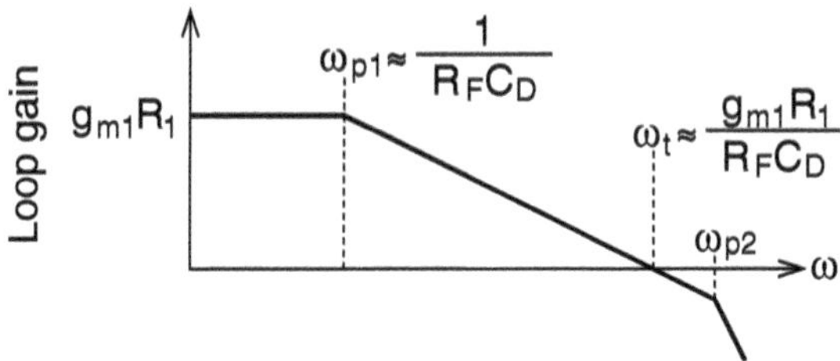

Fig. 6.134. Frequency plot of loop gain [331]

ω_{p2}. To avoid this, ω_t should track changes of R_F. Especially when resistor R_1 could track R_F (see eq. (6.97)), ω_t would remain constant and stability would be guarranteed. The bandwidth BW of the transimpdeance amplifiers is approximately

$$BW \approx \frac{1 + A}{R_F C_D} \approx \frac{g_{m1} R_1}{R_F C_D} \approx \omega_t \tag{6.98}$$

R_1 following R_F is also the condition to maintain a constant bandwidth. This tracking is difficult in the traditional circuit [331,335,336]. The structure shown in Fig. 6.133b, however, allows this tracking because R_1 is used as a shunt feedback resistor similar to R_F. Both resistors have essentially the same terminal voltages, with one terminal biased at $V_{GS1,2}$ and the other terminal connected to the output. Both, R_1 and R_F, therefore, can be realized as MOSFETs of the same type. Although R_1 and R_F can be designed to track to any fixed ratio, the design in [331] was simplified by making $R_1 = R_F$.

The optimization of the bandwidth makes a small-signal analysis necessary. The small-signal model is shown in Fig. 6.135. C_f is the shunt feedback capacitance across the gate and drain of transistor M2, C_i and C_o are the total capacitances at the internal node and output node, respectively.

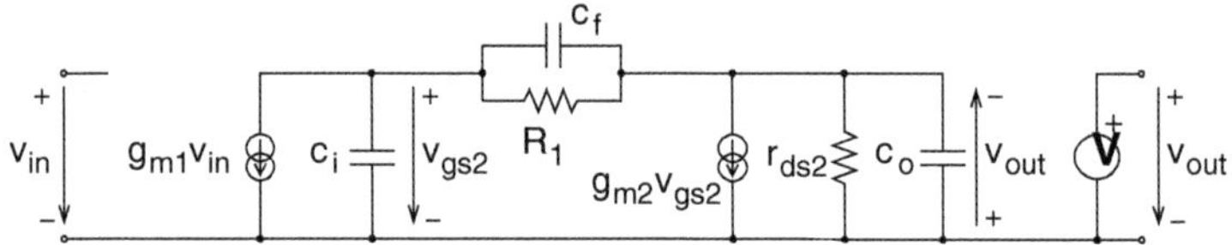

Fig. 6.135. Small-signal model of 3-stage amplifier [331]

Due to the capacitances C_i and C_o, the second stage exhibits a second order response. Assuming $g_{m2} \gg 1/r_{ds2}$ and $1/R_F$, and $C_F \ll C_o$ the frequency response is approximately [331]:

$$A(s) = \frac{v_{out}}{v_{in}}(s) \approx -\frac{g_{m1}}{C_i C_o} \frac{sC_f + (1/R_1 - g_{m2})}{s^2 + \frac{g_{m2} C_f}{C_i C_o} s + \frac{g_{m2}}{C_i C_o R_1}}. \tag{6.99}$$

By chosing C_f correctly, the poles of this transfer function can be made complex (see Fig. 6.136), leading to the following design equations:

$$\omega_0 \approx \sqrt{\frac{g_{m2}}{C_i C_o R_1}} \tag{6.100}$$

$$Q \approx \frac{1}{C_f} \sqrt{\frac{C_i C_o}{g_{m2} R_1}} \tag{6.101}$$

Commonly, a variable-gain transimpedance amplifier is optimized for a desired bandwidth at its maximum gain. At lower gain settings, only the stability of the circuit has to be confirmed. As one possible way to design the circuit in Fig. 6.133b, the start with the second stage was described in [331]. For a given bandwidth, the maximum value of R_1 can be determined from eq. (6.100) by estimating the values of g_{m2}, C_i, and C_o, based on device geometry and power dissipation, and by choosing a nondominant pole frequency ω_0 sufficiently higher than ω_t (e.g. $\omega_0 \approx 2\omega_t$) [331]. The value of C_f corresponding to a desired Q-factor (e.g. $1/\sqrt{2} = 0.71$ for a flat response [343])

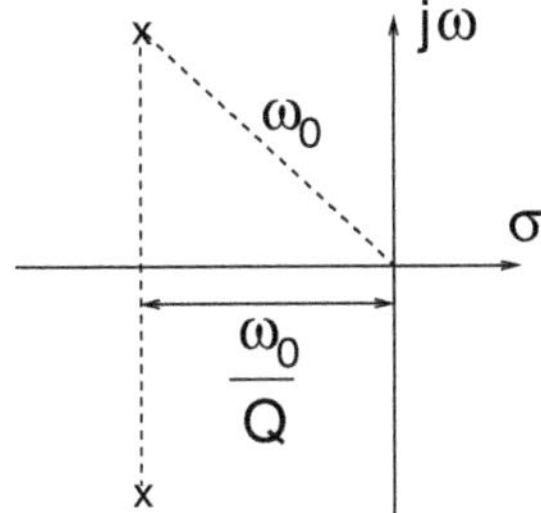

Fig. 6.136. Illustration of complex poles [331]

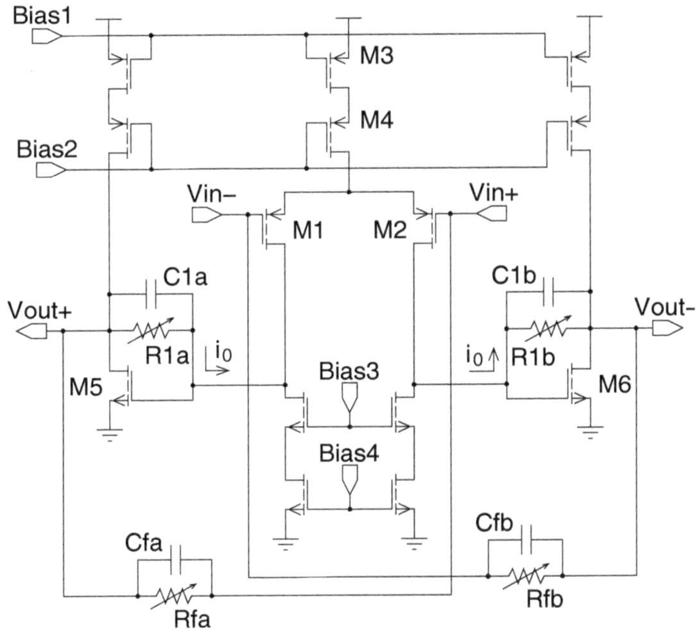

Fig. 6.137. Fully differential CMOS variable-transimpedance amplifier [331]

then can be found using (6.101). For a given photodiode capacitance C_D and estimating g_{m1}, eq. 6.97 then determines the maximum value of R_F.

Figure 6.137 shows the differential version of the variable-gain transimpedance amplifier. The transconductance first stage is here realized by a P-channel differential pair M1/M2, allowing to place the input transistors into an isolated N well to reduce substrate noise at the expense of a slightly larger thermal noise [309]. The NMOS transistors M_5 and M_6 bias each output at about 1 V. The inverting buffer shown in Fig. 6.133b in the single-ended design is eliminated in the differential version. Small-signal inversion is here achieved by simply cross coupling the differential outputs.

The dynamic range is the ratio of the noise floor of the amplifier to its maximum input current. The maximum signal current is limited to half of the tail current of the input differential pair.

A test chip with a bandwidth of 70 MHz for demonstration of a 100 Mb/s link was implemented in a 0.35 μm CMOS technology [331]. An input capacitance of 5 pF being typical for large photodiodes used in IR wireless receivers was assumed. The maximum possible photocurrent due to ambient light was 30 μA. This value corresponds to a 10 mm^2 silicon photodiode in direct sunlight. The maximum high-pass cutoff frequency was 1 MHz. A value of 1 MHz is low enough for a 100 Mb/s non-return-to-zero (NRZ) signal with a certain run-length limiting to prevent the transmission of long strings of 1's or 0's [331]. The tail current of the transimpedance amplifier's input differential pair was set to 800 μA to keep the power dissipation below 10 mW. The maximum input current, therefore, was 400 μA. The gains of the variable-transimpedance amplifier ranged from 20 kΩ (86 dBΩ) down to 500 Ω (54 dBΩ) corresponding to a gain interval of 32 dB [331].

The four variable resistors of the differential transimpedance amplifier were identical. To realize the mentioned gain range, each resistor consisted of three pairs of pass transistors (see Fig. 6.138). Three differently scaled pass transistors with three settings each were implemented, i. e. $3^3 = 27$ different resistance values were possible. To maximize the linearity, complementary N- and P-channel transistors were used, and the N wells of the P-channel transistors were tied to the outputs of the array to eliminate the back-gate effect. A speed penalty of about 40 % was mentioned in [331] due to the N-well capacitance. The transistor M_{tap} acts as a series resistor. By connecting

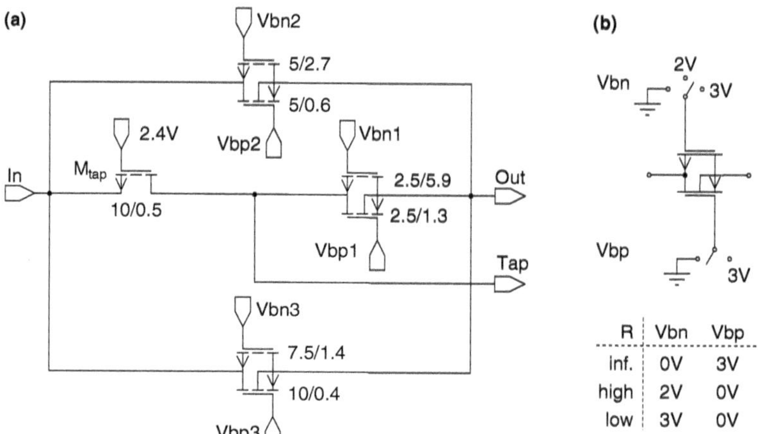

Fig. 6.138. Pass-transistor array **(a)** used to implement variable resistors and detail of pass transistor **(b)**. The W/L dimensions are given in units of micrometers [331]

the error amplifier to the middle tap terminal (see Fig. 6.138a) instead of from the output, the loop gain of the ambient photocurrent rejection circuit was reduced, thereby lowering the high-pass cutoff frequency [331].

The test chip occupied an area of 1.44 mm^2. The error amplifier was a common 2-stage CMOS OPAMP with a 5 pF internal compensation capacitor [331]. Its simulated dc gain was 94 dB and it had a dominant pole at 150 Hz. Measurements showed a maximum transimpedance gain of 19 kΩ and a gain range of 31 dB. Over the gain range, the measured bandwidth varied from 85 to only 103 MHz. A total input-referred noise current of 56 nA was present at the maximum gain. With this noise floor and a maximum input current of 400 μA, the preamplifier had a dynamic range of 77 dB (38.5 dB optical) [331]. The preamplifier consumed an electrical power of 8 mW at 3 V.

6.4.12 Hybrid Receivers

The optical absorption coefficient of GaAs and several related compounds is much higher than that of silicon in the visible and near infrared spectral range. III-V photodetectors, therefore, can be thinner and faster than silicon photodetectors with the same quantum efficiency. Consequently, III-V photodiodes hybrid integrated on silicon circuits are highly desirable. Polyimide bonding described in section 3.3 as a wafer-scale integration of photonic devices was applied to integrate GaAs PIN photodiodes with 0.8 μm CMOS transimpedance amplifier circuits [344]. The photodiode had a 3 μm thick intrinsic GaAs absorption layer realized in a mesa with 40 μm diameter. The capacitance of the hybrid integrated photodiode was 50 fF including almost no parasitics. The responsivity of this photodiode was 0.33 A/W. The dark current was smaller than 120 pA below 5 V [344].

Figure 6.139 shows the CMOS receiver circuit consisting of two amplifier stages. Two active MOS resistors are implemented as feedback elements in the transimpedance input stage. An NMOS transistor is always present.

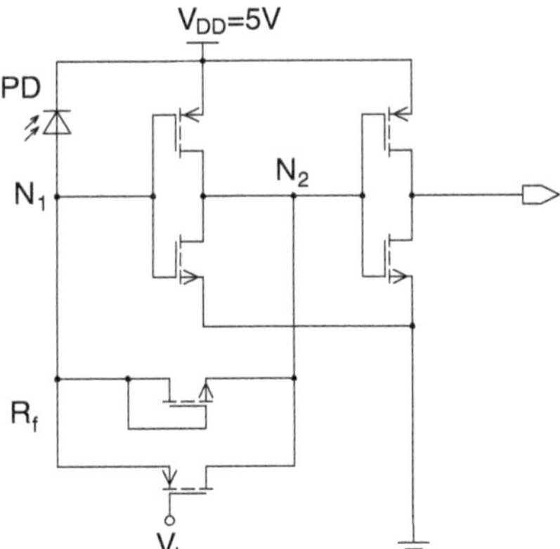

Fig. 6.139. Circuit of CMOS transimpedance receiver used in hybride OEIC with III/V photodiode [344]

The resistance of the PMOS transistor connected in parallel to the NMOS resistor can be controlled by the bias V_t. The second stage consists of an inverter as a voltage amplifier. For a feedback resistance $R_f = 5\,k\Omega$ a -3 dB bandwidth of 660 MHz was simulated for $C_D = 200\,fF$. For $C_D = 50\,fF$ a bandwidth of 1120 MHz was obtained. A reduction of the photodiode capacitance to 50 fF, which was in the range of the amplifier input capacitance, was an effective way to increase the bandwidth [344]. At the same time, the transistor widths could be chosen small for capacitive matching to C_D. In such a way, the low photodiode capacitance led to a high bandwidth and to a low power consumption. A power consumption between 7.5 and 10 mW at 5 V was measured finally.

According to the nominal value of 5 kΩ for R_f and a voltage gain of 10 for the second amplifier stage, a transimpedance gain of 95 dBΩ was achieved at the output of the polyimide bonded OEIC. A photocurrent change of 53 μA caused an output voltage swing of 3 V [344]. This results in a sensitivity large enough for chip-to-chip interconnects since VCSELs can emit even up to several milliwatts of optical power.

The receiver operated up to 800 Mb/s. At a BER of 10^{-9}, the measured sensitivities were -18.2, -16.1, -9.2, and -8.5 dBm for bit rates of 155, 311, 622, and 800 Mb/s, respectively [344].

The polyimide bonding allows a variety of device combinations. InGaAs photodiodes for wavelengths up to 1550 nm also can be hybrid bonded to silicon circuits. The same group improved their results with GaAs and InGaAs photodiodes for 850 and 1550 nm, respectively, bonded to 0.5 μm CMOS receiver circuits [345]. The GaAs and InGaAs photodiodes with a diameter of 30 μm both had an intrinsic layer thickness of 3 μm. Without antireflection coating their responsivities were 0.42 and 0.8 A/W, respectively. The dark currents were 2.5 pA and 20 nA, respectively, for a 5 V reverse bias [345].

The circuit working at a supply voltage of 3.3 V and shown in Fig. 6.140 is small and simple. The amplifier was designed with a transimpedance input stage and an inverter as voltage amplifier as well as two inverters with resistive feedback as an output buffer. This buffer had a bandwidth of more than 1 GHz, a voltage gain of 1, and an output impedance of 50 Ω [345]. The voltage supply lines were divided into four sections to avoid feeding back amplified signals to the sensitive input part through these lines. The receiver had a high sensitivity and a high bandwidth due to the direct attachment of III-V photodiodes and due to the absence of any parasitic capacitance. A bandwidth of about 1 GHz was possible for $R_f = 5\,k\Omega$ [345].

Due to the high R_f value optoelectrical conversion efficiencies of 10,900 and 15,900 V/W were obtained for 850 and 1550 nm, respectively. For the GaAs-CMOS receiver (850 nm) and a BER of 10^{-9} the sensitivities were -30.1 and -27.4 dBm for bit rates of 622 and 1000 Mb/s, respectively. Wide dynamic ranges of 22.2 and 19.5 dB were observed for the respective bit rates. For the InGaAs-CMOS receiver (1550 nm) and a BER of 10^{-9} the sensitivities were

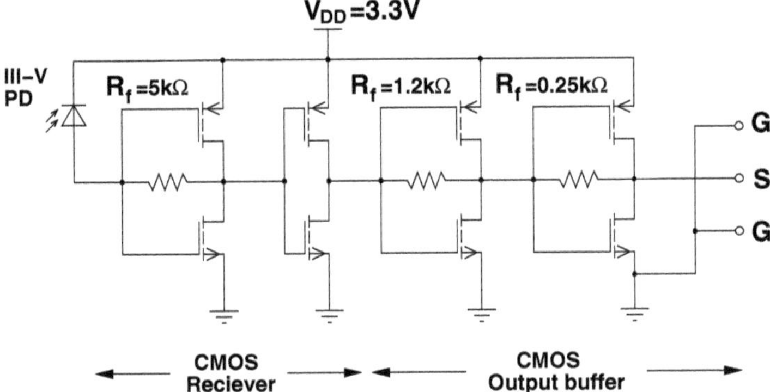

Fig. 6.140. Circuit diagram of III-V CMOS hybrid receiver [345]

-31.4 and -28 dBm with dynamic ranges of 22.4 and 18.0 dB. The receivers operated up to 1.3 Gb/s with a total power consumption below 30 mW [345].

6.4.13 Speed Enhancement Techniques

g_m Enhancement Technique

High-frequency circuits in CMOS technology generally suffer from a lower transconductance and larger parasitic capacitances compared to their HBT or BJT counterparts [346, 347]. Although higher transconductances can be obtained at the expense of a higher power consumption and larger device size, i. e. larger transistor width, the parasitic capacitances finally limit the signal bandwidth. To increase the bandwidth and the gain, i. e. the sensitivity, of a CMOS transimpedance amplifier, a g_m enhancement technique as well as an inductive peaking technique, however, can be applied. Let us have a look on the g_m enhancement technique first.

Typical high-frequency circuits adopt a common-gate input stage. For an amplifier with a common-gate input configuration, the conversion gain can be boosted by increasing the transconductance g_m of the input transistor M1. Compared to the conventional common gate input stage (Fig. 6.141a) the effective transconductance is increased by the feedback amplifier gain (-A) for the circuit shown in Fig. 6.141b.

On the other hand, the input impedance of the transimpedance amplifier with a g_m-enhanced common-gate input stage as shown in Fig. 6.141 is lowered by the factor A. This is especially advantageous for a high input node capacitance introduced by large-area photodiodes or external PIN photodiodes where also bondpad capacitances are involved. The inverting amplifier has to be designed for a unity gain bandwidth larger than that of the transimpedance amplifier for gain and bandwidth enhancement. Let us now describe the second technique applied in a 2.5 GHz CMOS transimpedance amplifier.

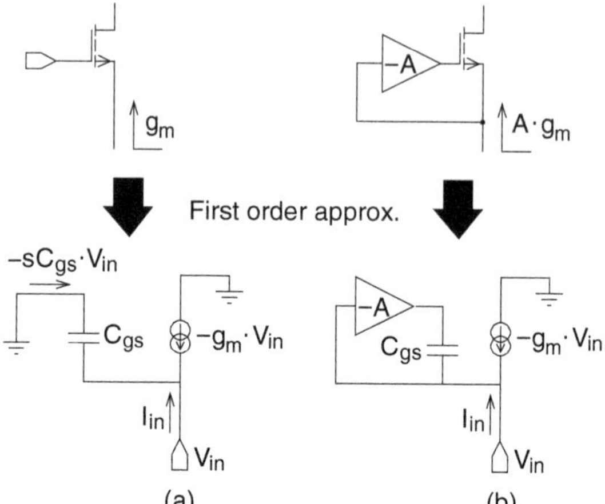

Fig. 6.141. Input impedance analysis without **(a)** and with **(b)** local feedback loop [348]

Active Inductors

Inductors instead of resistive load elements can be implemented in order to eliminate their parasitic capacitance effect and their noise contribution. In [349] spiral inductors were integrated needing a large area, however. It is much better for a small chip area to use active inductors. An active inductor can be formed by an NMOS transistor and a resistor (see Fig. 6.142).

The NMOS transistor is operated in the saturation region. The resistor can be realized by a PMOS transistor operating in the triode region. The output impedance looking from the source of the NMOS transistor can be approximated by [348]:

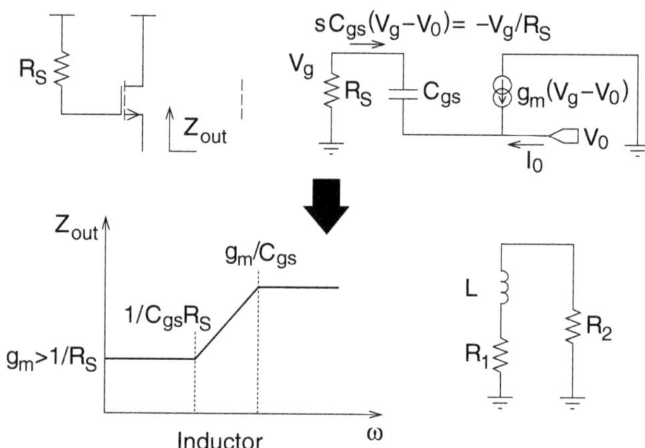

Fig. 6.142. Principle of active inductors [348]

$$Z_{\text{out}} = \frac{1}{g_{\text{m}}} \frac{1 + sR_{\text{s}}C_{\text{gs}}}{1 + s\frac{1}{g\text{m}}C_{\text{gs}}} \tag{6.102}$$

with a zero at $\omega_{\text{zero}} = -1/(R_{\text{s}}C_{\text{gs}})$ and a pole at $\omega_{\text{pole}} = -g_{\text{m}}/C_{\text{gs}}$. For Z_{out} to behave as an inductor, it is necessary that $g_{\text{m}} > 1/R_{\text{s}}$. Z_{out} can be modelled as an ideal inductor L in series with a passive resistor R_1 and with another resistor R_2 in parallel as shown in Fig. 6.142 in the inductive region $\omega_{\text{zero}} < \omega < \omega_{\text{pole}}$. It is possible to tune the inductive region and the inductance by adjusting the locations of the pole and zero. For a better gain and boosted bandwidth of a transimpedance amplifier, the inductive region was designed to cover the desired -3dB bandwidth [348].

The schematic of the transimpedance amplifier proposed in [348] combining the g_{m} enhancement technique with active inductors is shown in Fig. 6.143. R_{s1} and M_{r1} form the active inductor load of the input transistor M1 operating in common-gate configuration. M2 operates in common-source configuration with the active inductor load formed with R_{s2} and M_{r2}. A current-injection technique is also implemented. To increase the current in M1 and thereby its g_{m}, the current through transistor Mb is added to the current flowing through the active inductor load of M1. In such a way the dc voltage drop across the active inductor can be kept low and the width of transistor M_{r1} can be reduced. In such a way, the self-resonance frequency of the active inductor can be increased and the voltage headroom required by the active inductor can be reduced.

The open-loop transimpedance of the circuit shown in Fig. 6.143 is approximately [348]:

$$T_{\text{ol}}(0) = Z_1 g_{\text{m2}} Z_2 \tag{6.103}$$

where Z_1 and Z_2 are the inductances of the first and second active inductor, respectively. A value of 62.8 dBΩ was reported for T_{ol}. The common-source gain stage provided only unity gain to avoid the Miller effect [348]. The

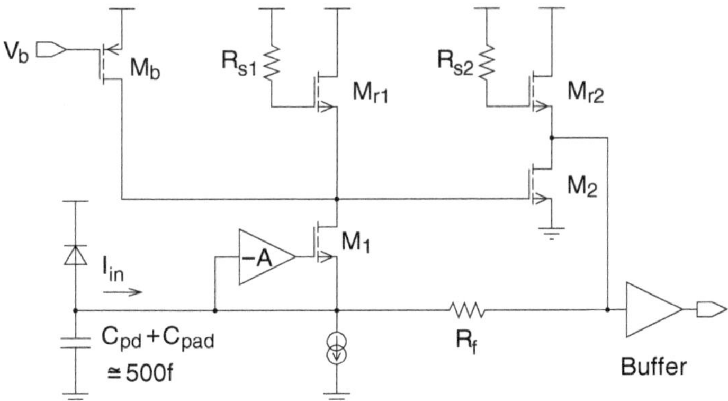

Fig. 6.143. Circuit schematic of TIA with active inductor loads [348]

feedback resistor R_f was chosen to be 3 kΩ for effective transimpedance and bandwidth optimization. The closed-loop transimpedance

$$T_{cl}(0) = \frac{T_{ol(0)}}{1 + T_{ol(0)} \frac{1}{Rf}} \tag{6.104}$$

therefore was about 60 dBΩ [348].

For comparison simulated results were given in [348] for a conventional transimpedance amplifier with resistive loads and without g_m enhancement. For such a conventional transimpedance amplifier (TIA) the dominant pole is at the source of M1 with a -3dB frequency of $f_{-3dB} = \frac{1}{2\pi} \frac{gm1}{Cin} \approx 630$ MHz. A conversion gain of 45 dBΩ belongs to this conventional TIA [348].

With current injection increasing g_{m1} a conversion gain of 54 dBΩ and a -3dB bandwidth of 1.2 GHz were achieved. When the feedback amplifier with A = 15 dB and a -3dB bandwidth of 5 GHz for g_m enhancement was added, the pole at the source of M1 became $A\frac{1}{2\pi} \frac{gm1}{Cin} \approx 3.7$ GHz [348]. This input pole is far away from the signal band of interest and the -3dB bandwidth is dominated by the pole at the input of the common-source gain stage, which is given by $f_{-3dB} = \frac{1}{2\pi} \frac{1}{Z1C1} \approx 2.6$ GHz [348], where C_1 is the parasitic capacitance at the drain of transistor M1. The effect of the active inductors when also added in the simulations was rather small. The -3dB bandwidth was extended to 2.7 GHz with a transimpedance gain of 54 dBΩ [348].

A transimpedance gain of 54.5 dBΩ and a bandwidth of 2.5 GHz were measured for the transimpedance amplifier fabricated in TSMC 0.35 μm 1P4M digital CMOS process. In this technology the core area of the transimpedance amplifier occupied only 45 × 55 μm^2. The measured RMS input noise current was 800 nA in a 2 GHz bandwidth, which is a rather high value. The power consumption was 22.5 mW with a 3 V supply [348].

Electric-Field Enhancement Technique

Optoelectronic integrated circuits usually suffer from slow carrier diffusion of photogenerated carriers in the integrated photodiodes due to light penetration depth larger than the width of the space-charge/drift region caused by doping concentrations of 10^{15} cm^{-3} and higher as well as by limited chip supply voltages. Bandwidths of many integrated photodiodes are lower than several tens of MHz [62, 64]. Base-collector photodiodes in standard-buried-collector based bipolar and BiCMOS processes achieve several hundred MHz, however, with a low quantum efficiency [171, 172, 182]. Double photodiodes perform similarly concerning speed partially with a higher responsivity [67, 327]. Integration of PIN photodiodes with low-doped intrinsic regions allows wider drift regions and high quantum efficiencies in CMOS technology with little or no additional process complexity [77, 320, 329]. In bipolar and BiCMOS processes larger process modifications are necessary for PIN photodiode integration [35, 305]. Lateral photodiodes in a SOI film with a thickness of the order of one micrometer eliminate or reduce the slow-diffusion effect, however, suffer from a low quantum efficiency [350]. Another possibility to speed up integrated photodiodes is to use a second higher supply voltage for the photodiode to increase the electric field, i. e. the drift velocity, and the width of the space-charge region. A second supply voltage of 30 or 24 V was also necessary for lateral PIN photodiodes on SOI [188,315]. Due to the thickness of the SOI layer of 2 µm the quantum efficiency was only about 10 % at 850 nm.

Many applications, especially consumer applications, however, do not allow to use a second, higher supply voltage for the photodiode due to cost reasons. With steadily reducing structure sizes of new silicon technologies towards the nanometer regime, chip supply voltages reduce steadily, leading to reduced drift velocity and increasing drift times, i. e. rise and fall times of the integrated photodiodes. Therefore, an innovative OEIC containing an integrated voltage-up-converter (VUC) to generate the high photodiode supply voltage from the circuit supply voltage on chip was suggested [351]. The VUC-OEIC, therefore, only needs one external supply voltage for the circuits and for speed enhancement of the integrated photodiode.

In [351] a PIN photodiode allowing the use of a higher reverse bias than the circuit supply voltage [35] with a large I-layer thickness was implemented leading to a quantum efficiency of 89 % for a wavelength of 670 nm and 50 % at 850 nm. Due to autodoping problems in epitaxy [35] it is difficult to achieve a low doping concentration throughout the whole intrinsic region. Here the VUC also helps to relax demands on epitaxy and to achieve a high speed of the integrated photodiodes nevertheless. The VUC actually allowed to extend the data rate for a given technology from about 50 Mb/s to 1.5 Gb/s [351].

The OEIC consisted of a PIN photodiode, an amplifier for the photocurrent and a charge pump which acts as a VUC. The photodiode had a capacitance of 160 fF at a reverse bias of 15 V and a responsivity of 0.48 A/W for 670 nm. The VUC shown in Fig. 6.144 is based on a 4-stage Dickson charge pump [352]. It consists of an inverter as a predriver which produces the necessary phase shift of 180 ° between the clock signals and uses three additional inverters. The clock signal was connected from outside the chip to allow for a high flexibility to characterize the test chip. The capacitors C0, C1, C2, and C3 had values of 10 pF. C_{G1} was equal to 20 pF. To reduce the voltage ripple of the supply VCG for the photodiode there was a filter implemented formed by $R_G = 10\,\mathrm{k}\Omega$ and $C_{G2} = 20\,\mathrm{pF}$ [351]. The photodiode was connected to the node VPD (see Fig. 6.144). The rectifier diodes were implemented as base/emitter diodes of a bipolar NPN transistor. The charge pump had a current consumption of about 2 mA at a clock frequency of 20 MHz.

Figure 6.145 shows the whole amplifier which consists of a transimpedance preamplifier with a feedback resistor $R_1 = 1.5\ \mathrm{k}\Omega$ followed by two voltage amplifier stages. This results in an overall transimpedance of 11 kΩ. To achieve a high bandwidth only NMOS transistors with minimum gate lengths (0.6 µm)

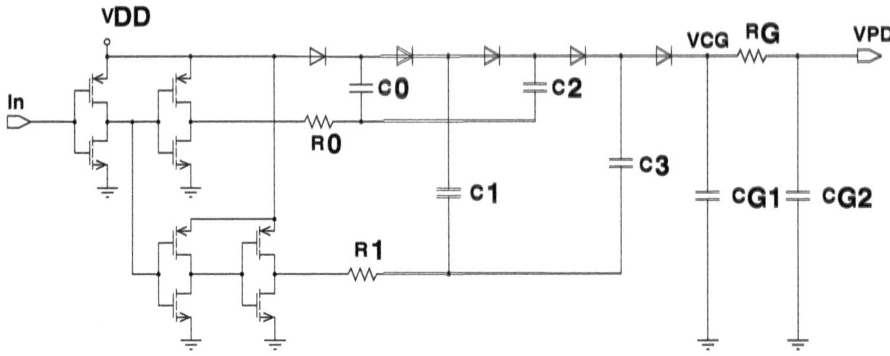

Fig. 6.144. Circuit schematic of the VUC [351]

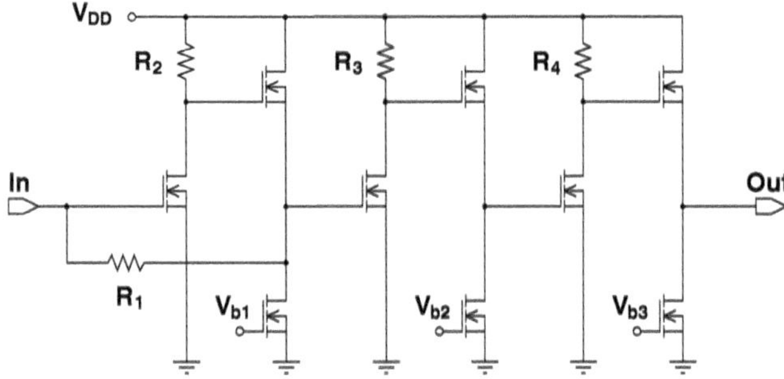

Fig. 6.145. Circuit schematic of the CMOS TIA implemented in the VUC-OEIC [351]

were used. The amplifier drew a current of 33 mA from a 5 V supply. The OEIC had separate analog and digital supplies for the amplifier and the charge pump to suppress unwanted digital noise. The power consumption of the whole chip at a supply voltage of 5 V was 175 mW. Figure 6.146 shows the micrograph of the VUC-OEIC chip with an area of 1240 μm × 960 μm. The chip was manufactured in a 0.6 μm BiCMOS process [351].

For measurements, a 670 nm laser diode was modulated via a bias-T by a HP8753E network analyzer or a digital 2.5-Gb/s ECL pattern generator. A Tektronix communication analyzer CSA8000 was used to measure eye diagrams and determine sensitivity values of the VUC OEIC. The output signal of the OEIC was measured with a 3-GHz picoprobe having an input capacitance of 100 fF.

The improvement of the frequency response due to the VUC can be seen in Fig. 6.147. With the charge pump a cutoff frequency of 655 MHz was obtained compared to only 22 MHz without VUC [351]. The rise and fall times measured at the OEIC output without VUC were 13.3 ns and 12.1 ns. With the VUC, these values reduced to 0.55 and 0.57 ns, respectively [351].

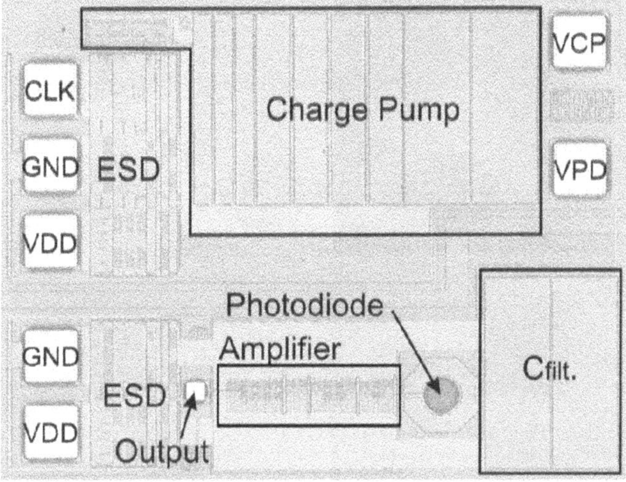

Fig. 6.146. Chipfoto of VUC-OEIC [353]

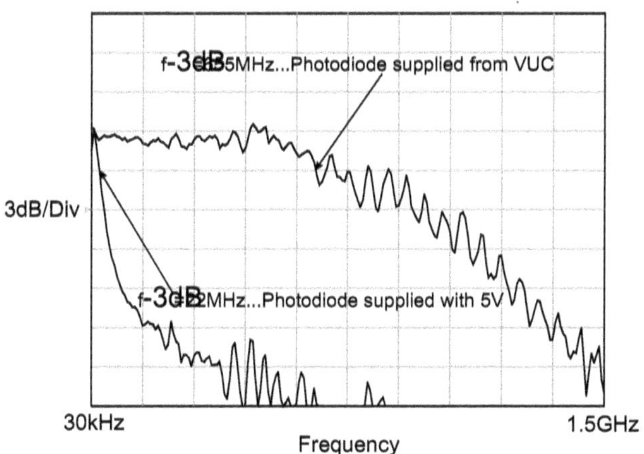

Fig. 6.147. Frequency response of the VUC-OEIC [351]

The maximum achievable data rate is 1.5 Gb/s (see Fig. 6.148) with VUC compared to only about 50 Mb/s without VUC.

Table 6.10 lists the sensitivity results of the VUC-OEIC for $\lambda = 670$ nm and a bit error rate of 10^{-9}. The chip was wire-bonded to a PCB. Both analog and digital pads were connected to the same 5 V power supply. The bias voltage of the photodiode UPD = 16.3 V was measured at the VPD pad. For comparison, for the photodiode biased with 16.3 V from an external power supply sensitivities for the OEIC of -17.5, -17.5, -17, and -16.8 dBm were found for data rates of 0.622, 1.0, 1.25, and 1.5 Gb/s, respectively [353]. These values were not corrected for about 1.5 dB due to the picoprobe noise and were

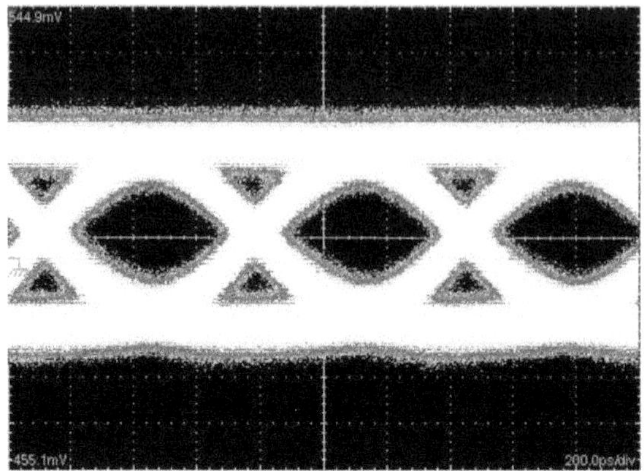

Fig. 6.148. Eye diagram of the VUC-OEIC at 1.5 Gb/s, $\lambda = 670$ nm, BER $= 10^{-9}$ [353]

Table 6.10. Comparison of sensitivity for external supply UPD=16.3V and for VUC ($\lambda = 670$ nm, BER $= 10^{-9}$) [353]

Data rate (Mb/s)	Sensitivity with UPD = 16.3V (dBm)	Sensitivity with VUC (dBm)
622	-17.5	-14.3
1000	-17.5	-14.1
1250	-17	-14.1
1500	-16.8	-13.1

limited by the non-optimized MOS amplifier. The corresponding sensitivity values of the VUC-OEIC were -14.3, -14.1, -14.1, and -13.1 dBm, indicating that the VUC reduces the sensitivity by up to about 3.5 dB. For the VUC the sensitivity for decreasing data rates remains almost constant, indicating that there was some cross-talk or substrate noise from the VUC to the amplifier or photodiode. An influence of the VUC output voltage ripple could be excluded due to the R-C-filtering of the photodiode bias. The problem is to be solved in a redesign by an improved layout to reduce coupling from VUC to substrate and from there into the photodiode [351]. The amplifier noise can also be reduced in a redesign by an optimized MOS or BiCMOS amplifier.

In another investigation, the bandwidth of the photodiode in an optoelectronic integrated circuit with a single 5 V supply was extended from 20 MHz to 771 MHz by an innovative on-chip voltage-up-converter [354]. In this approach, a shunt-regulator keeps the die area small and exploits breakdown voltages of available devices best. Rise and fall times below 0.5ns were achieved allowing operation at data rates in excess of 1 Gbit/s. A possible receiver sensitivity of -22.5 dBm was obtained for a wavelength of 670 nm.

In [354] a monolithically integrated OEIC, which incorporates the same photodiode as in [351], was presented. The slow carrier-diffusion effect in the photodiode was significantly reduced by extending the space-charge region with the help of the innovative voltage-up-converter. The OEIC in 0.6 μm BiCMOS technology implemented a new design concept consisting of a charge pump with regulator and a photodiode with a transimpedance amplifier followed by a buffer to drive the 50 Ω measurement equipment [354]. The charge pump generated the high voltage (14.5 V) needed for the supply of the photodiode from the circuit supply voltage of VDD = 5 V. It consisted of a Dickson-type [352] voltage multiplier (Fig. 6.149) with six stages and a shunt regulator at its output.

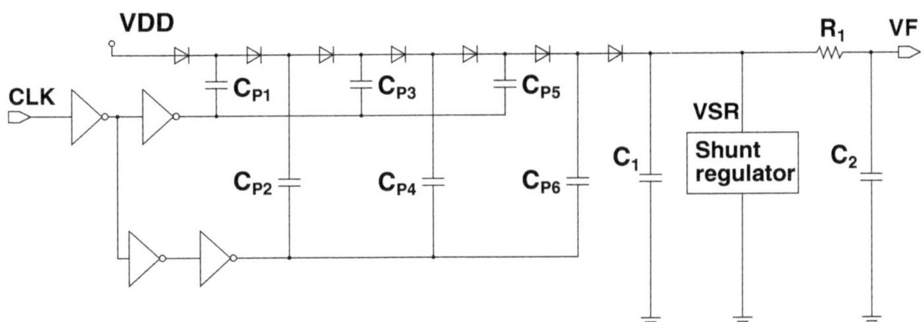

Fig. 6.149. Circuit schematic of the VUC with regulator [354]

The shunt regulator (Fig. 6.150) helps to minimize the influence of the variable load current (the photodiode's current, i. e. the photocurrent) on the output voltage in order to guarantee high photodiode speed for all light intensities and to reduce the output ripple present at C1 (5 pF). The output ripple voltage is further reduced with the help of an RC filter (R1 and C2 in Fig. 6.149). The clock signal could be in the frequency range from 25 MHz to 100 MHz coming from an external source [354]. At 50 MHz, the pump could deliver 100 μA.

With the help of several inverting stages the pump signal and an inverted pump signal is generated from the clock (CLK) signal. The clock is applied externally for flexibility of the test chip. Without regulator, the value of the unloaded output voltage of the charge pump VSR would be about two times the value of the regulated voltage (14.5 V) [354]. Under this condition, the maximum output power for a given clock frequency and a fixed pump-capacitor area is available. A shunt regulator was chosen to keep the output voltage at the lower fixed level. The unloaded voltage condition for the charge pump never occurs (as it would with a series regulator). The current difference between the maximum deliverable charge-pump current and the photocurrent is flowing through the shunt regulator. Therefore, the pump itself has always the same load, independent of photocurrent, and has constant power consumption, too.

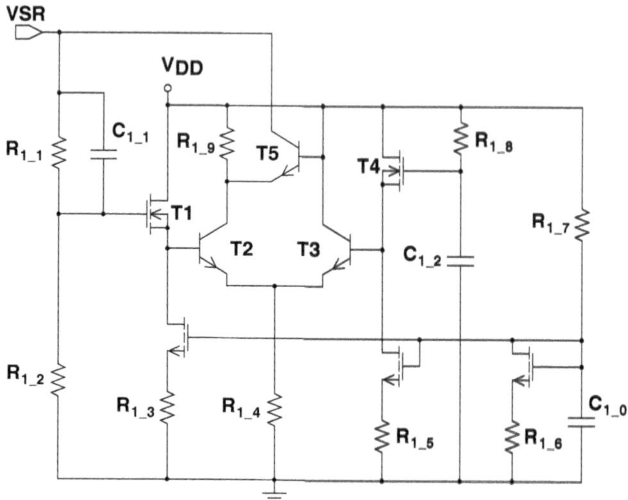

Fig. 6.150. Circuit schematic of the shunt regulator [354]

Keeping the free running (but never occuring) voltage of the charge pump higher and using more stages has a further advantage: The maximum reverse voltage the rectifier diodes have to withstand is lowered. In this case, the reverse voltage across the diodes is about 5 V. The diodes were realized with bipolar transistors. With the low reverse voltage it was possible to use the more efficient base-emitter diode instead of the collector-base diode, which is slower according to simulations. In such a way, the high frequencies mentioned above were applicable. This reduced the pump capacitors C_{P1}–C_{P6} to 1.2 pF only (compared to 10 pF in [351]) and, therefore, saved chip area [354].

As mentioned before, the shunt regulator keeps the pump output voltage fixed allowing low breakdown voltages of the poly-poly capacitors in the charge pump. Therefore, shunt regulators are preferred against series regulators. Fig. 6.150 shows the regulator used. It consists of a FET-input bipolar differential stage for high gain and a high voltage cascode transistor T5 with a biasing resistor R1_9 in parallel. The FET-input stage is necessary because of the high-resistance voltage divider needed. The regulator was designed to sink at least 1 mA and takes its reference directly from the supply voltage VDD [354]. The RC filter consisting of R1_8 and C1_2 serves to smoothen the reference voltage. Due to its structure and due to C1_1, fast responses to load current changes are accomplished.

The receiver shown in Fig. 6.151 consists of a photodiode (PD) with a light-sensitive diameter of 50 μm followed by a transimpedance amplifier formed by T1, R3, T3, T5, C1, and R1. Together with a dummy amplifier formed by T2, R4, T4, T6, C2 and R2 the voltage difference is amplified with a differential amplifier (T7, T8 and R5) and buffered with two emitter followers (T9 and T10) for driving the 50 Ω output. The usage of a nearly

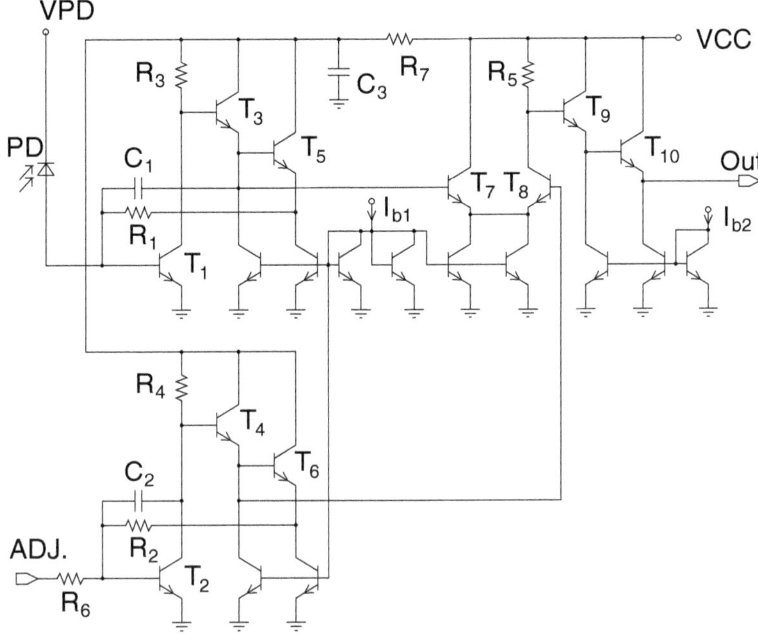

Fig. 6.151. Circuit schematic of the receiver [354]

identical dummy amplifier helps to reduce the VUC switching noise coupled through the substrate [354]. The dummy amplifier has an additional input pad (ADJ) only for evaluation purposes, in a final chip this bond pad is not necessary.

Figure 6.152 shows the resulting chip with the strictly separated charge pump (left side) and optical receiver (right side). The active diameter of the photodiode is 50 µm, but it has an overall dimension of 130 µm which includes a light shield and the connection for the cathode. The active area of the amplifier itself is only 80 µm × 160 µm. The whole chip occupies an area of 1080 µm × 870 µm.

The complete OEIC was bonded on a PCB. The output of the receiver was directly connected via a 50 Ω micro-strip line to a SMA-connector. The clock for the charge pump was externally generated from an impulse generator. For the comparison of the speed enhancement, the photodiode was first connected to 5 V and then to the regulated output of the charge pump. The photodiode supply VPD (see Fig. 6.151) was connected to the output of the charge pump VF (see Fig. 6.149) and to an external chip capacitor which acts together with the internal 10 kΩ resistor R1 as a low pass. The value of this supply blocking capacitor in parallel to the on-chip capacitor C2 (only 2 pF) was 10 nF and was realized as a small SMD-chip capacitor. The charge pump generated a voltage of 14.5 V and was clocked at a speed of 50 MHz, for which it can

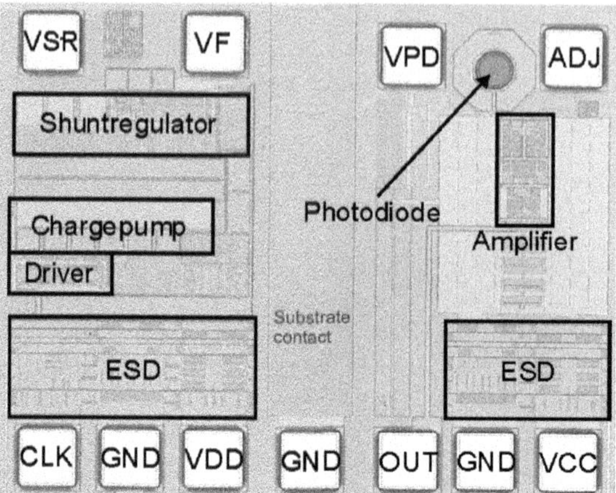

Fig. 6.152. Chipphoto of receiver with VUC and shunt regulator [354]

supply a current of 100 μA. The current consumption of the charge pump is low compared to that of the receiver and has the value of 2 mA [354].

The photodiode was stimulated by a 670 nm laser to which the test signals were fed via a bias-T. The output of the receiver was connected to a Tektronix Oscilloscope CSA8000 for the measurement of transient behavior. A 2.5 Gbit/s pattern generator modulated the laser via the bias-T. In Figs. 6.153 and 6.154 the step response can be seen for the case that the photodiode was supplied from VDD = 5 V and from the charge pump, respectively. In both figures, 256 measurements were averaged to reduce noise for rise/fall time determination. In Fig. 6.154 the asynchronous charge pump noise is also averaged over 256 measurements. The rise and fall times are reduced from above 25 ns by a factor of approximately 50 to below 0.5 ns, because the higher photo-

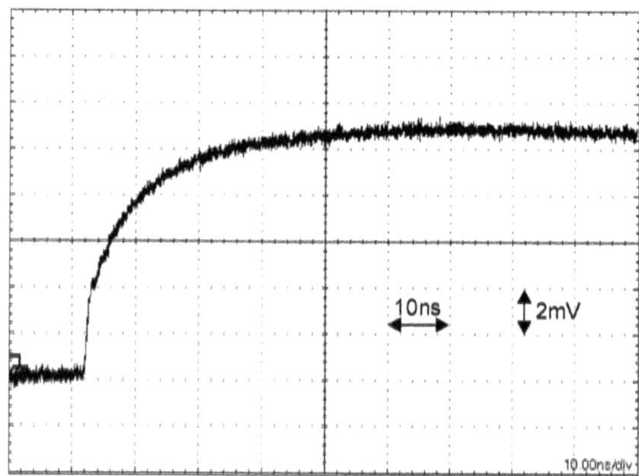

Fig. 6.153. Step response of receiver without VUC, i. e. for VPD = 5 V [354]

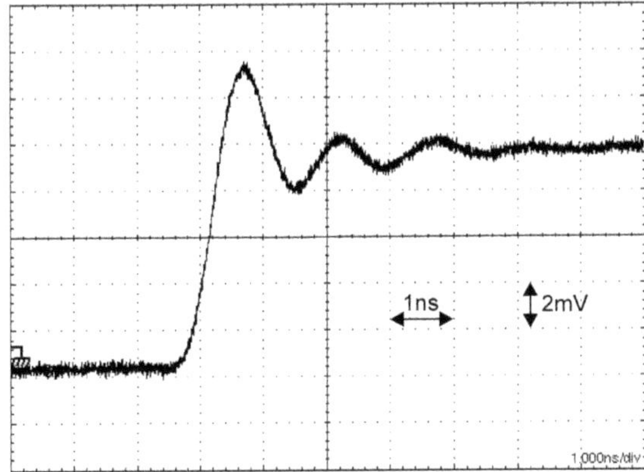

Fig. 6.154. Step response of receiver with VUC and shunt regulator (VPD = 14.5 V) [354]

diode supply voltage generated by the voltage-up-converter (VUC) increases the width of the space-charge region and thereby eliminates slow diffusion of photogenerated carriers. A HP8753 network analyzer was used to measure the frequency response of the OEIC. The bandwidth enhancement due to the voltage-up-converter is from 20 MHz to 771 MHz indicating an improvement by a factor of 38. In Table 6.11 the results are summarized.

Due to the switching function of the charge pump, a small amount of cross talk of about 12 mVpp was observed at the output of the amplifier. The receiver without charge pump produced an output noise voltage of 2.2 mVeff. Without the SMD capacitor, but with a 10 kΩ resistor between VF and VPD (activating an additional RC filter together with a 5 pF blocking capacitor integrated close to the photodiode), the cross talk signal at the output was 22 mVpp. After a redesign, i. e. the optimization of the integrated RC filter, neither the external SMD capacitor nor the additional 10 kΩ resistor will be necessary [354]. When this OEIC is used for data communication at a BER of 10^{-9}, the estimated sensitivity for ISI-free data rates is −22.5 dBm with a laser source which has an infinite extinction ratio [354]. The power penalty due to the switching noise of the charge pump can be determined by about

Table 6.11. Summary of speed enhancement ($\lambda = 670$ nm, VDD = VCC = 5 V) [354]

	OEIC with VUC (VPD = 14.5 V)	OEIC without VUC (VPD = 5 V)
Bandwidth (MHz)	771	20
Rise time (ps)	492	27500
Fall time (ps)	481	25800

1.5 dB considering that the sensitivity of the receiver without the noise of the charge pump is −24 dBm [354]. The current consumption of the receiver alone was 17 mA, which means a power dissipation of 85 mW.

Let us summarize. The OEIC described in [354] used a charge pump to improve the bandwidth of integrated photodiodes from 20 MHz to 771 MHz with a single supply of 5 V. This bandwidth is sufficient for 1.25 Gbit/s receiver operation. The charge pump produced a voltage of 14.5 V with a current capability of 100 μA at a clock frequency of 50 MHz. The estimated upper bound for the optical power penalty due to the switching noise was only 1.5 dB. After a redesign for reduction of overshoot, the sensitivity of −22.5 dBm can be obtained at a data rate of 1.25 Gbit/s [354], which is more than 8 dB better compared to [351], where no regulator was implemented.

A current consumption of 95 mW was achieved for the complete chip including charge pump, shunt regulator, and amplifier. Compared with [351] the current consumption was halved, the bandwidth higher, the sensitivity was much better and the chip area was smaller. Chip size can be reduced further in a redesign to about $920 \times 770\,\mu\text{m}^2$ by eliminating the external connections from the charge pump to the photodiode, integrating a clock generator and reducing the spacing between VUC and receiver [354].

6.4.14 Plastic-Optical-Fiber Receivers

Many applications require fast and sensitive optical sensors with a large light-sensitive area. Industrial automation and optical data transmission via plastic optical fibers (POFs) in automobiles and local area networks as well as optical interconnects are examples for such applications. Large-area sensors relax demands on mechanical adjustment and POFs have rather large core diameters of up to 1 mm. Therefore, optical sensors with large-area photodetectors are advantageous or even required. But the high junction capacitance of a photodiode with a large diameter limits both the speed and the sensitivity of transimpedance amplifiers (TIAs) connected to the photodiode (Figure 6.155a). A photoreceiver using a photodiode with a diameter of 700 μm, having a capacitance of 20 pF, achieved a maximum data rate of 20 Mb/s and a sensitivity of -20 dBm [355]. According to [356], the -3dB bandwidth of a transimpedance amplifier can be estimated as follows

$$f_{3\text{dB}} = \frac{A+1}{2\pi R_{\text{F}} C_{\text{T}}} \tag{6.105}$$

where A is the open-loop voltage gain of the inverting amplifier, R_{F} is the resistance of the feedback resistor, and C_{T} is the sum of the photodiode capacitance C_{D} and the input capacitance C_{in} of the amplifier. To achieve higher data rates, it is important to reduce the photodiode capacitance C_{D}. However, even with a low-capacitance PIN structure offering a capacitance of only $0.011\,\text{fF}/\mu\text{m}^2$ [77, 357], an octagonal photodiode with a diameter of 400 μm

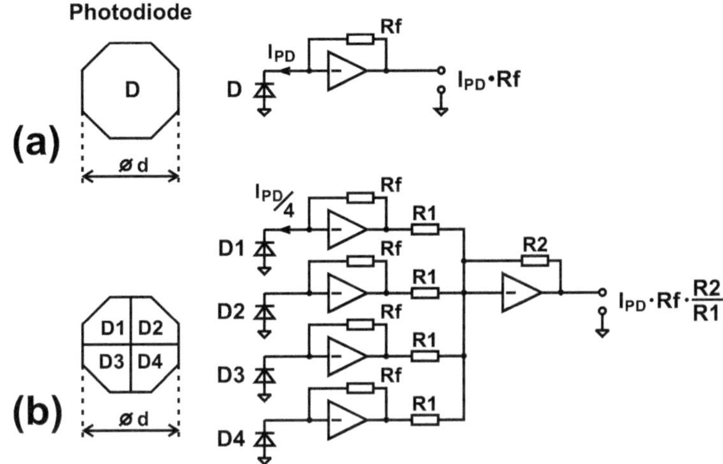

Fig. 6.155. Block diagrams of the optical sensors: transimpedance amplifier connected to a large-area photodiode **(a)** and four transimpedance amplifiers connected to four quarters of a large-area photodiode and a summation amplifier **(b)** [358]

shows a capacitance of 1.6 pF. Decreasing the resistance of the feedback resistor would increase the bandwidth at the cost of a reduced sensitivity. To further enhance the bandwidth of the optical receiver without reducing sensitivity, therefore, the photodiode was divided into four sections [358]. Each of the four-quarter photodiodes was connected to a TIA. The output voltages of the four TIAs were combined with a summation amplifier (Fig. 6.155b). The capacitance of each of the four-quarter photodiodes is only one fourth of the capacitance of the unseparated photodiode, therefore the bandwidth of the four TIAs can be almost four times (if C_{in} is small) as high as the bandwidth of the receiver shown in Fig. 6.155a.

All amplifiers shown in Figures 6.155a and b consist of three inverter stages (Fig. 6.156), because inverters are very area-efficient. The PMOS diodes P2, P4 and P6 reduce the gain of the amplifier, increase its bandwidth and ensure stability of the TIA [309]. The feedback resistor of the TIA shown in Fig. 6.155a as well as the feedback resistors of the four TIAs shown in Fig. 6.155b are formed by identical PMOS transistors P7, which have a small-signal resistance of 80 kΩ. P7 is not present in the summation amplifier. The resistors R1 and R2 are formed by polysilicon resistors. The summation amplifier (Fig. 6.155b) uses the same amplifier as the TIAs, but shows a higher bandwidth because the parasitic capacitance on the summation node between R1 and R2 is much smaller than the photodiode capacitances.

PIN photodiodes, similar to that described in [77, 357], but with an area 60 times as large were applied. An antireflection coating (ARC) was used to achieve a high photodiode quantum efficiency $\eta = 96.5\,\%$ at $\lambda = 670\,\text{nm}$. Rise times of 610 ps and fall times of 515 ps were measured for the photodiode.

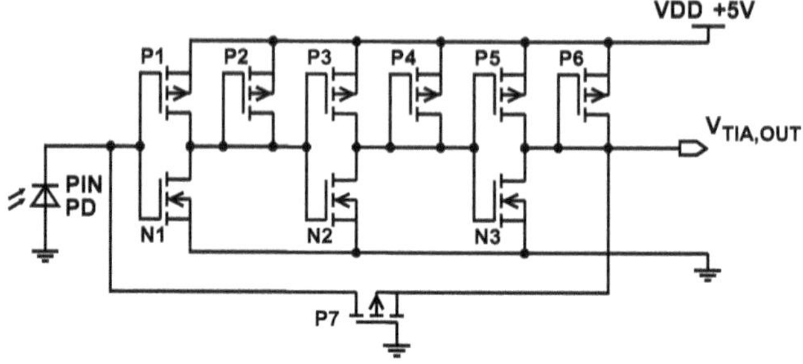

Fig. 6.156. Schematic of transimpedance amplifiers [358]

The unseparated photodiode D of the optical receiver shown in Fig. 6.155a has a diameter of 400 μm and a capacitance of 1.6 pF at a reverse voltage of 2.5 V. A second photodiode with a diameter of 400 μm was divided into four sections. These four-quarter photodiodes, forming the photodiodes D1 to D4 shown in Fig. 6.155b, have a capacitance of 415 fF at 2.5 V each [358].

The chip was fabricated using a 0.6 μm CMOS process [358]. Both optical sensors occupied a die area of $700 \times 780\,\mu m^2$. The receivers were measured on wafer level. For all measurements, the supply voltage VDD was 5 V. The light was coupled into the photodiodes via an optical fiber adjusted on a wafer prober. The light source was a laser diode with an optical wavelength of 670 nm modulated by a 2.5 Gb/s generator via a bias-tee. The output voltages were measured with a picoprobe (input capacitance 0.1 pF, bandwidth 3 GHz). The transient responses were measured with a digital storage oscilloscope Tektronix CSA8000. Figure 6.157 shows the transient response of the optical sensor with the unseparated photodiode.

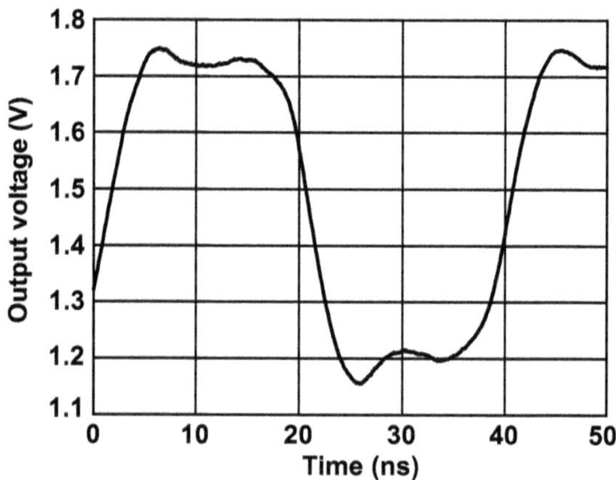

Fig. 6.157. Transient response of the optical sensor with the unseparated photodiode for $\lambda = 670$nm, $P_{opt} = 10\,\mu W$ [358]

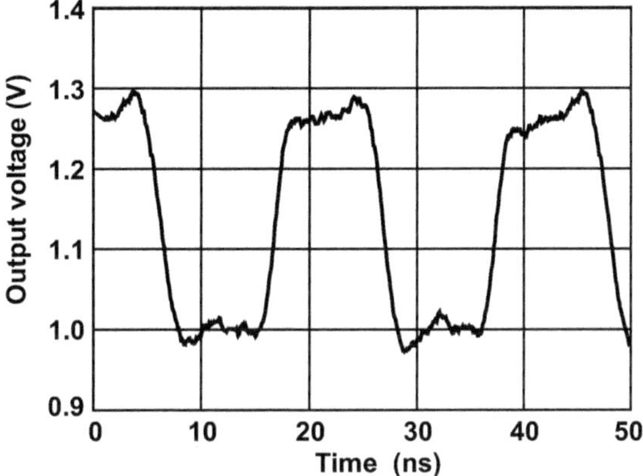

Fig. 6.158. Transient response of the optical sensor with the quartered photodiode for $\lambda = 670\text{nm}$, $P_{\text{opt}} = 2.5\,\mu\text{W}$ [358]

Rise times of 5.11 ns and fall times of 4.55 ns were measured for the receiver with the unseparated photodiode [358]. The transient response of the optical receiver with the quartered photodiode is shown in Fig. 6.158. Rise times of 1.7 ns and fall times of 1.73 ns were measured for the receiver with the quartered photodiode. This means that the sensor with the quartered photodiode is about three times faster than the receiver with the unseparated photodiode. The maximum data rate DR can be estimated from the measured rise times and fall times by $\text{DR} = 1/(t_{\text{r}} + t_{\text{f}})$. Therefore maximum data rates of 104 Mb/s for the receiver with the unseparated photodiode as well as 292 Mb/s for the receiver with the quartered photodiode can be achieved. For 300 Mb/s and a bit error rate of 10^{-9}, an average optical input power of $1.48\,\mu\text{W}$ was determined. Further results of both optical receivers are compared in Table 6.12. DC photo-sensitivities of $36.8\,\text{mV}/\mu\text{W}$ as well as $75\,\text{mV}/\mu\text{W}$ were measured. The photo-sensitivity of the receiver with the quartered photodiode was 2.04 times higher than the photo-sensitivity of the standard receiver due to the resistance ratio R2/R1, which defines the amplification of the summation amplifier (Fig. 6.155b). The simulated values of the input-referred spectral noise densities at 100 MHz are shown in Table 6.12, which could not be measured because of bad noise behavior of the picoprobe. The lower noise current density of the receiver with the quartered photodiode is the result of the lower capacitance of each of the quarter photodiodes. The power consumption of each amplifier was 5.2 mW. Since the power consumption of the output driver consisting of a source-follower stage (not shown in Fig. 6.155a and b) was 18.8 mW, the values of 24 mW and 44.7 mW resulted for the two different sensors [358].

An optical sensor using a quartered photodiode, four transimpedance amplifiers and a summation amplifier was presented. This new concept allowed speed and sensitivity enhancement of optical sensors with large-area photode-

Table 6.12. Comparison of the optical sensors with unseparated and quartered photodiode ($\lambda = 670$ nm, VDD = 5 V) [358]

	Unseparated photodiode	Quartered photodiode
Photo-sensitivity (mV/μW)	36.8	75.0
Bandwidth (MHz)	79.1	223
Rise time (ns)	5.11	1.70
Fall time (ns)	4.55	1.73
Input-referred spectral noise current density at 100 MHz ($pA/\sqrt{Hz}$)	5.35	3.29
Power consumption (mW)	24.0	44.7

tectors [358]. With a PIN photodiode of 400 μm in diameter, a POF receiver achieved a maximum data rate of 300 Mb/s with a sensitivity of -28.3 dBm compared to values of 20 Mb/s and -20 dBm, respectively, reported in [355].

6.4.15 Burst-Mode Optical Receivers

The demand for broadband access networks has been rapidly increasing in the last years. An asynchronous transfer mode passive optical network (ATM-PON) is one of the best means to satisfy this demand in terms of bandwidth and cost [359]. Figure 6.159 shows an ATM-PON system in which several Optical Network Units (ONUs) are connected to an Optical Line Terminal (OLT) through optical fiber and a 1:N star coupler. The downstream transmission is in continuous mode whereas the upstream transmission is in cell-based Time Division Multiple Access (TDMA) burst mode. In downstream direction each ONU receives packages with constant optical power. There can be large differences in received optical power between different ONUs, however, depending on the fiber length between individual ONUs and the star coupler. As a consequence of the different fiber length resulting in different attenuation of the signals sent by the ONUs, the OLT receives packages with different optical power. The sensitivity of the OLT receiver, therefore, has to be changed very fast between two cells. The name burst mode stems from the different signal height of the cells received by the OLT.

According to the global standard G.983.1 class C announced by ITU-T in 1998, the burst cells at an OLT can have a very large power difference [360]. In addition, the laser diodes are biased above threshold to achieve a high data rate and a low jitter. The receiver in an OLT, therefore, must realize high sensitivity, wide dynamic range (more than 30 dB difference between strong and weak cells), and it must receive signals with an extinction ratio as low as 10 dB.

Conventional preamplifiers, which control their transimpedance bit by bit have problems to receive large signals with a low extinction ratio be-

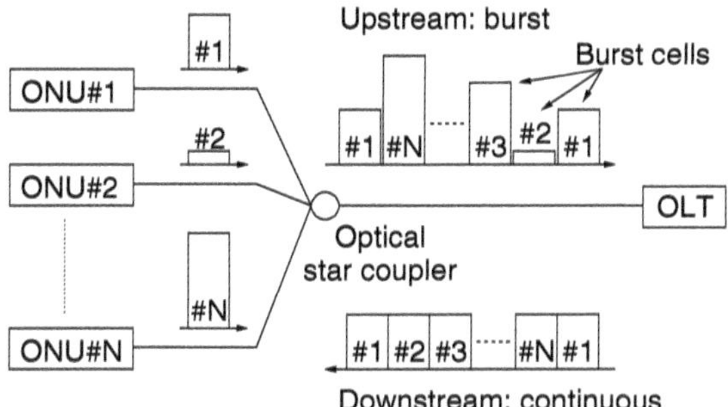

Fig. 6.159. Principle of optical burst-mode transmission [360]

cause of the nature of logarithmic amplifier operation [361–364]. In these preamplifiers, a current-bypass diode or current-bypass circuit is connected in parallel with the feedback resistor. When the output voltage of the transimpedance preamplifier exceeds the turn-on voltage of the bypass circuit, current flows through this circuit lowering the transimpedance. This type of transimpedance control was called bit-AGC (bit automatic gain control) in [360].

If a large input signal with a low extinction ratio is applied to such a bit-AGC preamplifier, the output waveform has a large dc part (see Fig. 6.160). The ac amplitude of the output signal is therefore reduced and it is difficult to decide between "0" and "1" properly. To solve the problem of output signal reduction, a novel circuit controlling the transimpedance cell by cell was proposed [360]. This type of control was called cell-AGC. The cell-AGC (see Fig. 6.161) is based on a bottom level detector (BLD), a gain control

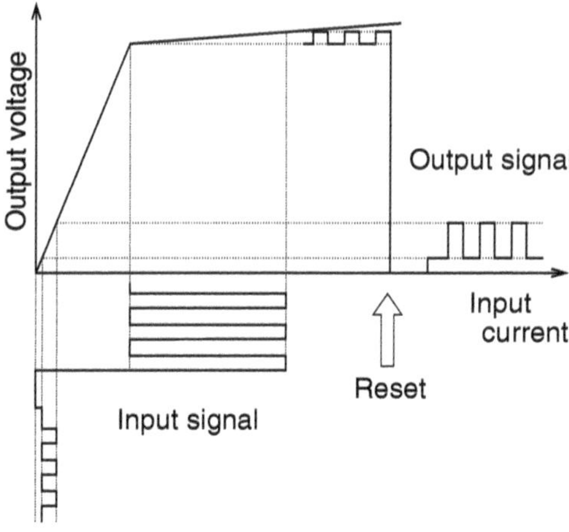

Fig. 6.160. Operation of bit-AGC [360]

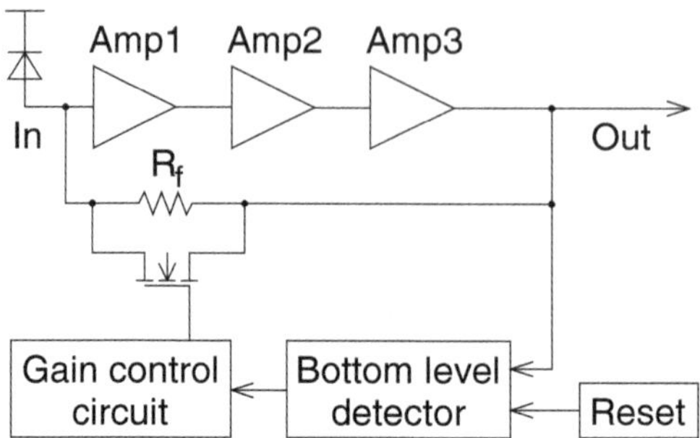

Fig. 6.161. Block diagram of 3-stage burst-mode receiver [360]

circuit (GCC), a reset circuit, and a MOSFET connected in parallel with the feedback resistor.

BLD has quickly to detect the output signal of the three-stage transimpedance amplifier. A capacitor in BLD holds this bottom level. Depending on this level, the GCC generates a constant voltage during operation in a cell and determines the resistance of the MOSFET connected in parallel to R_{f} via its gate source voltage V_{GS}.

Between two cells, the reset signal is applied to BLD, the hold capacitor is charged, and the output of GCC returns to its initial state [360]. To increase transmission quality, a fast response from BLD and GCC is necessary during the the first bit in the new cell. The response of cell-AGC to input signals with different power levels is illustrated in Fig. 6.162. For a large-signal cell, the transimpedance is set to G1 instead of to G2 for a weak-signal cell. Because the output voltage is proportional to the input current, the dc background level for the "0" level is not as large as that in bit-AGC. With the help of this

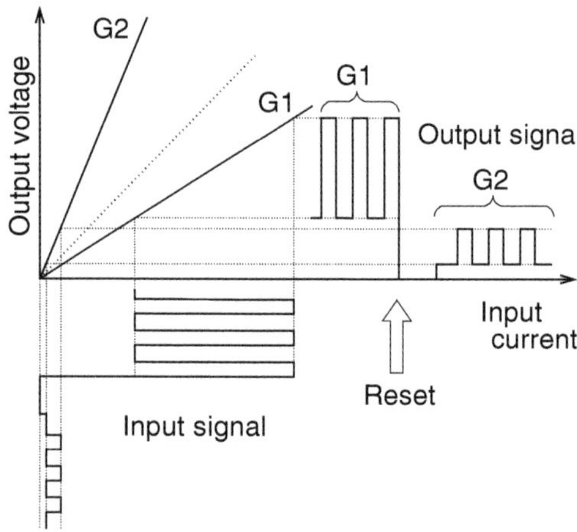

Fig. 6.162. Operation of cell-AGC [360]

curcuit, therefore, a decision between "0" and "1" can be performed properly and burst cells with a low extinction ratio can be received.

A 0.25 µm CMOS technology was selected for the realization of the cell-AGC burst-mode receiver aiming at noise reduction for a high sensitivity. A feedback resistor with a value of 40 kΩ was implemented in the trans-impedance preamplifier. The transconductance g_{m} of the input MOSFET was set to about 50 mS. Stability is the major problem to be solved for three-stage transimpedance amplifiers. Therefore, the influence of changes in g_m with process deviations on the stability has to be suppressed and the value of g_{m} has to be kept close to the minimum for a low power consumption. This requirement can be met with the series MOSFET load configuration shown in Fig. 6.163. Each of the three amplifier stages in Fig. 6.161 consists of the circuit shown in Fig. 6.163.

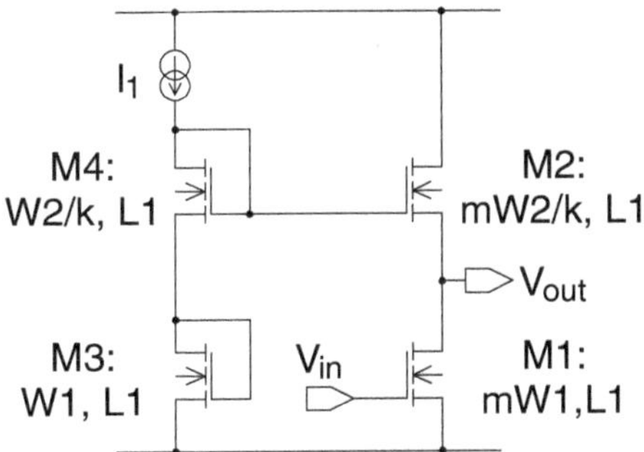

Fig. 6.163. Configuration of g_{m}-over-g_{m} amplifier stage [360]

Due to the condition of $V_{\mathrm{in}} = V_{\mathrm{out}}$ for negative feedback, V_{GS} of M1 is equal to V_{GS} of M3 and V_{GS} of M2 is equal to V_{GS} of M4. The transconductances are given by:

$$\begin{aligned} g_{\mathrm{m1}} &= \sqrt{2\mu_{\mathrm{n}} C_{\mathrm{ox}} \frac{mW_1}{L_1} I_1} \\ g_{\mathrm{m2}} &= \sqrt{2\mu_{\mathrm{n}} C_{\mathrm{ox}} \frac{mW_1}{kL_1} I_1} \end{aligned} \tag{6.106}$$

where μ_{n} is the electron mobility, C_{ox} is the gate oxide capacitance per area, I_1 is a reference current, and k is the ratio of the gate widths. The gain of this amplifier is equal to the ratio of the transconductances. This amplifier therefore is called g_{m}-over-g_{m} amplifier.

$$G = \frac{g_{\mathrm{m1}}}{g_{\mathrm{m2}}} = \sqrt{k} \tag{6.107}$$

The gain G of such an amplifier stage only depends on the parameter k. It is completely independent from process deviations and operating temperature. The current flowing through M1 is $m \times I_1$, such that the gain G

and g_m can be stabilized also under power supply deviations. As G is stabilized, the bandwidth f_{3dB} is kept stable, i. e. constant, due to the relation $f_{3dB} = \frac{1+G}{2\pi R_f C_f}$. The bandwidth — important for constant integrated noise — therefore can be designed to be stable around the optimum bandwidth.

The preamplifier in a 0.25 µm CMOS technology with an chip size of $1.3 \times 1.12\,mm^2$ was characterized at a supply voltage of $2.5\,V \pm 5\,\%$ with an external photodiode having a responsivity of 0.9 A/W ($\lambda = 1.3\,\mu m$) and a capacitance of 0.6 pF [360]. The bandwidth varied from 122 to 164 MHz in the operating temperature range from −40 to +85 °C and also due to remaining R_f and C_f deviations. The power consumption was 60–71.4 mW at 2.5 V supply voltage. A sensitivity of −39.3 dBm for a BER of 10^{-10} at a data rate of 156 Mb/s was reported [360]. The maximum optical input power was -6 dBm.

6.4.16 Deep-sub-µm Receivers

Optical communications is rapidly expanding into applications in short distance ranges. Low-cost applications like infrared wireless links and fiber-to-the-home (FTTH) are needed. CMOS implementations of optical receivers, therefore, are attractive with respect to costs and system size especially when systems-on-a-chip solutions are considered. But the trend towards lower supply voltages in CMOS poses new problems for the design of transimpedance amplifier front-ends. Traditionally, low-voltage operation has not been a requirement. Below 2 V the performance of conventional transimpedance amplifiers, however, is seriously degraded. One design using low-threshold devices achieved 1.2 V operation at a burst-mode data rate of 50 Mb/s [365]. In [366] an optical front-end for 1 V operation needing no special devices was described.

The difficulties in 1 V operation lie in a wide dynamic range and that the photodiode is provided with a sufficiently high reverse bias voltage in addition to maximizing the gain and the sensitivity for a given bandwidth. The wide-swing low-voltage transimpedance amplifier shown in Fig. 6.164 enables a bias voltage of 0.8 V for the photodiode [366]. The circuit consists of a sub-1 V current amplifier [367] in transimpedance configuration. It achieves a closed-loop gain approaching $-R_f$ for large current gains. The common-gate input stage with M1 sets the input voltage to $V_{bias} - V_{GS1}$ which can be adjusted and reduced to the saturation voltage of transistor M_2, V_{DSsat2}, to keep M_2 working as a constant current source. The resulting photodiode reverse voltage is $V_{DD} - V_{DSsat2}$ which is about 80 % of the 1 V supply voltage [366]. The output is biased at $V_{GS3} = V_{Th,n} + V_{DSsat3}$ and it can swing down to V_{DSsat3}. The amplifier, therefore, has a wide output swing equal to the threshold voltage $V_{Th,n}$ which was about 60 % of the 1 V supply voltage in a 0.35 µm CMOS process with a typical threshold voltage of 0.6 V for the NMOS transistor and -0.65 V for the PMOS transistor [366].

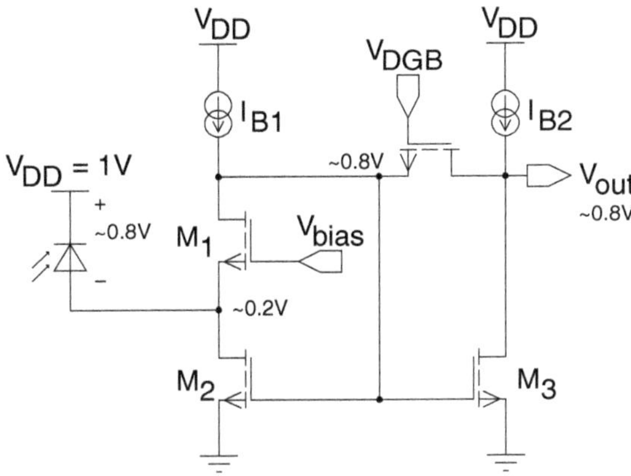

Fig. 6.164. Transimpedance amplifier for opearation at 1 V [366]

By adapting the transimpedance gain in accordance to the signal strength, the dynamic range of the front-end can be further enhanced in addition to a wide output swing. Variable-gain transimpedance amplifiers, however, have been a challenge to stabilize [331]. This problem is overcome in the design described in [366] by replacing the voltage amplifier by a current amplifier [368]. As a consequence, a constant bandwidth being ideally independent of the transimpedance gain is obtained. Out-of-band noise, therefore, is not a matter for this design. An NMOS feedback resistor instead of a PMOS device is preferred because of its lower resistance for strong photocurrents. The NMOS feedback resistor realizes soft limiting further enhancing the dynamic range [366]. The source and drain potentials of the NMOS feedback transistor lie near the supply. For a 1 V supply and without low-threshold devices, a charge pump is needed to bias the gate of the NMOS feedback transistor above the supply voltage. For a constant sensitivity of the receiver, the gate bias voltage has to be stable and free of ripple. By driving the gate directly from the output of the charge pump (called dynamic gate biasing: DGB), no current is drawn from the charge pump and the output ripple is minimized. The charge pump is shown in Fig. 6.165.

The charge pump is a voltage doubler with two unique characteristics. First, the voltage doubler has a separate input, V_{IN}, not to double spikes on the front-end supply voltage. Second, in order to overcome the limitation that the charge pump supply voltage must be at least 0.5 V above the threshold voltage [145], the standard charge pump had to be improved to take advantage of the full supply voltage when driving the switches. The improved voltage doubler implements three tightly coupled charge pumps. The inner-most charge pump consisting of the devices M_1 and M_2 generates level-shifted clock signals with the full supply swing. These chlock signals are used to drive the outer-most charge pump which performs the actual doubling of the bias voltage using the transistors M_3 and M_4. The clock signals $\Phi_{1\mathrm{Vin}}$

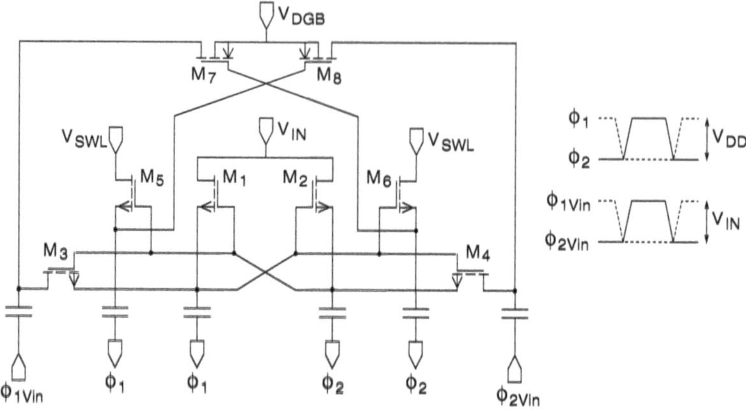

Fig. 6.165. Sub-1 V bias voltage doubler [366]

and $\Phi_{2\mathrm{Vin}}$ have a reduced voltage swing which is equal to the input voltage V_{IN}. The final charge pump with transistors M_5 and M_6 generates full-swing clock signals, but here the low level is shifted to V_{SWL} which is optimized for driving the PMOS output switches M_7 and M_8. The full-swing clock signals Φ_1 and Φ_2 were generated from an integrated, non-overlapping, two-phase clock generator.

The architecture of the optical front-end test chip fabricated in a 0.35 μm CMOS process is shown in Fig. 6.166. The transimpedance amplifier is followed by two gain stages that reuse the optimized transimpedance design in a transconductance-transimpedance topology, i. e. the output voltage of a

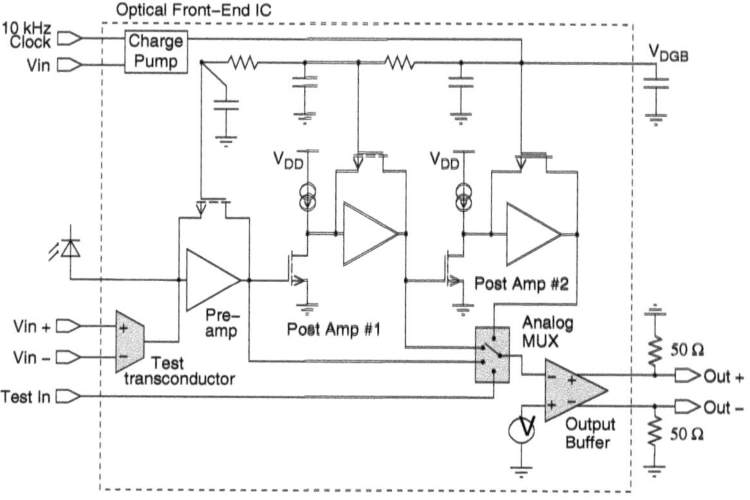

Fig. 6.166. Block diagram of 1 V optical receiver front-end [366]

transimpedance stage is converted to a current signal by an NMOS transistor in common-source configuration with a constant current load to obtain a current input signal for the following transimpedance amplifier. RC-filtering of the bias voltage V_{DGB} reduces ripple due to charge injection from the switches and isolates the gain stages in addition. In the test chip, additional circuitry as the test transconductor, an analog multiplexer, and a 50 Ω output buffer were incorporated. The preamplifier alone achieved a transimpedance gain of 2.4 kΩ with a bandwidth of 45 MHz, whereas the complete front-end provided an effective transimpedance gain of 210 kΩ with a bandwidth of 50 MHz [366]. The overall gain of the front-end was adjustable from 210 kΩ down to 19 kΩ by changing the gate bias voltage from 1.68 V to 2 V [366]. At the maximum gain the input-referred noise current density was 11 $pA/\sqrt{Hz}$. Operation at 75 Mb/s was demonstrated with a photodiode having a capacitance of 1 pF [366]. A maximum photocurrent of 40 μA was reported. The active chip area was 0.13 mm^2 and the total test chip area was 1.6 mm^2. The power dissipation was 1 mW at 1 V.

6.4.17 Comparison of Fiber Receivers

It is worthwhile to compare published results on silicon OEICs in Table 6.13. Polyimide bonding was applied to integrate GaAs PIN photodiodes with 0.8 μm CMOS transimpedance amplifier circuits resulting in data rates up to 800 Mb/s [344]. The same group achieved high sensitivities at data rates up to 1.3 Gb/s with GaAs and InGaAs photodiodes for 850 and 1550 nm, respectively, hybridized to 0.5 μm CMOS receiver circuits [345]. A hybrid receiver-transmitter circuit consisting of a 0.8 μm CMOS amplifier and of a flip-chip bonded GaAs-AlGaAs multi-quantum-well modulator, which also could be used as a PIN photodiode, has been reported to operate at 625 Mb/s [369]. For a hybrid OEIC with a GaAs photodetector and an amplifier in a 0.25 μm CMOS technology, finally a data rate of 1.25 Gb/s has been obtained [370]. A SiGe OEIC with a PIN photodiode and heterojunction bipolar transistors achieved a bandwidth of 460 MHz corresponding to a bit rate of about 690 Mb/s with a photodiode bias of 9 V [231]. A lateral PIN photodiode was used in an NMOS receiver OEIC, which operated at a bit rate of 300 Mb/s [313], when biased at VDD = 8 V and with a photodiode bias V_{PD} = 30 V. In a redesigned version, the NMOS OEIC achieved a data rate of 1 Gb/s at a sensitivity of −9.3 dBm [314]. Furthermore, a receiver OEIC on high-resistivity bulk silicon exhibited a sensitivity of −31.9 dBm at 622 Mb/s with a wavelength of 850 nm, with VDD = 3.5 V, and V_{PIN} = 30 V [315]. With a lateral PIN photodiode in a 3.0 μm thick N-type (100)-oriented SOI layer and an NMOS transimpedance amplifier with the circuit topology of Fig. 6.91 a sensitivity of -15.3 dBm at 622 Mb/s and at a BER of 10^{-9} was obtained for VDD = 3 V and V_{PIN} = 3 V with λ = 850 nm [315]. With higher supply voltages of VDD = 5 V and V_{PIN} = 5 V a sensitivity of -12.0 dBm at a data rate of 1.5 Gb/s was found. A sensitivity of -12.2 dBm finally was reported with

Table 6.13. Comparison of silicon receiver OEICs (BR = bit rate; PR = photoreceiver; PD = photodiode; *: with ARC)

Process (μm)	VDD (V)	λ (nm)	BR_{PR} ($\frac{Mb}{s}$)	S_{PR} (dBm)	V_{PD} (V)	BR_{PD} or f_{3dB}	R_{PD} ($\frac{A}{W}$)	η_{PD} (%)
0.8 CMOS-GaAs [344]	5	850	155	-18.2	2.5	-	0.33	47
0.8 CMOS-GaAs [344]	5	850	311	-16.1	2.5	-	0.33	47
0.8 CMOS-GaAs [344]	5	850	622	-9.2	2.5	-	0.33	47
0.8 CMOS-GaAs [344]	5	850	800	-8.5	2.5	-	0.33	47
0.5 CMOS-GaAs [345]	3.3	850	622	-30.1	1.7	-	0.42	60
0.5 CMOS-GaAs [345]	3.3	850	1000	-27.4	1.7	-	0.42	60
0.5 CMOS-GaAs [345]	3.3	850	1300	-18.5	1.7	-	0.42	60
0.5 CMOS-InGaAs [345]	3.3	1550	622	-31.4	1.7	-	0.8	64
0.5 CMOS-InGaAs [345]	3.3	1550	1000	-28.0	1.7	-	0.8	64
0.5 CMOS-InGaAs [345]	3.3	1550	1300	-22.8	1.7	-	0.8	64
0.8 CMOS-MQW [369]	5	850	625		2.5	-	0.5	75
0.25 CMOS-MQW [370]	3.3	850	1250		-	-	-	-
Bipolar SiGe [231]	6	850	690		9	460 MHz	0.3*	43*
1.0 NMOS [313]	8	870	300		30	500 Mb/s	0.52*	74*
1.0 NMOS [314]	1.8	850	1000	-9.3	30	1.0 Gb/s	0.54*	80*
1.0 NMOS [315]	3.5	850	622	-31.9	30	234 MHz	0.57*	83*
1.0 NMOS [315]	3.5	850	1000	-23.2	30	234 MHz	0.57*	83*
1.0 NMOS [315]	3.5	850	622	-23.2	10	124 MHz		
1.0 NMOS SOI [315]	3	850	622	-15.3	3		0.20*	29*
1.0 NMOS SOI [315]	5	850	1500	-12.0	5	900 MHz		
1.0 NMOS SOI [315]	5	850	2000	-12.2	20	2.8 GHz		
0.13 CMOS SOI [188]	3.2	850	2000	-16.6	24.0	5 GHz	0.07	10
0.13 CMOS SOI [188]	3.2	850	3125	-15.4	24.0	5 GHz	0.07	10
0.13 CMOS SOI [188]	3.2	850	5000	-10.9	24.0	5 GHz	0.07	10
1.5 Bipolar [37]	5	850	150		4.2	150 Mb/s	0.21	30
Bipolar [35]	5	780	50		3.0	300 MHz	0.35	56
Bipolar [305]	5	830	150		3.0	280 MHz	0.5*	75*
0.6 BiCMOS [171]	3.3	850	531	-13.1	2.5	700 MHz	0.07	10
0.6 BiCMOS [172]	3.3	670	622	-16.6	2.5	700 MHz	0.16	29
0.8 BiCMOS [316]	5	638	531		3.3	531 Mb/s	0.49*	95*
0.6 BiCMOS [317]	5	670	1250	-22.7	5	1.25 Gb/s	0.40	74
0.6 BiCMOS [317]	5	670	622	-24.5	17	1.5 Gb/s	0.40	74
0.6 BiCMOS [317]	5	670	1000	-24.3	17	1.5 Gb/s	0.40	74
0.6 BiCMOS [317]	5	670	1250	-24.1	17	1.5 Gb/s	0.40	74
0.6 BiCMOS [317]	5	670	1500	-22.1	17	1.5 Gb/s	0.40	74
0.6 BiCMOS [351]	5	670	1250	-14.1	VUC	1.5 Gb/s	0.40	74
0.6 BiCMOS [351]	5	670	1500	-13.1	VUC	1.5 Gb/s	0.40	74
0.6 BiCMOS [354]	5	670	1250	-22.5	VUC	1.5 Gb/s	0.40	74
0.35 CMOS [69]	3.3	850	1000	-6.3	10	1.0 Gb/s	0.04	5.9
0.35 CMOS [69]	3.3	850	1000	-8.6	10	1.0 Gb/s	0.04	5.9
0.25 CMOS [322]	3.3	860	700	-18.7	1.7	0.7 Gb/s	0.1	15
0.25 CMOS SOI [192]	2.0	850	1500		1.5	1.0 GHz	0.4	59
1.0 CMOS [77]	5	638	622	-14	3	1.7 GHz	0.25	49
1.0 CMOS [77]	3.3	638	622		1.8	1.4 GHz	0.48*	94*
1.0 CMOS [329]	5	638	1000	-15.4	3	1.7 GHz	0.25	49

VDD = 5 V and V_{PIN} = 20 V for a data rate of 2.0 Gb/s. These impressive results are due to the optimized SOI layer thickness allowing the supression of the slow drift in the depth between the N^+- and P^+-fingers, which is present in the lateral PIN photodiode on high-resistivity bulk silicon.

The same research group finally switched over to IBM's 130 nm CMOS process. With this process on a 2 μm thick SOI material, allowing for a quantum efficiency of about 10 % for 850 nm, a photodetector speed of 5 GHz with a reverse bias of -20 V [188] was achieved. With a 1 GHz design, a receiver sensitivity of -19 dBm was reported. With a faster 3 GHz circuit, sensitivities of -16.6, -15.4, and -10.9 dBm for data rates of 2, 3.125, and 5 Gb/s, respectively, were obtained for a BER of 10^{-9} [188].

With a bipolar OEIC [37] using the BEST-process of AT&T with 1.5 μm design rules [371], a data rate of 150 Mb/s was achieved. An N^+/P-substrate photodiode limited the speed to this data rate. In [35] and [305], vertical PIN photodiodes have been integrated with bipolar transistor technology. The optical receivers described in these two articles demonstrated a data rate of 50 Mb/s [35] and a frequency response of 147 MHz [305] at a supply voltage of 5 V although the PIN photodiodes showed bandwidths of ≈300 MHz at a bias of 3 V.

The thin P^+/N-collector/N^+-buried collector PIN photodiode in a 0.6 μm BiCMOS process was used together with a MOS amplifier in [171, 172] with a rather low responsivity of the photodiode. A double photodiode in a standard 0.8 μm BiCMOS technology enabled a bit rate of the OEIC with a bipolar amplifier in excess of 531 Mb/s [67, 316]. A 0.6 μm BiCMOS OEIC with a photodiode similar to that described in [35] achieved a sensitivity of −22.7 dBm at 1.25 Gb/s for BER = 10^{-9} with $\lambda = 670$ nm at a single 5 V supply [317]. With a second (external) supply of 17 V for the photodiode, sensitivities of −24.5, −24.3, and −24.1 dBm were obtained at bit rates of 622 Mb/s, 1.0 Gb/s, and 1.25 Gb/s, respetively. The sensitivity for 1.5 Gb/s operation finally was −22.7 dBm. With a photodiode bias of 16.3 V internally generated on the chip from VDD = 5 V with a voltage-up-converter (VUC), i. e. without an external supply voltage for the photodiode, sensitivities of −14.1 dBm at 1.25 Gb/s and of −13.1 dBm at 1.5 Gb/s were achieved [351]. With an improved approach for the VUC, a sensitivity of -22.5dBm was obtained at 1.25 Gb/s finally [354].

A very low responsivity of 0.04 A/W has been reported for the P^+/N-well photodiode of a 1 Gb/s OEIC in a 0.35 μm CMOS technology [69]. The most sophisticated approach with a spatially-modulated-light detector (SML detector) [70] in standard CMOS 0.6 μm and 0.25 μm [322] technology enabled data rates of 250 Mb/s and 700 Mb/s, with sensitivities of −18 dBm and −18.7 dBm, respectively, at 860 nm and a BER of 10^{-9}. The SML detector, however, suffers from a low quantum efficiency of 15 % [70], because half of the detector area is covered by metal and because carriers photogenerated below the N-well/P-substrate space-charge region do not contribute.

This low detector quantum efficiency finally limits the sensitivity of SML receivers. The power consumption of the 0.25 µm receiver with 2.5 mW per channel [322], however, was remarkably low. An avalanche photodiode in a SOI CMOS technology enabled an OEIC with a bandwidth of 1 GHz corresponding to a bit rate of about 1.5 Gb/s at a supply voltage of only 2 V [192].

Summarizing, a low responsivity has been present in order to achieve a high data rate in the investigations [37,69,171,172], or a high voltage for the photodiode was needed [188, 231, 313, 314, 317], or a high additional process complexity has been needed for the integration of fast and highly efficient photodiodes as in [35, 305]. Compared to the sophisticated detectors in [69] and [192, 322], the vertically integrated PIN photodiode developed in the CMOS DVD project and described in [77] achieved a higher speed and a much higher quantum efficiency with and without an antireflection coating. The CMOS-integrated vertical PIN photodiode, although slower than the lateral SOI PIN photodiodes, shows the largest speed-responsivity product of all integrated silicon photodiodes reported so far whereby standard or near-standard processes have been used.

It can be concluded that with CMOS receiver OEICs comparable or even better data rates can be obtained than with the bipolar OEICs described in [35,305]. In contrast to the integration of PIN photodiodes in bipolar circuits, the integration of vertical PIN photodiodes in CMOS circuits requires little additional process complexity. The vertical PIN photodiodes combine a high data rate and a high quantum efficiency at a low reverse bias with a single power supply voltage for the OEIC. Low cost PIN-CMOS receiver OEICs for optical data transmission and for optical interconnects on boards and between boards via optical backplanes seem feasible.

6.4.18 Special Circuits

In [372] a new opto-electrical phase-locked-loop (OE-PLL) was introduced. This OE-PLL is based on the CMOS photonic mixer device (CMOS PMD) described in [157]. The PMD combines the detection of the light and the mixing of this optical signal with an electrical signal in itself [157, 275]. The CMOS PMD, therefore, can be used as a phase detector in a phase-locked loop (PLL). An OE-PLL consequently can be manufactured with the complete readout electronics for the retiming and reshaping in low-cost CMOS technology. The OE-PLL, for instance, can be applied in optical fiber and wireless communication.

Figure 6.167 shows the block diagram of the OE-PLL. The CMOS PMD receives light pulses in a return-to-zero (RZ) mode from an optical fiber or via free space. This optical input signal is mixed in the PMD with the differential electrical output signal of the voltage-controlled oscillator (VCO). The static frequency f_{center} of the VCO is located close to the known value of the clock rate f_{bit} of the incoming binary signal carrying information. The difference signal PD_{out} is used to adjust the frequency and the phase of the VCO.

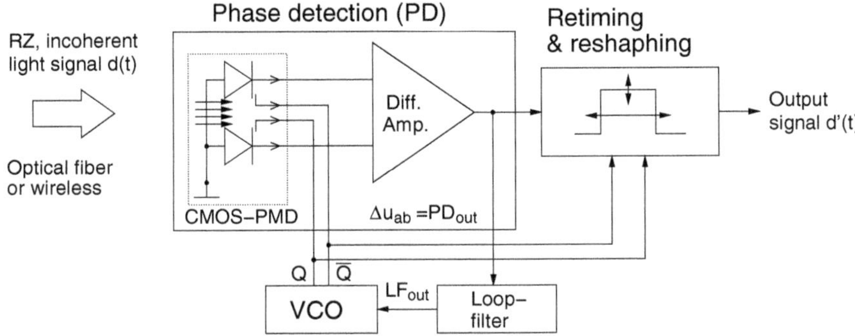

Fig. 6.167. Block diagram of an optoelectronic phase-locked loop [372]

Therefore, the difference signal PD_{out} is low-pass filtered allowing LF_{out} to be used as the input voltage of the VCO as it is done in conventional electrical PLLs. In order to allow for a safe locking of the PLL, a synchronization sequence preceedes the data stream. A locking time of 450 ns was reported in [372] for a jump of the data rate from 60 to 65 Mb/s.

The PMD readout circuit is depicted in Fig. 6.168. The charges separated in the PMD to its outputs a and b according to the modulation inputs u_{am}

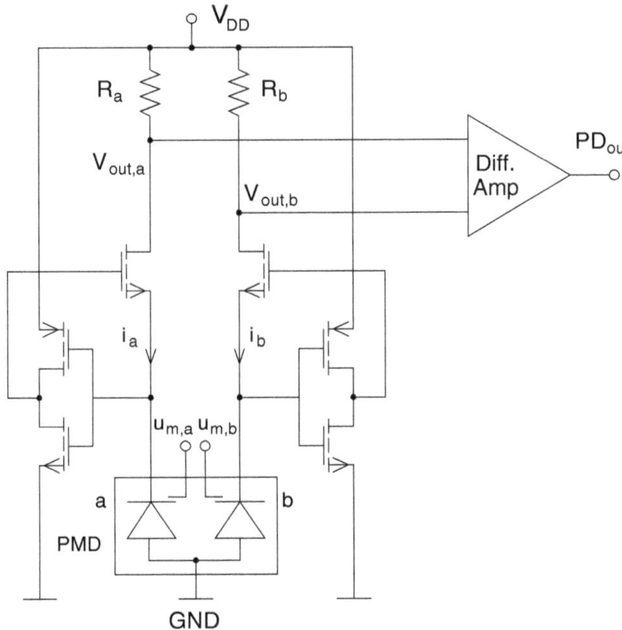

Fig. 6.168. PMD readout circuit with differential amplifier as a phase detector [372]

and u_{bm} lead to the photocurrents i_{a} and i_{b}. The low-impedance readout circuit consists of an inverter and a control transistor. The PMD output voltages are amplified by the inverters and fed back via the source follower control transistors. Due to this feedback circuit the cathode voltage of the PMD is adjusted to the inverter threshold of about 2.5 V independent of the drain voltage of the control transistor. The photocurrents are converted to voltages across the resistors R_{a} and R_{b}.

The difference between the two voltages $v_{\mathrm{out,a}}$ and $v_{\mathrm{out,b}}$ is a measure for the phase-shift between the light signal and the signal at the input gates u_{am} and u_{bm} of the PMD. This difference is amplified with a gain of 250, whereby the output of the amplifier referred to 2.5 V for identical input voltages [372]. It should be mentioned that a high CMRR of the amplifier is essential for a good suppression of background light. Also an output driver is necessary in the amplifier to obtain a low output impedance being necessary to drive the loop filter and the following VCO.

A common oscillator topology in monolithic PLLs is the ring oscillator, in which a cascade of M gain stages with a total phase shift of 180° is placed in a feedback loop. This phase requirement can be achieved by implementing an odd number of CMOS inverters. The loop oscillates with a period $T = 2MT_{\mathrm{d}}$, where T_{d} is the delay of each stage with a fan-out of one. To vary the frequency of oscillation, the delay of each stage has to change. This can be realized with the current-starved VCO shown in Fig. 6.169. An NMOS and a PMOS current source is added to each inverter to limit and to control the current available to the inverters. The current for the PMOS sources has to be the same as for the NMOS sources. This is achieved by mirroring the NMOS source current from the control NMOS device at the input to the PMOS sources. This current I_{D} determines the oscillation frequency f_{osc}:

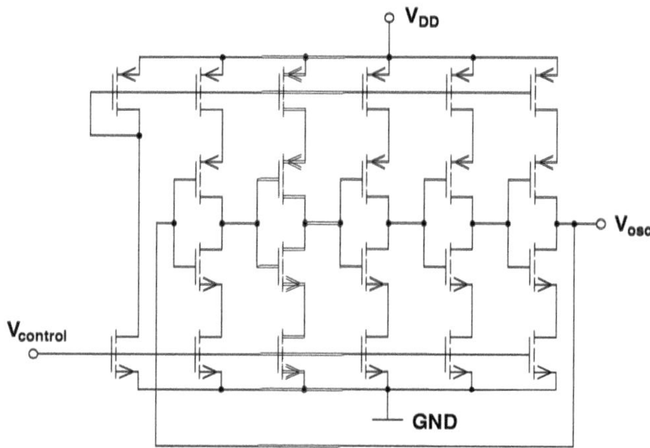

Fig. 6.169. Five-stage current-starved voltage-controlled oscillator [372]

$$f_{\mathrm{osc}} = \frac{I_{\mathrm{D}}}{M C_{\mathrm{tot}} V D D}. \qquad (6.108)$$

M is the number of inverter stages, C_{tot} is the inverter output capacitance. For the use in a PLL, a threshold switch is placed to the VCO output. The complementary signal is obtained by an inverter stage. The tuning range of the VCO was approximately 15–95 MHz in the input-voltage range 1.5–3.5 V [372].

The output signal of the amplifier PD_{out} goes from 2.5 V via 5 V back to 2.5 V and then to 0 V within one period of the RZ signal for a transmitted "1" [372]. For a logical "0", PD_{out} stays at 2.5 V, however. Therefore, a circuit is needed, which delivers a high output level for an input signal close to 0 V and close to 5 V, whereas its output level has to be low for an input signal around 2.5 V. Such a circuit is illustrated in Fig. 6.170. The inverters 1 and 2 have different switching points at 1.7 V and 3.2 V, respectively (Fig. 6.171). The two inverters, therefore, hand over three different logical states to the NAND gate with one inverted input.

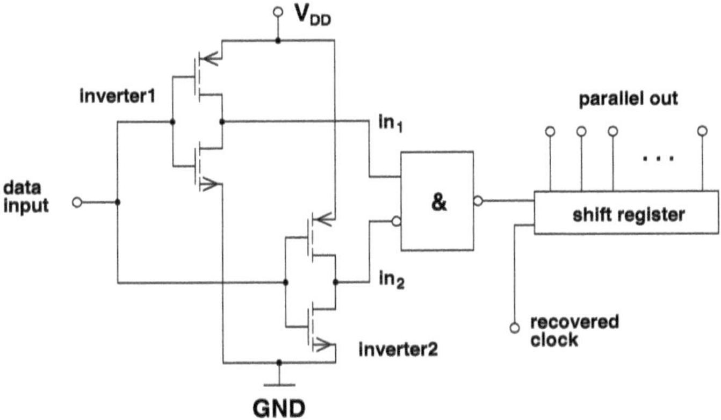

Fig. 6.170. Data recovery circuit [372]

The truth table of this NAND gate is included in Fig. 6.171. The whole circuit causes a 5 V data output for data input voltages from 0 to 1.7 V and from 3.2 to 5 V. For data input voltages between 1.7 and 3.2 V the data output level becomes low (0 V) [372]. Therefore, each data bit "1" (light on) which reaches the optical input of the PLL causes a high level at the data output of the recovery circuit. For a "0" (light off), the data output becomes low. This characteristics fulfills exactly the function of a data recovery circuit [372].

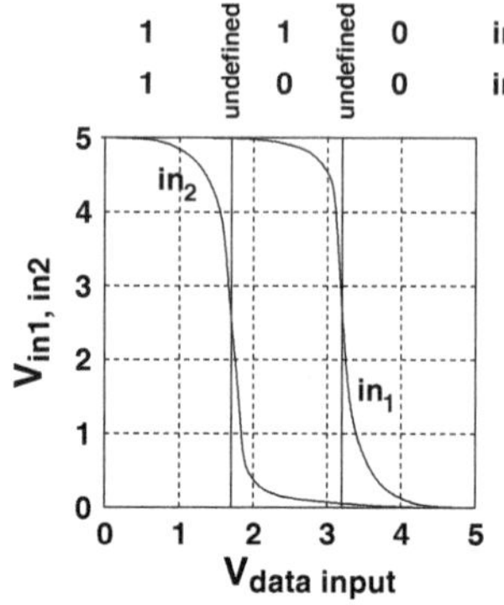

in1	in2	data output
0	0	1
0	1	1
1	0	0
1	1	1

Fig. 6.171. DC characteristics and truth table of the NAND gate with the inverted input in2 in the data recovery circuit [372]

6.5 Summary

OEICs span over a huge variety of circuit topologies and applications. Digital optical CMOS receivers based on sense-amplifier flip-flops, which require small die areas and are therefore very interesting for the application in massively parallel optical interconnects, were described. An innovative SML detector was combined with a sense amplifier. OEICs with current mirror amplifiers and current comparators were shown to be appropriate for the wafer-level test of digital CMOS and BiCMOS circuits with considerably increased frequencies.

Many analog photoreceiver circuits were included in this chapter. For instance, a bipolar circuit for optical flame detection with a very high transimpedance was described. The pixel circuits of an amorphous-silicon-CMOS imager with a dynamical range of 100 dB and of adaptive cameras-on-a-chip as well as complex smart-pixel circuits were described. Several approaches towards optical distance measuring chips and 3D cameras were discussed.

The most important results for the development of OEICs for optical storage systems are: a PIN CMOS OEIC requiring a very small chip area for DVD applications with a bandwidth of 147 MHz was realized in full custom design. This bandwidth is the same as that of former BiCMOS OEICs.

The CMOS-integrated PIN photodiodes can handle much higher data rates (> 1 Gb/s) than are necessary for DVD applications. Therefore, a 1.0 µm PIN CMOS OEIC for fiber receiver applications with a non-return-to-zero (NRZ) data rate of 1.0 Gb/s was demonstrated. This data rate was limited by the amplifier in 1.0 µm CMOS technology. OEICs with data rates in excess of 1 Gb/s, therefore, are possible with submicrometer CMOS processes. 0.6 µm BiCMOS OEICs with PIN photodiodes achieved a data rate of 1.5 Gb/s. Possible applications for PIN CMOS and PIN BiCMOS OEICs seem to be Local Area Networks (LANs), such as the Gigabit Ethernet or the Fiber Channel with 850 nm light.

References

1. K. J. Ebeling, *Integrated optoelectronics*, Springer, Berlin, Heidelberg, 1993.
2. S. Selberherr, *Analysis and simulation of semiconductor devices*, Springer, Wien, 1984.
3. E. D. Palik, *Handbook of optical constants of solids*, pp. 547–569, Academic Press, Inc., Orlando, 1985.
4. D. E. Aspnes and A. A. Studna, Phys. Rev. B **27**(2), pp. 985–1009, 1983.
5. Technology Modeling Association, Inc., Palo Alto, CA, *MEDICI manual*, 1994.
6. D. M. Caughey and R. E. Thomas, "Carrier mobilities in silicon empirically related to doping and field", Proc. IEEE **55**, pp. 2192–2193, 1967.
7. S. Selberherr, "Process and device modeling for VLSI", Microelectron. Reliability **24**(2), pp. 225–257, 1984.
8. S. M. Sze, *Physics of semiconductor devices*, Wiley, New York, 1981.
9. J. M. Senior, *Optical fiber communications*, p. 436, Prentice Hall, New York, 1992.
10. J. Muller, "Photodiode for optical communication", Advances in Electronics and Electron Physics **55**, pp. 192–216, 1981.
11. S. M. Sze, *Physics of semiconductor devices*, p. 758, Wiley, New York, 1981.
12. H. Zimmermann, "Improved CMOS-integrated photodiodes and their application in OEICs", in *IEEE Int. Workshop on High Performance Electron Devices for Microwave & Optoelectronic Applications*, pp. 346–351, 1997.
13. R. N. Hall, "Electron-hole recombination in germanium", Phys. Rev. **87**, p. 387, 1952.
14. W. Shockley and W. T. Read, "Statistics of the recombination of holes and electrons", Phys. Rev. **87**, p. 835, 1952.
15. C. T. Sah, R. N. Noyce, and W. Shockley, "Carrier generation and recombination in p-n junction and p-n junction characteristics", Proc. IRE **45**, p. 1228, 1957.
16. D. M. Chapin, C. S. Fuller, and G. L. Pearson, "A new silicon p-n junction photocell for converting solar radiation into electrical power", J. Appl. Phys. **25**, pp. 676–677, 1954.
17. A. Goetzberger, J. Knobloch, and B. Voss, *Crystalline silicon solar cells*, Wiley, Chichester, New York, 1998.
18. J. A. Mazer, *Solar cells: an introduction to crystalline photovoltaic technology*, Kluwer Academic Publishers, Boston, 1997.
19. L. Partain, *Solar cells and their applications*, Wiley, New York, 1995.
20. H. J. Möller, *Semiconductors for solar cells*, Artech House, Boston, 1993.
21. P. A. H. Hart, "Bipolar transistors and integrated circuits", in C. Hilsum, ed., *Handbook on Semiconductors*, vol. 4, (Device Physics), pp. 99–264, North-Holland, Amsterdam, 1993.

22. J. M. Senior, *Optical fiber communications*, p. 437, Prentice Hall, New York, 1992.
23. G. Winstel and C. Weyrich, *Optoelektronik II*, p. 76, Springer, Berlin, Heidelberg, 1986.
24. A. Ghazi, *Entwurf und Realisierung von integrierten Photodetektoren und CMOS-Schaltkreisen für optische Speichersysteme*, Wissenschaftsverlag Mainz, Aachen, 2000.
25. Landolt-Börnstein, *Numerical data and functional relationships in science and technology*, vol. 17c, p. 474, Springer, Berlin, 1984.
26. M. Takagi, K. Nakayama, C. Terada, and H. Kamoka, Proc. 4th Conf. Solid-State Devices, Suppl. Jpn. Soc. Appl. Phys. **42**, p. 101, 1973.
27. D. D. Tang, T. H. Ning, R. D. Isaac, G. C. Feth, S. K. Wiedmann, and H.-N. Yu, "Subnanosecond self-aligned I^2L/MTL Circuits", IEEE J. Solid-Sate Circuits **15**(4), pp. 444–449, 1980.
28. E. F. Labuda and J. T. Clemens, "Integrated circuit technology", in R. E. Kirk and D. F. Othmer, eds., *Encyclopedia of Chemical Technology*, Wiley, New York, 1980.
29. J. A. Appels, E. Kooi, M. M. Paffen, J. J. H. Schlorje, and W. H. C. G. Verkuylen, "Local oxidation of silicon and its application in semiconductor technology", Philips Res. Rep. **25**, p. 118, 1970.
30. M. Grossman, "Recessed-oxide isolation hikes IBM's LSI density and speed", Electron Design **12**(6), pp. 26–28, 1979.
31. R. J. Blumberg and S. Brenner, "A 1500 gate, random logic, large-scale integrated (LSI) masterslice", IEEE J. Solid-State Circuits **14**(5), pp. 818–822, 1979.
32. D. D. Tang, P. M. Soloman, T. H. Ning, R. D. Issac, and R. E. Burger, "1.25–μm deep-groove-isolated self-aligned bipolar circuits", IEEE J. Solid-Sate Circuits **17**(5), pp. 925–931, 1982.
33. A. Hayasaka, Y. Takami, M. Kawamura, K. Ogiue, and S. Ohwaki, "U-groove isolation technique for high speed bipolar VLSI's", in *IEDM Digest Technical Papers*, pp. 62–65, 1982.
34. H. Goto, T. Takada, R. Abe, Y. Kawabe, K. Oami, and M. Tanaka, "An isolation technology for high performance bipolar memories: IOP-II", in *IEDM Digest Technical Papers*, pp. 58–61, 1982.
35. M. Yamamoto, M. Kubo, and K. Nakao, "Si-OEIC with a built-in PIN-photodiode", IEEE Transactions on Electron Devices **42**(1), pp. 58–63, 1995.
36. H.-M. Rein and R. Ranfft, *Integrierte Bipolarschaltungen*, p. 50, Springer, Berlin, Heidelberg, 1987.
37. H. H. Kim, R. G. Swartz, Y. Ota, T. K. Woodward, M. D. Feuer, and W. L. Wilson, "Prospects for silicon monolithic opto-electronics with polymer light emitting diodes", IEEE J. Lightwave Technology **12**(12), pp. 2114–2121, 1994.
38. J. Popp and H. v. Philipsborn, "10 Gbit/s on-chip photodetection with self-aligned silicon bipolar transistors", in *ESSDERC*, pp. 571–574, 1990.
39. J. Wieland, H. Duran, and A. Felder, "Two-channel 5Gbit/s silicon bipolar monolithic receiver for parallel optical interconnects", Electronics Letters **30**(4), p. 358, 1994.
40. H. Kabza, K. Ehinger, T. F. Meister, H.-W. Meul, P. Weger, I. Kerner, M. Miura-Mattausch, R. Schreiter, D. Hartwig, M. Reisch, M. Ohnemus,

R. Köpl, J. Weng, H. Klose, H. Schaber, and L. Treitinger, "A 1-μm polysilicon self-aligned bipolar process for low-power high-speed integrated circuits", IEEE Electron Device Letters **10**(8), pp. 344–346, 1989.
41. J. Wieland, H. Melchior, M. Q. Kearley, C. R. Morris, A. M. Moseley, M. J. Goodwin, and R. C. Goodfellow, "Optical receiver array in silicon bipolar technology with selfaligned, low parasitic III/V detectors for DC-1 Gbit/s parallel links", Electronics Letters **24**(27), pp. 2211–2213, 1994.
42. G. Winstel and C. Weyrich, *Optoelektronik II*, p. 97, Springer, Berlin, Heidelberg, 1986.
43. J. Lindemayer and C. Y. Wrigley, "Beta cuttoff frequencies of junction transistors", Proc. IRE **50**, pp. 194–198, 1962.
44. D. Bolliger, P. Malcovati, A. Häberli, H. Baltes, P. Sarro, and F. Maloberti, "Integrated ultraviolet sensor system with on-chip 1 GΩ transimpedance amplifier", in *ISSCC*, pp. 328–329, 1996.
45. D. Bolliger, R. S. Popovic, and H. Baltes, "Integration of a smart selective UV detector", in *Transducers'95 and Eurosensors IX, Digest of Technical Papers 2 (8th Int. Conf. Solid-State Sensors and Actuators)*, pp. 144–147, 1995.
46. L. K. Nanver, E. J. G. Goudena, and H. W. van Zeijl, "DIMES-01, a baseline BIFET process for smart sensor experimentation", Sensors and Actuators A **36**, pp. 139–147, 1993.
47. C. T. Kirk, "A theory of transistor cutoff frequency (f_T) falloff at high current densities", IRE Trans. Electron Devices **ED-9**, pp. 164–174, 1962.
48. L. A. Hahn, "The saturation characteristics of high-voltage transistors", Proc. IEEE **55**(8), pp. 1384–1388, 1967.
49. J. R. A. Beale and J. A. G. Slatter, "The equivalent circuit of a transistor with a lightly doped collector operating in saturation", Solid-State Electron. **11**, pp. 241–252, 1968.
50. J. A. Pals and H. C. de Graaff, "On the behaviour of the base-collector junction of a transistor at high collector current densities", Philips Res. Rep. **24**, pp. 53–69, 1969.
51. R. J. Whittier and D. A. Tremere, "Current gain and cutoff frequency falloff at high currents", IEEE Trans. Electron Devices **ED-16**(1), pp. 39–57, 1969.
52. H. C. Poon and H. K. Gummel, "Modeling of emitter capacitance", Proc. IRE **57**, pp. 2181–2182, 1969.
53. H. C. de Graaff, "Collector models for bipolar transistors", Solid-State Electron. **16**, pp. 587–600, 1973.
54. G. Rey and J. P. Bailbe, "Some aspects of current gain variations in bipolar transistors", Solid-State Electron. **17**, pp. 1045–1057, 1974.
55. T. H. Ning, D. D. Tang, and P. M. Solomon, "Scaling properties of bipolar devices", in *IEEE Int. Electron Device Meeting*, pp. 61–64, Wash., D.C., 1980.
56. G. Schumicki and P. Seegebrecht, *Prozesstechnologie*, pp. 370–380, Springer, Berlin, Heidelberg, 1991.
57. L. C. Parrillo, R. S. Payne, R. E. Davies, G. W. Rentlinger, and R. L. Field, "Twin-tub CMOS - a technology for VLSI circuits", in *IEDM Digest Technical Papers*, pp. 752–755, 1980.
58. T. Hori, *Gate dielectrics and MOS ULSIs*, Springer, New York, 1997.
59. C.-Y. Lu, J. J. Sung, H. C. Kirsch, N.-S. Tsai, R. Liu, A. S. Manocha, and S. J. Hillenius, "High-perfomance salicide shallow-junction CMOS devices for submicrometer VLSI application in twin-tub VI", IEEE Trans. Electron Devices **36**(11), pp. 2530–2536, 1989.

60. R. A. Chapman, C. C. Wei, D. A. Bell, S. Aur, G. A. Brown, and R. A. Haken, "0.5 micron CMOS for high performance at 3.3 V", in *IEDM Digest Technical Papers*, pp. 52–55, 1988.
61. B. Davari, W. H. Chang, M. Wordeman, C. S. Oh, Y. Taur, K. E. Petrillo, D. Moy, J. J. Bucchignano, H. Y. Ng, M. G. Rosenfield, F. J. Hohn, and M. D. Rodriguez, "A high performance 0.25 μm CMOS technology", in *IEDM Digest Technical Papers*, pp. 56–59, 1988.
62. E. Braß, U. Hilleringmann, and K. Schumacher, "System integration of optical devices and analog CMOS amplifiers", IEEE J. Solid-State Circuits **29**(8), pp. 1006–1010, 1994.
63. U. Hilleringmann and K. Goser, "Optoelectronic system integration on silicon: waveguides, photodetectors, and VLSI CMOS circuits on one chip", IEEE Trans. Electron Devices **42**(5), pp. 841–846, 1995.
64. E. Fullin, G. Voirin, M. Chevroulet, A. Lagos, and J.-M. Moret, "CMOS-based technology for integrated optoelectronics: a modular approach", in *IEDM Digest Technical Papers*, pp. 527–530, 1994.
65. R. Kauert, W. Budde, and A. Kalz, "A monolithic field segment photo sensor system", IEEE J. Solid-State Circuits **30**(7), pp. 807–811, 1995.
66. P. Lee, A. Simoni, A. Sartori, and G. Torelli, "A photosensor array for spectrophotometry", Sensors and Actuators A **46–47**, pp. 449–452, 1995.
67. H. Zimmermann, K. Kieschnick, T. Heide, and A. Ghazi, "Integrated high-speed, high-responsivity photodiodes in CMOS and BiCMOS technology", in *Proc. 29th European Solid-State Device Conference (ESSDERC)*, pp. 332–335, 1999.
68. M. L. Simpson, M. N. Ericson, G. E. Jellison, W. B. Dress, A. L. Wintenberg, and M. B. Bobrek, "Application specific spectral response with CMOS compatible photodiodes", IEEE Trans. Electron Devices **46**(5), pp. 905–913, 1999.
69. T. K. Woodward and A. V. Krishnamoorthy, "1 Gbit/s CMOS photoreceiver with integrated detector operating at 850 nm", Electronics Letters **34**(12), pp. 1252–1253, 1998.
70. C. Rooman, D. Coppee, and M. Kuijk, "Asynchronous 250-Mb/s optical receivers with integrated detector in standard CMOS technology for optocoupler applications", IEEE J. Solid-State Circuits **35**(7), pp. 953–957, 2000.
71. M. Kuijk, D. Coppee, and R. Vounckx, "Spatially modulated light detector in CMOS with sense-amplifier receiver operating at 180 Mb/s for optical data link applications and parallel optical interconnects between chips", IEEE J. Selected Topics in Quantum Electronics **4**(6), pp. 1040–1045, 1998.
72. L. D. Garrett, J. Qi, C. L. Schow, and J. C. Campbell, "A silicon-based integrated NMOS-p-i-n photoreceiver", IEEE Trans. Electron Devices **43**(3), pp. 411–416, 1996.
73. S. He, L. D. Garrett, K.-H. Lee, and J. C. Campbell, "Monolithic integrated silicon NMOS PIN photoreceiver", Electronics Letters **30**(22), pp. 1887–1888, 1994.
74. S. M. Sze, *VLSI Technology*, McGraw-Hill, New York, 1988.
75. T. Hori, J. Hirase, Y. Odake, and T. Yasui, "Deep-submicrometer large-angle-tilt implanted drain (LATID) technology", IEEE Trans. Electron Devices **39**(10), pp. 2312–2324, 1992.

76. H. Zimmermann, "Monolithic Bipolar-, CMOS-, and BiCMOS-receiver OE-ICs", in *Proc. Int. Semicond. Conference (CAS'96)*, pp. 31–40, Sinaia, Romania, 1996.
77. H. Zimmermann, T. Heide, and A. Ghazi, "Monolithic high-speed CMOS-photoreceiver", IEEE Photonics Technology Letters **11**(2), pp. 254–256, 1999.
78. H. Zimmermann, U. Müller, R. Buchner, and P. Seegebrecht, "Optoelectronic receiver circuits in CMOS-technology", in *Mikroelektronik'97, GMM-Fachbericht 17*, pp. 195–202, VDE-Verlag, Berlin, Offenbach, 1997.
79. H. Zimmermann, T. Heide, A. Ghazi, and K. Kieschnick, "PIN-CMOS-receivers for optical interconnects", in *Ext. Abstr. 2nd IEEE Workshop on Signal Propagation on Interconnects, Travemünde, Germany*, pp. 88–89, 1998.
80. H. Zimmermann, A. Ghazi, T. Heide, R. Popp, and R. Buchner, "Advanced photo integrated circuits in CMOS technology", in *Proc. 49th Electronic Components and Technology Conference (ECTC)*, pp. 1030–1035, 1999.
81. R. A. Chapman, R. A. Haken, D. A. Bell, C. C. Wei, R. H. Havemann, T. E. Tang, T. C. Holloway, and R. J. Gale, "An 0.8 μm CMOS technology for high performance logic applications", in *IEDM Digest Technical Papers*, pp. 362–365, 1987.
82. P. Pavan, G. Spiazzi, E. Zanoni, M. Muschitiello, and M. Cecchetti, "Latch-up DC triggering and holding characteristics of n-well, twin-tub and epitaxial CMOS technologies", IEE Proceedings-G **138**(5), pp. 605–612, 1991.
83. H. Zimmermann, T. Heide, A. Ghazi, and P. Seegebrecht, "PIN-CMOS-receivers for optical interconnects", in H. Grabinski, ed., *Signal Propagation on Interconnects*, vol. II, Kluwer, Amsterdam, 1999.
84. A. Ghazi, T. Heide, and H. Zimmermann, "PIN CMOS OEIC for DVD systems", in *Proc. 43rd Int. Scientific Colloquium, TU Ilmena, Germany*, vol. 2, pp. 380–385, 1998.
85. H. Zimmermann, *Integrated Silicon Optoelectronics*, Springer, Berlin, Heidelberg, 2000.
86. W. S. Boyle and G. E. Smith, "Charge coupled semiconductor devices", Bell Syst. Tech. **49**(4), pp. 587–593, 1970.
87. E. G. Stevens, B. C. Burkey, D. N. Nicols, Y. S. Yee, D. L. Losee, T.-H. Lee, T. J. Tredwell, and R. P. Khosla, "A 1-megapixel, progressive-scan image sensor with antiblooming control and lag-free operation", IEEE Transactions on Electron Devices **38**(5), pp. 981–988, 1991.
88. T. Kuriyama, H. Kodama, T. Kozono, Y. Kitahama, Y. Morita, and Y. Hiroshima, "A 1/3-in 270000 pixel CCD image sensor", IEEE Transactions on Electron Devices **38**(5), pp. 949–953, 1991.
89. S. Kawai, N. Mutoh, and N. Teranishi, "Thermionic-emission-based barrier height analysis for precise estimation of charge handling capacity in CCD registers", IEEE Trans. Electron Devices **44**(10), pp. 1588–1592, 1997.
90. J. P. Lavine and E. K. Banghart, "The effect of potential obstacles on charge transfer in image sensors", IEEE Trans. Electron Devices **44**(10), pp. 1593–1598, 1997.
91. R. Miyagawa and T. Kanade, "CCD-based range-finding sensor", IEEE Transactions on Electron Devices **44**(10), pp. 1648–1652, 1997.
92. T. Spirig, M. Marley, and P. Seitz, "The multitap lock-in CCD with offset subtraction", IEEE Transactions on Electron Devices **44**(10), pp. 1643–1647, 1997.

93. B. E. Burke, J. A. Gregory, M. W. Bautz, G. Y. Prigozhin, S. E. Kissel, B. B. Kosiki, A. H. Loomis, and D. J. Young, "Soft-X-ray CCD imagers for AXAF", IEEE Transactions on Electron Devices **44**(10), pp. 1633–1642, 1997.
94. K. Itakura, T. Nobusada, Y. Saitou, N. Kokusenya, R. Nagayoshi, M. Ozaki, Y. Sugawara, K. Mitani, and Y. Fujita, "A 2/3-in 2.0 M-pixel CCD imager with an advanced M-FIT architecture capable of progressive scan", IEEE Transactions on Electron Devices **44**(10), pp. 1625–1632, 1997.
95. K. Tachikawa, T. Umeda, Y. Oda, and T. Kuroda, "Device design with automatic simulation system for basic CCD characteristics", IEEE Transactions on Electron Devices **44**(10), pp. 1611–1616, 1997.
96. T. Yamada, Y. Kawakami, T. Nakano, N. Mutoh, K. Orihara, and N. Teranishi, "Driving voltage reduction in a two-phase CCD by suppression of potential pockets in inter-electrode gaps systems", IEEE Trans. Electron Devices **44**(10), pp. 1580–1587, 1997.
97. B. E. Burke, R. K. Reich, J. A. Gregory, W. H. McGonagle, A. M. Waxman, E. D. Savoye, and B. B. Kosiki, "640 $\times$ 480 back-illuminated CCD imager with improved blooming control for night vision", in *IEDM Digest Technical Papers*, pp. 33–36, 1998.
98. J. T. Bosiers, A. C. Kleinman, A. van der Sijde, L. Korthout, D. W. Verbugt, H. L. Peek, E. Roks, A. Heringa, F. F. Vledder, and P. Opmeer, "A 2/3" 2-M pixel progressive scan FT-CCD for digital still camera applications", in *IEDM Digest Technical Papers*, pp. 37–40, 1998.
99. A. Tanabe, Y. Kudoh, Y. Kawakami, K. Masubuchi, S. Kawai, T. Yamada, M. Morimoto, K. Arai, K. Hatano, M. Furumiya, Y. Nakashiba, N. Mutoh, K. Orihara, and N. Teranishi, "Dynamic range improvement by narrow-channel effect suppression and smear reduction technologies in small pixel IT-CCD image sensors", in *IEDM Digest Technical Papers*, pp. 41–44, 1998.
100. H. Fiedler and K. Knupfer, "Market overview: charge-coupled devices", in H. Baltes, W. Göpel, and J. Hesse, eds., *Sensors update*, vol. 1, pp. 223–271, VCH, Weinheim, 1996.
101. P. Seitz and K. Knop, "Image sensors", in H. Baltes, W. Göpel, and J. Hesse, eds., *Sensors update*, vol. 2, pp. 85–103, VCH, Weinheim, 1996.
102. Y. Hagiwara, "High-density and high-quality frame transfer CCD imager with very low smear, low dark current, and very high blue sensitivity", IEEE Trans. Electron Devices **43**(12), pp. 2122–2130, 1996.
103. J. Hynecek, "Low-noise and high-speed charge detection in high-resolution CCD image sensors", IEEE Transactions on Electron Devices **44**(10), pp. 1679–1688, 1997.
104. N. Tanaka, N. Nakamura, Y. Matsunaga, S. Manabe, H. Tango, and O. Yoshida, "A low driving voltage CCD with single layer electrode structure for area image sensor", IEEE Transactions on Electron Devices **44**(11), pp. 1869–1874, 1997.
105. S. M. Sze, *Physics of semiconductor devices*, p. 412, Wiley, New York, 1981.
106. R. H. Walden, R. H. Krambeck, R. J. Strain, J. McKenna, N. L. Schryer, and G. E. Smith, "The buried channel charge coupled device", Bell Syst. Tech. **51**, pp. 1635–1640, 1972.
107. M. M. Blouke, J. R. Janesick, J. E. Hall, M. W. Cowens, and P. J. May, "800 $\times$ 800 charge coupled device image sensor", Opt. Eng. **22**(5), pp. 607–614, 1983.

108. Y. Ishihara, E. Oda, H. Tanigawa, A. Kohno, N. Teranishi, E.-I. Takeuchi, I. Akiyama, and T. Kamata, "Interline CCD image sensor with an antiblooming structure", IEEE Trans. Electron Devices **31**(1), pp. 83–88, 1984.
109. N. Teranishi and Y. Ishihara, "Smear reduction in interline CCD image sensor", IEEE Trans. Electron Devices **34**(5), pp. 1052–1056, 1987.
110. E. Oda, K. Orihara, T. Tanaka, T. Kamata, and Y. Ishihara, "1/2-in 768(H) × 492(V) pixel CCD image sensor", IEEE Trans. Electron Devices **36**(1), pp. 46–53, 1989.
111. K. Knop, "Image sensors", in W. Göpel, J. Hesse, and J. N. Zemel, eds., *Optical sensors*, pp. 233–252, VCH, Weinheim, 1992.
112. P. Centen, "CCD on-chip amplifiers: noise performance versus MOS transistor dimensions", IEEE Transactions on Electron Devices **38**(5), pp. 1206–1216, 1991.
113. J. Hojo, Y. Naito, H. Mori, K. Fujikawa, N. Kato, T. Wakayama, E. Komatsu, and M. Itasaka, "A 1/3-in 510(H) × 492(V) CCD image sensor with mirror image function", IEEE Trans. Electron Devices **38**(5), pp. 954–959, 1991.
114. N. Mutoh, "Simulation for 3-D optical and electrical analysis of CCD", IEEE Trans. Electron Devices **44**(10), pp. 1604–1610, 1997.
115. M. Yamagishi, M. Negishi, H. Yamada, T. Tsunakawa, K. Shinohara, T. Ishimaru, Y. Kamide, Y. Yamazaki, H. Abe, H. Kanbe, Y. Tomiya, K. Yonemoto, T. Iizuka, S. Nakamura, K. Harada, and K. Wada, "A 2 million pixel FIT-CCD image sensor for HDTV camera systems", IEEE Trans. Electron Devices **38**(5), pp. 976–980, 1991.
116. G. Williams and J. Janesick, "Cameras with CCD's capture new markets", Laser Focus World, Detector Handbook pp. S5–S9, 1996.
117. A. L. Lattes, S. C. Munroe, and M. M. Seaver, "Ultrafast shallow-buried-channel CCD's with built-in drift fields", IEEE Electron Device Lett. **12**(2), pp. 104–107, 1991.
118. T. Satoh, N. Mutoh, M. Furumiya, I. Murakami, S. Suwazono, C. Ogawa, K. Hatano, H. Utsumi, S. Kawai, K. Arai, M. Morimoto, K. Orihara, T. Tamura, N. Teranishi, and Y. Hokari, "Optical limitations to cell size reduction in IT-CCD image sensors", IEEE Trans. Electron Devices **44**(10), pp. 1599–1603, 1997.
119. R. Dawson, J. Preisig, J. Carnes, and J. Pridgen, "CMOS/buried-N-channel CCD compatible process for analog signal processing applications", RCA Review **38**, pp. 406–435, 1977.
120. D. Ong, "An all-implanted CCD/CMOS process", IEEE Transactions on Electron Devices **28**(1), pp. 6–12, 1981.
121. E. R. Fossum, "CMOS image sensors: electronic camera-on-a-chip", IEEE Transactions on Electron Devices **44**(10), pp. 1689–1698, 1997.
122. A. Simoni, A. Sartori, M. Gottardi, and A. Zorat, "A digital vision sensor", Sensors and Actuators A **46–47**, pp. 439–443, 1995.
123. P. Noble, "Self-scanned silicon image detector arrays", IEEE Transactions on Electron Devices **15**(4), pp. 202–209, 1968.
124. F. Andoh, K. Taketoshi, J. Yamazaki, M. Sugawara, Y. Fujita, F. Mitani, Y. Matuzawa, K. Miyata, and S. Araki, "A 250000 pixel image sensor with FET amplification at each pixel for high-speed television cameras", in *ISSCC Dig. Tech. Papers*, pp. 212–213, 1990.

125. H. Kawashima, F. Andoh, N. Murata, K. Tanaka, M. Yamawaki, and K. Taketoshi, "A 1/4 inch format 250000 pixel amplifier MOS image sensor using CMOS process", in *IEDM Digest Technical Papers*, pp. 575–578, 1993.
126. M. Sugawara, H. Kawashima, F. Andoh, N. Murata, Y. Fujita, and M. Yamawaki, "An amplified MOS imager suited for image processing", in *ISSCC Dig. Tech. Papers*, pp. 228–229, 1994.
127. E. Oba, K. Mabuchi, Y. Iida, N. Nakamura, and H. Miura, "A 1/4 inch 330 k square pixel progressive scan CMOS active pixel image sensor", in *ISSCC Dig. Tech. Papers*, pp. 180–181, 1997.
128. C. Aw and B. Wooley, "A 128 × 128 pixel standard CMOS image sensor with electronic shutter", IEEE J. Solid-State Circuits **31**(12), pp. 1922–1930, 1996.
129. R. H. Nixon, S. E. Kemeny, B. Pain, C. O. Staller, and E. R. Fossum, "256 × 256 CMOS active pixel sensor camera-on-a-chip", IEEE J. Solid-State Circuits **31**(12), pp. 2046–2050, 1996.
130. R. H. Nixon, S. E. Kemeny, R. C. Gee, B. Pain, Q. Kim, and E. R. Fossum, "CMOS active pixel image sensors for highly integrated imaging systems", IEEE J. Solid-State Circuits **32**(2 (Feb.)), pp. 187–197, 1997.
131. O. Yadid-Pecht and E. R. Fossum, "Wide intrascene dynamic range CMOS APS using dual sampling", IEEE Trans. Electron Devices **44**(10), pp. 1721–1723, 1997.
132. Z. Zhou, B. Pain, and E. R. Fossum, "CMOS active pixel sensor with on-chip successive approximation analog-to-digital converter", IEEE Trans. Electron Devices **44**(10), pp. 1759–1763, 1997.
133. Z. Zhou, B. Pain, and E. R. Fossum, "Frame-transfer CMOS active pixel sensor with pixel binning", IEEE Trans. Electron Devices **44**(10), pp. 1764–1768, 1997.
134. G. Yang, O. Yadid-Pecht, C. Wrigley, and B. Pain, "A snap-shot CMOS active pixel imager for low-noise, high-speed imaging", in *IEDM Digest Technical Papers*, pp. 45–48, 1998.
135. D. Scheffer, B. Dierickx, and G. Meynants, "Random addressable 2048 × 2048 active pixel image sensor", IEEE Trans. Electron Devices **44**(10), pp. 1716–1720, 1997.
136. F. Pardo, B. Dierickx, and D. Scheffer, "CMOS foveated image sensor: signal scaling and small geometry effects", IEEE Trans. Electron Devices **44**(10), pp. 1731–1737, 1997.
137. A. Dickinson, B. Auckland, E.-S. Eid, D. Inglis, and E. R. Fossum, "A 256x256 CMOS active pixel image sensor with motion detection", in *ISSCC*, pp. 226–227, 1995.
138. R. M. Guidash, T.-H. Lee, P. P. K. Lee, D. H. Sackett, C. I. Drowley, M. S. Swenson, L. Arbaugh, R. Hollstein, F. Shapiro, and S. Domer, "A 0.6 μm CMOS pinned photodiode color imager technology", in *IEDM Digest Technical Papers*, pp. 927–929, 1997.
139. S. Mendis, S. E. Kemeny, and E. R. Fossum, "CMOS active pixel image sensor", IEEE Trans. Electron. Devices **41**, pp. 452–453, 1994.
140. M.-H. Chi, "Technologies for high performance CMOS active pixel imaging system-on-a-chip", in *Proc. 5th Int. Conf. on Solid-State and Integrated-Circuit Technology*, pp. 180–183, 1998.
141. M. Kunii, K. Hasegawa, H. Oka, Y. Nakazawa, T. Takeshita, and H. Kurihara, "Performance of a high-resolution contact-type linear image sensor with a-

Si:H/a-SiC:H heterojunction photodiodes", IEEE Transactions on Electron Devices **36**(12), pp. 2877–2881, 1989.
142. H. Kakinuma, M. Sakamoto, Y. Kasuya, and H. Sawai, "Characterisitics of Cr Schottky amorphous silicon photodiodes and their application in linear image sensors", IEEE Transactions on Electron Devices **37**(1), pp. 128–133, 1990.
143. L. E. Antonuk, J. Boudry, Y. El-Mohri, W. Huang, J. Siewerdsen, J. Yorkston, and R. A. Street, "A high-resolution, high frame rate flatpanel TFT array for digital X-Ray imaging", SPIE, Phys. Med. Imag. **2163**, pp. 118–127, 1994.
144. N. C. Bird, C. J. Curling, and C. van Berkel, "Large-area image sensing using amorphous silicon NIP diodes", Sensors and Actuators **46–47**, pp. 444–448, 1995.
145. X. D. Wu, R. A. Street, R. Weisfield, S. Ready, and S. Nelson, "Page sized a-Si:H two-dimensional array as imaging devices", in *Proc. 4th Int. Conf. Solid-State and Integrated-Circuit Technology*, pp. 724–726, 1995.
146. E. D. Palik, *Handbook of optical constants of solids*, pp. 571–586, Academic Press, Inc., Orlando, 1985.
147. R. H. Bube, "Solar cells", in C. Hilsum, ed., *Handbook on Semiconductors*, vol. 4, (Device Physics), pp. 825–826, North-Holland, Amsterdam, 1993.
148. M. Böhm, F. Blecher, A. Eckhardt, B. Schneider, S. Benthien, H. Keller, T. Lule, P. Rieve, M. Sommer, R. C. Lind, L. Humm, M. Daniels, and N. Wu, "High dynamic range image sensors in thin film on ASIC - technology for automotive applications", in D. E. Ricken and W. Gessner, eds., *Advanced Microsystems for Automotive Applications*, pp. 157–172, Springer, Berlin, Heidelberg, 1998.
149. B. Schneider, H. Fischer, S. Benthien, H. Keller, T. Lule, P. Rieve, M. Sommer, J. Schulte, and M. Böhm, "TFA image sensors: from the one transistor cell to a locally adaptive high dynamic range sensor", in *IEDM Digest Technical Papers*, pp. 209–212, 1997.
150. H. Fischer, J. Schulte, P. Rieve, and M. Böhm, "Technology and performance of TFA (Thin Film on ASIC)-sensors", in *Mat. Res. Soc. Symp. Proc.*, vol. 336, pp. 867–872, 1994.
151. J. Schulte, H. Fischer, T. Lule, Q. Zhu, and M. Böhm, "Properties of TFA (Thin Film on ASIC) sensors", in H. Reichl and A. Heuberger, eds., *Micro System Technologies*, pp. 783–790, 1994.
152. M. P. Vidal, M. Bafleur, J. Buxo, and G. Sarrabayrouse, "A bipolar photodetector compatible with standard CMOS technology", Solid-State Electron. **34**(8), pp. 809–814, 1991.
153. E. A. Vittoz, "MOS transistors operated in the lateral bipolar mode and their application in CMOS technology", IEEE J. Solid-State Circuits **18**(6), pp. 273–279, 1983.
154. R. W. Sandage and J. A. Connelly, "A fingerprint opto-detector using lateral bipolar phototransistors in a standard CMOS process", in *IEDM Digest Technical Papers*, pp. 171–174, 1995.
155. W. T. Holman and J. A. Connelly, "A compact low-noise operational amplifier for a 1.2 μm digital CMOS technology", IEEE J. Solid-State Circuits **30**(6), pp. 710–714, 1995.
156. H. Beneking, "Gain and bandwidth of fast near-infrared photodetectors: a comparison of diodes, phototransistors, and photoconductive devices", IEEE Trans. Electron Devices **29**(9), pp. 1420–1430, 1982.

157. R. Schwarte, Z. Xu, H. Heinol, J. Olk, R. Klein, B. Buxbaum, H. Fischer, and J. Schulte, "A new electrooptical mixing and correlating sensor: facilities and applications of the photonic mixer device (PMD)", in *Proc. SPIE — Sensors, sensor systems, and sensor data processing*, vol. 3100, pp. 245–253, 1997.
158. R. Schwarte, "Dynamic 3D-vision", in *Proc. IEEE Int. Symp. on Electron Devices for Microwave and Optoelectronic Applications*, pp. 241–248, 2001.
159. S. Wolf, *Silicon processing for the VLSI era, Vol. 2 – Process integration*, Lattice Press, Sunset Beach, 1990.
160. R. H. Havemann and R. H. Eklund, "Process integration issues for submicron BiCMOS technology", Solid State Technology (6), pp. 71–76, 1992.
161. T. Ikeda, A. Watanabe, Y. Nishio, I. Masuda, N. Tamba, M. Odaka, and K. Ogiue, "High-speed BiCMOS technology with a buried twin well structure", IEEE Trans. Electron Devices **34**(6), pp. 1304–1310, 1987.
162. S. S. Ahmed, W. W. Asakawa, M. T. Bohr, S. S. Chambers, T. Deeter, M. Denham, J. K. Greason, W. W. Holt, R. R. Taylor, and I. Young, "A triple diffused approach for high performance 0.8 µm BiCMOS technology", Solid State Technology (10), pp. 33–40, 1992.
163. R. A. Chapman, D. A. Bell, R. H. Eklund, R. H. Havemann, M. G. Harward, and R. A. Haken, "Submicrometer BiCMOS well design for optimum circuit performance", in *IEDM Digest Technical Papers*, pp. 756–759, 1988.
164. H. Iwai, G. Sasaki, Y. Unno, Y. Niitsu, M. Norishima, Y. Sugimoto, and K. Kanzaki, "0.8-µm Bi-CMOS technology with high f_T ion-implanted emitter bipolar transistor", in *IEDM Digest Technical Papers*, pp. 28–31, 1987.
165. T.-Y. Chiu, G. M. Chin, M. Y. Lau, R. C. Hanson, M. D. Morris, K. F. Lee, A. M. Voshchenkov, R. G. Swartz, V. D. Archer, M. T. Y. Liu, S. N. Finegan, and M. D. Feuer, "Non-overlapping super self-aligned BiCMOS with 87 ps low power ECL", in *IEDM Digest Technical Papers*, pp. 752–755, 1988.
166. W. R. Burger, C. Lage, T. Davies, M. DeLong, D. Haueisen, J. Small, G. Huglin, A. Landau, F. Whitwer, and B. Bastani, "An advanced self-aligned BICMOS technology for high performance 1-megabit ECL i/O SRAMs", in *IEDM Digest Technical Papers*, pp. 421–424, 1989.
167. Y. Kobayashi, C. Yamaguchi, Kobayashi, Y. Amemiya, and T. Sakai, "High perfomance LSI process technology: SST CBi-CMOS", in *IEDM Digest Technical Papers*, pp. 760–763, 1988.
168. K. Sakaue, Y. Shobatake, M. Motoyama, Y. Kumaki, S. Takatsuka, S. Tanaka, H. Hara, K. Matsuda, S. Kitaoka, M. Noda, Y. Niitsu, M. Norishima, H. Momose, K. Maeguchi, M. Ishibe, S. Shimizu, and T. Kodama, "A 0.8-µm BiCMOS ATM switch on an 800-Mb/s asynchronous buffered banyan network", IEEE J. Solid-State Circuits **26**(8), pp. 1133–1144, 1991.
169. M. El-Diwany, J. Borland, J. Chen, S. Hu, P. v. Wijnen, C. Vorst, V. Akylas, M. Brassington, and R. Razouk, "An advanced BiCMOS process utilizing ultra-thin silicon epitaxy over arsenic buried layers", in *IEDM Digest Technical Papers*, pp. 245–248, 1989.
170. M. Norishima, Y. Niitsu, G. Sasaki, H. Iwai, and K. Maeguchi, "Bipolar transistor design for low process-temperature 0.5 µm BI-CMOS", in *IEDM Digest Technical Papers*, pp. 237–240, 1989.
171. P. J.-W. Lim, A. Y. C. Tzeng, H. L. Chuang, and S. A. S. Onge, "A 3.3 V monolithic photodetector/CMOS preamplifier for 531 Mb/s optical data link applications", in *ISSCC*, pp. 96–97, 1993.

172. D. M. Kuchta, H. A. Ainspan, F. J. Canora, and R. P. Schneider, "Performance of fiber-optic data links using 670 nm CW VCSELs and a monolithic Si photodetector and CMOS preamplifier", IBM J. Res. Develop. **39**(1/2), pp. 63–72, 1995.
173. H. Zimmermann, K. Kieschnick, M. Heise, and H. Pless, "BiCMOS OEIC for optical storage systems", Electronics Letters **34**(19), pp. 1875–1876, 1998.
174. H. Zimmermann, "Full custom CMOS and BiCMOS OPTO-ASICs", in *Proc. 5th Int. Conf. on Solid-State and Integrated-Circuit Technology*, pp. 344–347, 1998.
175. K. Kieschnick, H. Zimmermann, and P. Seegebrecht, "Silicon-based optical receivers in BiCMOS technology for advanced optoelectronic integrated circuits", in *Proc. European Materials Research Society Meeting (E-MRS), Strasbourg, June 1–4*, 1999.
176. H. Zimmermann, K. Kieschnick, M. Heise, and H. Pless, "High-bandwidth BiCMOS OEIC for optical storage systems", in *IEEE Int. Solid-State Circuits Conference*, pp. 384–385, 1999.
177. K. Kieschnick, T. Heide, A. Ghazi, H. Zimmermann, and P. Seegebrecht, "High-speed photonic CMOS and BiCMOS receiver ICs", in *Proc. 25th European Solid-State Circuits Conference (ESSCIRC)*, pp. 398–401, 1999.
178. S. Groiss and J. Sturm, "Low-noise sampling system for photocurrent detection with monolithically integrated photodiodes", in *Proc. 27th European Solid-State Circuits Conference (ESSCIRC)*, pp. 180–183, 2001.
179. J. Sturm, S. Hainz, G. Langguth, and H. Zimmermann, "Integrated photodiodes in standard BiCMOS technology", in *Proc. SPIE*, vol. 4997B, 2003.
180. K. Kieschnick, H. Zimmermann, H. Pless, and P. Seegebrecht, "Integrated photodiodes for DVD and CD-ROM applications", in *Proc. 30th European Solid-State Device Conference (ESSDERC)*, pp. 252–255, 2000.
181. H. Zimmermann and K. Kieschnick, "Low-offset BiCMOS OEIC for Optical Storage Systems", Electronics Letters **36**(14), pp. 1223–1224, 2000.
182. G. W. de Jong, J. R. M. Bergervoet, J. H. A. Brekelmans, and J. F. P. van Mil, "A DC-to-250 MHz current pre-amplifier with integrated photodiodes in standard CBiMOS for optical storage systems", in *ISSCC*, pp. 362–363, 2002.
183. J.-P. Colinge, *Silicon-on-insulator technology: materials to VLSI*, Kluwer Academic Publishers, Boston, 1991.
184. J.-P. Colinge, "p-i-n photodiodes made in laser-recrystallized silicon-on-insulator", IEEE Transactions on Electron Devices **33**(2), pp. 203–205, 1986.
185. M. Ghioni, F. Zappa, V. P. Kesan, and J. Warnock, "A VLSI-compatible high-speed silicon photodetector for optical data link applications", IEEE Transactions on Electron Devices **43**(7), pp. 1054–1060, 1996.
186. C. L. Schow, R. Li, J. D. Schaub, and J. C. Campbell, "Design and implementation of high-speed planar Si photodiodes fabricated on SOI substrates", IEEE J. Quantum Electronics **35**(10), pp. 1478–1482, 1999.
187. S. M. Csutak, J. D. Schaub, W. E. Wu, R. Shimer, and J. C. Campbell, "CMOS-compatible high-speed planar silicon photodiodes fabricated on SOI substrates", IEEE J. Quantum Electronics **38**(2), pp. 193–196, 2002.
188. S. M. Csutak, J. D. Schaub, W. E. Wu, and J. C. Campbell, "High-speed monolithically integrated silicon optical receiver fabricated in 130 nm CMOS technology", IEEE Photonics Technology Letters **14**(4), pp. 516–518, 2002.

189. E. Chen and S. Y. Chou, "High-efficiency and high-speed metal-semiconductor-metal photodetectors operating in the infrared", Appl. Phys. Lett. **70**(6), pp. 753–755, 1997.
190. J. Y. L. Ho and K. S. Wong, "High-speed and high-sensitivity silicon-on-insulator metal-semiconductor-metal photodetector with trench structure", Appl. Phys. Lett. **69**(1), pp. 16–18, 1996.
191. B. F. Levine, J. D. Wynn, F. P. Klemens, and G. Sarusi, "1 Gb/s Si high quantum efficiency monolithically integrable $\lambda = 0.88\,\mu m$ detector", Appl. Phys. Lett. **66**(22), pp. 2984–2986, 1995.
192. T. Yoshida, Y. Ohtomo, and M. Shimaya, "A novel p-i-n photodetector fabricated on SIMOX for 1 GHz 2 V CMOS OEICs", in *IEDM Digest Technical Papers*, pp. 29–32, 1998.
193. W. N. Grant, "Electron and hole ionization rates in epitaxial silicon at high electric fields", Solid-State Electron. **16**, pp. 1189–1203, 1973.
194. A. Apsel and A. G. Andreou, "5 mW Gbit/s silicon on sapphire CMOS optical receiver", Electronics Letters **37**(19), pp. 1186–1188, 2001.
195. A. G. Andreou, Z. K. Kalayjian, A. Apsel, P. O. Pouliquen, R. A. Athale, G. Simonis, and R. Reedy, "Silicon on Sapphire CMOS for optoelectronic microsystems", IEEE Circuits&Systems Magazine **1**(3), pp. 22–30, 2001.
196. K. Bernstein and N. J. Rohrer, *SOI circuit design concepts*, Kluwer Academic Publishers, Boston, 1998.
197. J. B. Kuo and K.-W. Su, *CMOS VLSI engineering silicon-on-insulator (SOI)*, Kluwer Academic Publishers, Boston, 2000.
198. T. Nakahara, H. Tsuda, K. Tateno, S. Matsuo, and T. Kurokawa, "Hybrid integration of smart pixels by using polyimide bonding: demonstration of a GaAs p-i-n photodiode/CMOS receiver", IEEE J. Selected Topics in Quantum Electronics **5**(2), pp. 209–216, 1999.
199. E. Kasper, H. J. Herzog, and H. Kibbel, "A one-dimensional SiGe superlattice grown by UHV epitaxy", Appl. Phys. **8**, pp. 199–205, 1975.
200. H. M. Manasevit, I. S. Gergis, and A. B. Jones, "The properties of $Si/Si_{1-x}Ge_x$ films grown on Si substrates by chemical vapor deposition", J. Electron. Mater. **12**(4), pp. 637–651, 1983.
201. J. C. Bean, T. T. Sheng, L. C. Feldman, A. T. Fiory, and R. T. Lynch, "Pseudomorphic growth of Ge_xSi_{1-x} on silicon by molecular beam epitaxy", Appl. Phys. Lett. **44**(1), pp. 102–104, 1984.
202. R. People, J. C. Bean, D. V. Lang, A. M. Sergent, H. L. Stormer, K. W. Wecht, R. T. Lynch, and K. Baldwin, "Modulation doping in Ge_xSi_{1-x}/Si strained layer heterostructures", Appl. Phys. Lett. **45**(11), pp. 1231–1233, 1984.
203. R. People, "Indirect band gap of coherently strained Ge_xSi_{1-x} bulk alloys on <001> silicon substrates", Phys. Rev. B **32**(2), pp. 1405–1408, 1985.
204. D. C. Ahlgren, M. Gilbert, D. Greenberg, S. J. Jeng, J. Malinowski, D. Nguyen-Ngoc, K. Schonenberg, K. Stein, R. Groves, K. Walter, G. Hueckel, D. Colavito, G. Freeman, D. Sunderland, D. L. Harame, and B. Meyerson, "Manufacturability demonstration of an integrated SiGe HBT technology for the analog and wireless marketplace", in *IEEE Int. Electron Device Meeting*, pp. 859–862, Wash., D.C., 1996.
205. S. Subbanna, D. Ahlgren, D. Harame, and B. Meyerson, "How SiGe evolved into a manufacturable semiconductor production process", in *IEEE Int. Solid-State Circuits Conference*, pp. 66–67, 1999.

206. D. J. Paul, "Silicon-germanium strained layer materials in microelectronics", Advanced Materials **11**(3), pp. 191–204, 1999.
207. S. Luryi, T. P. Pearsall, H. Tempkin, and J. C. Bean, "Waveguide infrared photodetectors on a silicon chip", IEEE Electron Device Lett. **7**(2), pp. 104–107, 1986.
208. A. Splett, T. Zinke, K. Petermann, E. Kasper, H. Kibbel, H.-J. Herzog, and H. Presting, IEEE Photonics Technology Letters **6**, pp. 59–, 1994.
209. S. B. Samavedam, M. T. Currie, T. A. Langdo, and E. A. Fitzgerald, "High-quality germanium photodiodes integrated on silicon substrates using optimized relaxed graded buffers", Appl. Phys. Lett. **73**(15), pp. 2125–2127, 1998.
210. T. L. Lin and J. Maserjian, "Characterization of wafer bonded photodetectors fabricated using various annealing temperatures and ambients", Appl. Phys. Lett. **57**(14), pp. 1422–1424, 1990.
211. T. L. Lin, E. W. Jones, A. Ksendzov, S. M. Dejewski, R. W. Fathauer, T. N. Krabach, and J. Maserjian, "A novel Si-based LWIR detector: the SiGe/Si heterojunction internal photoemission detector", in *IEDM Digest Technical Papers*, pp. 641–644, 1990.
212. R. P. G. Karunasiri, J. S. Park, and K. L. Wang, "$Si_{1-x}Ge_x$/Si multiple quantum well infrared detector", Appl. Phys. Lett. **59**(20), pp. 2588–2590, 1991.
213. R. A. Soref, "Silicon-based photonic devices", in *ISSCC*, pp. 66–67, 1995.
214. Q. Mi, X. Xiao, J. C. Sturm, L. Lenchyshyn, and M. L. W. Thewalt, "Room-temperature 1.3 μm electroluminescence from strained $Si_{1-x}Ge_x$/Si quantum wells", Appl. Phys. Lett. **60**(25), pp. 3177–3179, 1992.
215. H. Presting, T. Zinke, A. Splett, H. Kibbel, and M. Jaros, Appl. Phys. Lett. **69**, pp. 2376–, 1996.
216. R. A. Soref, "Silicon-based optoelectronics", Proc. IEEE **81**(12), pp. 1687–1706, 1993.
217. G. Abstreiter, "Engineering the future of electronics", Phys. World **5**, pp. 36–39, 1992.
218. J. C. Sturm, "Advanced column-IV epitaxial materials for silicon-based optoelectronics", MRS Bulletin (April), pp. 60–64, 1998.
219. D. C. Houghton, "Strain relaxation kinetics in $Si_{1-x}Ge_x$/Si heterostructures", J. Appl. Phys. **70**(4), pp. 2136–2151, 1991.
220. D. C. Houghton, C. Gibbings, C. Tuppen, M. Lyons, and M. Halliwell, "Equilibrium critical thickness for $Si_{1-x}Ge_x$ strained layers on (100) Si", Appl. Phys. Lett. **56**(5), pp. 460–462, 1990.
221. F. K. LeGoues, B. S. Meyerson, and J. Morar, "Anomalous strain relaxation in SiGe thin films and superlattices", Phys. Rev. Lett. **66**(22), pp. 2903–2906, 1991.
222. Y. J. Mii, Y. H. Xie, E. A. Fitzgerald, D. Monroe, F. A. Thiel, B. E. Weir, and L. C. Feldman, "Extremely high electron mobility in Si/Ge_xSi_{1-x} structures grown by molecular beam epitaxy", Appl. Phys. Lett. **59**(13), pp. 1611–1613, 1991.
223. K. Ismail, S. Rishton, J. O. Chu, K. Chan, and B. S. Meyerson, "High-performance Si/SiGe n-type modulation-doped transistors", IEEE Electron Device Lett. **14**(7), pp. 348–350, 1993.
224. G. Kissinger, T. Morgenstern, G. Morgenstern, and H. Richter, "Stepwise equilibrated graded Ge_xSi_{1-x} buffer with very low threading dislocation density on Si(001)", Appl. Phys. Lett. **66**(16), pp. 2083–2085, 1995.

225. E. J. Prinz, P. M. Garone, P. V. Schwartz, X. Xiao, and J. C. Sturm, "The effects of base dopant outdiffusion and undoped $Si_{1-x}Ge_x$ junction spacer layers in $Si/Si_{1-x}Ge_x/Si$ heterojunction bipolar transistors", IEEE Electron Device Lett. **12**(2), pp. 42–44, 1991.
226. J. W. Slotboom, B. Streutker, A. Pruijimboom, , and D. J. Gravesteijn, "Parasitic energy barriers in SiGe HBTs", IEEE Electron Device Lett. **12**(9), pp. 486–488, 1991.
227. E. D. Palik, *Handbook of optical constants of solids II*, pp. 607–636, Academic Press, Inc., Boston, 1991.
228. T. Tashiro, T. Tatsumi, M. Sugiyama, T. Hashimoto, and T. Morikawa, "A selective epitaxial SiGe/Si planar photodetector for Si-based OEIC's", IEEE Trans. Electron Devices **44**(4), pp. 545–550, 1997.
229. R. T. Carline, D. A. O. Hope, V. Nayar, D. J. Robbins, and M. B. Stanaway, "A vertical cavity longwave infrared SiGe/Si photodetector using a buried silicide mirror", in *IEEE Int. Electron Device Meeting*, pp. 891–894, Wash., D.C., 1997.
230. A. Schüppen, U. Erben, A. Gruhle, H. Kibbel, H. Schumacher, and U. König, "Enhanced SiGe heterojunction bipolar transistors with 160 GHz-f_{max}", in *IEDM Digest Technical Papers*, pp. 743–746, 1995.
231. J.-S. Rieh, D. Klotzkin, O. Qasaimeh, L.-H. Lu, K. Yang, L. P. B. Katehi, P. Bhattacharya, and E. T. Croke, "Monolithically integrated SiGe-Si PIN-HBT front-end photoreceiver", IEEE Photonics Technology Letters **10**(3), pp. 415–417, 1998.
232. J. M. Rabaey, *Digital integrated circuits: a design perspective*, Prentice Hall, New York, 1996.
233. S.-M. Kang and Y. Leblebici, *CMOS digital integrated circuits: analysis and design*, McGraw-Hill, New York, 1996.
234. T. A. DeMassa and Z. Ciccone, *Digital integrated circuits*, Wiley, New York, 1996.
235. K. R. Laker and W. M. C. Sansen, *Design of analog integrated circuits and systems*, McGraw-Hill, New York, 1994.
236. Dept. Electrical Engineering and Computer Sciences, University of California, Berkely, CA 94720, *BSIM3v3 manual*, 1995.
237. H. Shichman and D. Hodges, "Modelling and simulation of insulated-gate field-effect transistor switching circuits", IEEE J. Solid-State Circuits **3**(3), pp. 285–289, 1968.
238. S. Liu and L. W. Nagel, "Small-signal MOSFET models for analog circuit design", IEEE J. Solid-State Circuits **17**(6), pp. 983–998, 1982.
239. P. E. Allen and D. R. Holberg, *CMOS analog circuit design*, Holt, Rinehart, and Winston, New York, 1987.
240. P. R. Gray and R. G. Meyer, *Analysis and design of analog integrated circuits*, p. 39, Wiley, New York, 1993.
241. Technology Modeling Association, Inc., Palo Alto, CA, *SUPREM-3 manual*, 1994.
242. Technology Modeling Association, Inc., Palo Alto, CA, *TSUPREM-4 manual*, 1994.
243. P. Hoppe, *Übertragungsverhalten analoger Schaltungen*, p. 161, B. G. Teubner, Stuttgart, 1994.

244. K. Kieschnick and H. Zimmermann, "High-sensitivity BiCMOS OEIC for optical storage systems", IEEE Journal of Solid-State Circuits **38**(4), pp. 579–584, 2003.
245. K. Kieschnick, H. Zimmermann, and P. Seegebrecht, "BiCMOS OEIC with Enhanced Sensitivity for DVD Systems", in *Proc. 27th European Solid-State Circuits Conference (ESSCIRC)*, pp. 184–187, 2001.
246. J. Knorr and H. Zimmermann, "A transmission-line approach for modeling feedback resistors in integrated transimpedance amplifiers", IEEE Trans. Circuits and Systems I: Fundamental Theory and Applications **50**(9), 2003.
247. P. P. Sofiriadis and Y. Tsividis, "Integrators using a single distributed RC element", in *IEEE Symposium on Circuits and Systems (ISCAS)*, vol. 2, pp. 21–24, 2002.
248. X. Qinwei and P. Mazumder, "Efficient macromodeling for on-chip interconnects", in *Proceedings of ASP-DAC 2002: 7th Asia and South Pacific and the 15th International Conference on VLSI Design (Design Automation Conference)*, pp. 561–566, 2002.
249. J. Qian, S. Pullela, and L. Pillage, "Modelling the effective capacitance for the RC interconnect of CMOS gates", IEEE Trans. Computer-Aided Design of Integrated Circuits and Systems **13**(12), pp. 1526–1535, 1994.
250. K. Janchitrapongvej, V. Kawejan, S. Seatia, C. Benjangkaprasert, P. Tangtisanon, and O. Saingaroon, "Capacitive double layers uniformly distributed RC line and its applications to active filters", in *Proc. TENCON*, vol. 2, pp. 23–25, 2000.
251. R. Mongia, I. Bahl, and P. Bhartia, *RF and Microwave Coupled-Line Circuits*, pp. 65–74 and 90–97, Artech House, Inc., Norwood, MA, 1999.
252. G. Gonzalez, *Microwave transistor amplifiers*, pp. 4–25, Prentice Hall, Englewood Cliffs, N.J., 1984.
253. R. C. Jaeger, *Microelectronic circuit design*, p. 993, McGraw-Hill, New York, 1997.
254. P. R. Gray and R. G. Meyer, *Analysis and design of analog integrated circuits*, Wiley, New York, 1993.
255. K. Schneider and H. Zimmermann, "Design of low-noise transimpedance frontends for systems-on-a-chip", in *Proc. Austrochip*, pp. 115–118, 2001.
256. K. Ayadi, M. Kuijk, P. Heremans, G. Bickel, G. Borghs, and R. Vounckx, "A monolithic optoelectronic receiver in standard 0.7 μm CMOS operating at 180 MHz and 176 fJ light input energy", IEEE Photonics Technology Letters **9**(1), pp. 88–90, 1997.
257. J. F. Heanue, M. C. Bashaw, and L. Hesselink, "Volume holographic storage and retrieval of digital data", Science **265**(8), pp. 749–752, 1994.
258. M. E. Schaffer and P. A. Mitkas, "Smart photodetector array for page-oriented optical memory in 0.35–μm CMOS", IEEE Photonics Technology Letters **10**(6), pp. 866–868, 1998.
259. F. Esfahani, K.-O. Hofacker, A. Benedix, and H. H. Berger, "Small area optical inputs for high speed CMOS circuits", in *9th Int. IEEE ASIC Conference*, pp. 7–10, Rochester, N.Y., 1996.
260. H. H. Berger, J. Sturm, F. Esfahani, A. Benedix, S. von Aichberger, B. Müller, and K.-O. Hofacker, "Optical signal injection for high-speed wafer level function test of integrated circuits", in *IEEE Int. Conf. on Microelectronic Test Structures*, pp. 39–42, IEEE, Monterey, CA, 1997.

261. A. V. Krishnamoorthy, L. M. F. Chirovsky, W. S. Hobson, R. E. Leibenguth, S. P. Hui, G. J. Zydzik, K. W. Goossen, J. D. Wynn, B. J. Tseng, J. Lopata, J. A. Walker, J. E. Cunningham, and L. A. D'Asaro, "Vertical-cavity surface-emitting lasers flip-chip bonded to gigabit-per-second CMOS circuits", IEEE Photonics Technology Letters **11**(1), pp. 128–130, 1999.
262. T. C. Banwell, A. C. V. Lehmen, and R. R. Cordell, "VCSE laser transmitters for parallel data links", IEEE J. Quantum Electron. **29**(2), pp. 635–644, 1993.
263. D. L. Mathine, R. Droopad, and G. N. Maracas, "A vertical-cavity surface-emitting laser applied to a 0.8 μm NMOS driver", IEEE Photonics Technology Letters **9**(7), pp. 869–871, 1997.
264. C. Stanescu, S. Porumbescu, A. Hanganu, S. Costea, I. Mirea, G. Aungurecei, M. Furis, and B. Mihalea, "Bipolar preamplifier for monolithic OEIC receiver", in *Proc. Int. Semicond. Conference (CAS'97)*, pp. 567–570, Sinaia, Romania, 1997.
265. M. Seifart, *Analoge Schaltungen*, pp. 315–316, Verlag Technik, Berlin, 1996.
266. T. Lule, H. Fischer, S. Benthien, H. Keller, M. Sommer, J. Schulte, P. Rieve, and M. Böhm, "Image sensor with per-pixel programmable sensitivity in TFA technology", in H. Reichl and A. Heuberger, eds., *Micro System Technologies*, p. 675, 1996.
267. T. Hamamoto and K. Aizawa, "A computational image sensor with adaptive pixel-based integration time", IEEE J. Solid-State Circuits **36**(4), pp. 580–585, 2001.
268. M. Loose, K. Meier, and J. Schemmel, "A self-calibrating single-chip CMOS camera with logarithmic response", IEEE J. Solid-State Circuits **36**(4), pp. 586–596, 2001.
269. I. Stuttart, "The high-dynamic-range CMOS evaluation camera (HDRC)", in *Institut für Mikroelektronik Stuttgart, Germany, Publicity material*, 1997.
270. G. F. Marshall and S. Collins, "A high-dynamic-range front-end for automatic image processing applications", in *Proc. SPIE, Advanced Focal Plane Arrays and Electronic Cameras*, vol. 3410, 1998.
271. S. Kavadias, B. Dierickx, D. Scheffer, A. Alaerts, D. Uwaerts, and J. Bogaerts, "A logarithmic response CMOS image sensor with on-chip calibration", IEEE J. Solid-State Circuits **35**(8), pp. 1146–1152, 2000.
272. R. L. Geiger, P. E. Allen, and N. R. Strader, *VLSI design techniques for analog and digital circuits*, McGraw-Hill, New York, 1990.
273. O. Schrey, J. Huppertz, G. Filimonovic, W. Brockherde, and B. J. Hosticka, "A 1k × 1k high dynamic range CMOS image sensor with on-chip programmable region of interest readout", in *Proc. 27th European Solid-State Circuits Conference (ESSCIRC)*, pp. 124–127, 2001.
274. L. Viarani, D. Stoppa, L. Gonzo, M. Gottardi, and A. Simoni, "A CMOS smart pixel for active 3D vision applications", in *IEEE Proc. Sensors*, pp. 11–14, 2002.
275. Z. Xu, R. Schwarte, H. Heinol, B. Buxbaum, and T. Ringbeck, "Smart pixel — photonic mixer device (PMD)", in *Proc. M^2 VIP'98 — Int. Conf. on Mechatronics and Machine Vision in Practice*, pp. 259–264, 1998.
276. R. Jeremias, W. Brockherde, G. Doemens, B. Hosticka, L. Listl, and P. Mengel, "A CMOS photosensor array for 3D imaging using pulsed laser", in *ISSCC Dig. Tech. Papers*, pp. 252–253, 2001.
277. W. Budde, "Multidecade linearity measurements on Si photodiodes", Applied Optics **18**(10), pp. 1555–1558, 1979.

278. P. Seitz, "Smart pixels", in *IEEE Int. Symposium on High Performance Electron Devices for Microwave & Optoelectronic Applications*, pp. 229–234, 2001.
279. K. Engelhardt and P. Seitz, "High-resolution optical position encoder with large mounting tolerances", Applied Optics **36**(13), pp. 2912–2916, 1997.
280. S. U. Ay, S. Barna, and E. R. Fossum, "Differential mode CMOS active pixel sensor (APS) for optically programmable gate array (OPGA)", in *Proc. IEEE Workshop on Charge-Coupled Devices and Advanced Image Sensors*, pp. 161–164, 2001.
281. A. F. Fercher, "Optical coherence tomography", J. Biomedical Optics **1**(2), pp. 157–173, 1996.
282. S. Bourquin, P. Seitz, and R. P. Salathe, "Optical coherence topography based on a two-dimensional smart detector array", Optics Letters **26**(8), pp. 512–514, 2001.
283. M. Barbaro, P.-Y. Burgi, A. Mortara, P. Nussbaum, and F. Heitger, "A 100×100 pixel silicon retina for gradient extraction with steering filter capabilities and temporal output coding", IEEE J. Solid-State Circuits **37**(2), pp. 160–172, 2002.
284. J. G. Harris, C. Koch, E. Staats, and J. Luo, "Analog hardware for detecting discontinuities in early vision", Int. J. Computer Vision **4**, pp. 211–223, 1990.
285. A. G. Andreou, R. C. Meitzler, K. Strohbehn, and K. A. Boahen, "Analog VLSI neuromophic image acquisition and pre-processing systems", Neural Network **8**, pp. 1323–1347, 1995.
286. C.-Y. Wu and C.-F. Chiu, "A new structure of the 2-D silicon retina", IEEE J. Solid-State Circuits **30**, pp. 890–897, 1995.
287. L. Raffo, S. P. Sabatini, G. M. Bo, and G. M. Bisio, "Analog VLSI circuits as physical structures for perception in early visual tasks", IEEE Trans. Neural Networks **9**, pp. 1483–1494, 1998.
288. M. Ishikawa, K. Ogawa, T. Komuro, and I. Ishii, "A CMOS vision chip with SIMD processing element array for 1 ms image processing", in *IEEE Int. Solid-State Circuits Conference*, pp. 206–207, 1999.
289. R. Etienne-Cummings, J. van der Spiegel, and P. Mueller, "A foveated silicon retina for two-dimensional tracking", IEEE Trans. Circuits Syst. I **47**, pp. 504–517, 2000.
290. W. T. Freeman and E. L. Adelson, "The design and use of steerable filters", IEEE Trans. Pattern Anal. Mach. Intell. **13**, pp. 891–906, 1991.
291. A. Mortara and E. A. Vittoz, "A 12-transistor PFM demodulator for analog neural networks communication", IEEE Trans. Neural Networks **6**, pp. 1280–1283, 1995.
292. E. A. Vittoz, "Analog VLSI signal processing: Why, where, and how?", J. VLSI Signal Processing **8**, pp. 27–44, 1994.
293. Y. Perelman and R. Ginosar, "A low-light-level sensor for medical diagnostic applications", IEEE J. Solid-State Circuits **36**(10), pp. 1553–1558, 2001.
294. B. J. Hosticka, "CMOS sensor systems", Sensors and Actuators **A 66**, pp. 335–341, 1998.
295. G. de Graaf and R. F. Wolffenbuttel, "Smart optical sensor systems in CMOS for measuring light intensity and color", Sensors and Actuators **A 67**, pp. 115–119, 1999.
296. B. Fowler and A. El-Gammal, "Techniques for pixel level analog to digital conversion", in *Proc. SPIE, Infrared Readout Electronics IV*, vol. 3360, pp. 2–12, 1998.

297. E. Säckinger and W. Guggenbuhl, "A high swing, high impedance MOS cascode circuit", IEEE J. Solid-State Circuits **25**(1), pp. 289–297, 1990.
298. S. R. Norsworthy, R. Schreier, and G. C. Temes, *Delta-sigma data converters: theory, design and simulation*, IEEE Press, New York, 1997.
299. D. Droste and J. Bille, "An ASIC for Hartmann-Shack wavefront detection", IEEE J. Solid-State Circuits **37**(2), pp. 173–182, 2002.
300. E. D. Malacara, *Optical Shop Testing*, Wiley, New York, 1991.
301. F. Zernike, "Beugungstheorie des Schneidenverfahrens und seine verbesserte Form der Phasenkontrastmethode", Physica **1**, p. 689, 1934.
302. J. L. et al., "Winner-take-all networks of O(N) complexity", in *Proc. Neural Information Processing Systems (NIPS)*, p. 703, 1989.
303. J. A. S. et al., "CMOS current mode winner-take-all circuit with both excitory and inhibitory feedback", Electronics Letters **29**(10), 1993.
304. M. Mansuripur and G. Sincerbox, Proc. IEEE **85**(11), pp. 1780–1796, 1997.
305. M. Kyomasu, "Development of an integrated high speed silicon PIN photodiode sensor", IEEE Transactions on Electron Devices **42**(6), pp. 1093–1099, 1995.
306. M. Förtsch and H. Zimmermann, "A low-offset low-area 147-MHz CMOS DVD OEIC", in *Proc. Austrochip*, 2003.
307. H. Zimmermann, T. Heide, and A. Ghazi, "DVD OEIC and 1 Gbit/s Fiber Receiver in CMOS Technology", in *IEEE Int. Workshop on High Performance Electron Devices for Microwave & Optoelectronic Applications*, pp. 224–229, 2000.
308. B.-E. Kim, M.-S. Jeong, D.-M. Cho, J.-K. KIM, J.-S. Lee, S.-K. KIM, and S.-W. KIM, "0.8 μm CMOS analog front-end processor for CD-ROM", IEEE Trans. Consumer Electronics **42**(3), pp. 826–831, 1996.
309. M. Ingels, G. V. D. Plas, J. Crols, and M. Steyart, "A CMOS 18 THzΩ 240 Mb/s transimpedance amplifier and 155 Mb/s LED-driver for low cost optical fiber links", IEEE J. Solid-State Circuits **29**(12), pp. 1552–1559, 1994.
310. M. Ingels and M. Steyart, "A 1-Gb/s, 0.7-μm CMOS optical receiver with full rail-to-rail output swing", IEEE J. Solid-State Circuits **34**(7), pp. 971–976, 1999.
311. T. Takimoto, N. Fukunaga, M. Kubo, , and N. Okabayashi, "High speed Si-OEIC(OPIC) for optical pickup", IEEE Trans. Consumer Electronics **44**(1), pp. 137–142, 1998.
312. P. R. Gray and R. G. Meyer, *Analysis and design of analog integrated circuits*, p. 456, Wiley, New York, 1993.
313. J. Qi, C. L. Schow, L. D. Garrett, and J. C. Campbell, "A silicon NMOS monolithically integrated optical receiver", IEEE Photonics Technology Letters **9**(5), pp. 663–665, 1997.
314. C. L. Schow, J. D. Schaub, R. Li, J. Qi, and J. C. Campbell, "A monolithically integrated 1-Gb/s silicon photoreceiver", IEEE Photonics Technology Letters **11**(1), pp. 120–121, 1999.
315. J. D. Schaub, R. Li, S. M. Csutak, and J. C. Campbell, "High-speed monolithic silicon photoreceivers on high resistivity and SOI substrates", IEEE J. Lightwave Technology **19**(2), pp. 272–278, 2001.
316. K. Kieschnick, H. Zimmermann, H. Pless, and P. Seegebrecht, "BiCMOS receiver OEIC for optical interconnects", in *Ext. Abstr. 3nd IEEE Workshop on Signal Propagation on Interconnects, Neustadt, Germany*, pp. 72–73, 1999.

317. R. Swoboda and H. Zimmermann, "A low-noise monolithically integrated 1.5 Gb/s optical receiver in 0.6 μm BiCMOS technology", **9**(2), 2003.
318. R. Li, J. D. Schaub, S. M. Csutak, and J. C. Campbell, "A high-speed monolithic silicon photoreceiver fabricated on SOI", IEEE Photonics Technology Letters **12**(8), pp. 1046–1048, 2000.
319. S. M. Csutak, J. D. Schaub, W. E. Wu, R. Shimer, and J. C. Campbell, "High-speed monolithically integrated silicon photoreceivers fabricated in 130-nm CMOS technology", J. Lightwave Technology **20**(9), pp. 1724–1729, 2002.
320. T. Heide and H. Zimmermann, "Investigation of optical interconnect receivers in standard micron and sub-micron MOS technology", Optical Enginneering **42**(3), pp. 773–786, 2003.
321. R. C. Jaeger, *Microelectronic circuit design*, p. 934, McGraw-Hill, New York, 1997.
322. C. Rooman, M. Kuijk, R. Windisch, R. Vounckx, G. Borghs, A. Plichta, M. Brinkmann, K. Gerstner, R. Strack, P. van Daele, W. Woittiez, R. Baets, and P. Heremans, "Inter-chip optical interconnects using imaging fiber bundles and integrated CMOS detectors", in *Proc. 27th European Conf. on Optical Communication (ECOC)*, pp. 296–297, 2001.
323. G. Williams, "Lightwave receivers", in L. Tingye, ed., *Topics in lightwave systems*, pp. 79–148, Academic Press, New York, 1991.
324. T. K. Woodward, A. V. Krishnamoorthy, R. G. Rozier, and A. L. Lentine, "Low-power, small-footprint gigabit Ethernet compatible optical receiver circuit in 0.25 μm CMOS", Electronics Letters **36**(17), pp. 1489–1490, 2000.
325. K. Phang and D. Johns, "A CMOS optical preamplifier for wireless infrared communications", IEEE Trans. Circuits Syst. **46**(7), pp. 852–859, 1999.
326. T. Heide, A. Ghazi, and H. Zimmermann, "High speed optical PIN-CMOS-receiverss", in *IEEE Int. Workshop on High Performance Electron Devices for Microwave & Optoelectronic Applications*, pp. 72–76, 1998.
327. H. Zimmermann, T. Heide, and H. Pless, "High-Performance Receivers for Optical Interconnects in Standard MOS Technology", in *Optoelectronic Interconnects VIII*, pp. 1–9, SPIE Proc. Vol. 4292, 2001.
328. T. Heide, *Monolithische Lasertreiber- und Empfängerschaltkreise in CMOS-Technologie für die optische Kurzstreckenübertragung*, Ph.D. Thesis, Christian-Albrechts University of Kiel, 2000.
329. H. Zimmermann and T. Heide, "A monolithically integrated 1-Gb/s optical receiver in 1-μm CMOS technology", IEEE Photonics Technology Letters **13**(7), pp. 711–713, 2001.
330. G. Agrawal, *Fiber-Optic Communication Systems*, Wiley, New York, 1997.
331. K. Phang and D. A. Johns, "A CMOS optical preamplifier for wireless infrared communications", IEEE Trans. Circuits and Systems–II: Analog and Digital Signal Processing **46**(7), pp. 852–859, 1999.
332. Infrared Data Association, [Online]: http://www.irda.org, *Serial Infrared Physical Layer Link Specification*, 1997.
333. L. A. D. van den Broeke and A. J. Nieuwkerk, "Wide-band integrated optical receiver with improved dynamic range using a current switch at the input", IEEE Journal of Solid-State Circuits **28**(7), pp. 862–864, 1993.
334. P. Palojarvi, T. Ruotsalainen, and J. Kostamovaara, "A variable gain transimpedance amplifier channel with a timing discriminator for a time-of-flight laser radar", in *Proc. European Solid-State Circuits Conference*, pp. 384–387, 1997.

335. R. G. Meyer and W. D. Mack, "A wide-band low-noise variable-gain BiCMOS transimpedance amplifier", IEEE Journal of Solid-State Circuits **29**(6), pp. 701–706, 1994.
336. H. Khorramabadi, L. D. Tzeng, and M. J. Tarsia, "A 1.06 Gb/s-31 dBm to 0 dBm BiCMOS optical preamplifier featuring adaptive transimpedance", in *IEEE Int. Solid-State Circuits Conference*, pp. 54–55, 1995.
337. M. B. Ritter, F. Gfeller, W. Hirt, D. Rogers, and S. Gowda, "Circuit and system challenges in IR wireless communication", in *ISSCC*, pp. 398–399, 1996.
338. C. Petri, S. Rocchi, and V. Vignoli, "High dynamic CMOS preamplifiers for QW diodes", Electronics Letters **34**(9), pp. 877–878, 1998.
339. A. Tanabe, M. Soda, Y. Nakahara, A. Furukawa, T. Tamura, and K. Yoshida, "A single-chip 2.4 Gb/s CMOS optical receiver IC with low substrate crosstalk preamplifier", in *ISSCC*, pp. 304–305, 1998.
340. R. G. Swartz, Y. Ota, M. J. Tarsia, and V. D. Archer, "A burst mode, packet receiver with precision reset and automatic dark level compensation for optical bus communications", in *Proc. Symp. VLSI Technology*, pp. 67–68, 1993.
341. J. L. Hullett and S. Moustakas, "Optimum transimpedance broadband optical preamplifier design", Opt. Quantum Electron. **13**, pp. 65–69, 1981.
342. R. Coppoolse, J. Verbeke, P. Lambrecht, J. Codenie, and J. Vandewege, "Comparison of a bipolar and a CMOS front end in broadband optical transimpedance amplifiers", in *Proc. 38th Midwest Symp. Circuits and Systems*, pp. 1026–1029, 1996.
343. A. S. Sedra and K. C. Smith, *Microelectronic Circuits*, pp. 718–722, Oxford Univ. Press, Oxford, 1997.
344. T. Nakahara, H. Tsuda, K. Tateno, S. Matsuo, and T. Kurokawa, "Hybride integration of GaAs pin-photodiodes with CMOS transimpedance amplifier circuits", Electronics Letters **34**(13), pp. 1352–1353, 1998.
345. T. Nakahara, H. Tsuda, N. Ishihara, K. Tateno, and C. Amano, "High-sensitivity 1 Gbit/s CMOS receiver integrated with GaAs- or InGaAs-photodiode by wafer-bonding", Electronics Letters **37**(12), pp. 781–783, 2001.
346. T. M. et al., "45 GHz transimpedance 32 dB limiting amplifier and 40 Gbps 1:4 high-sensitivity demultiplexer with decision circuit using SiGe HBTs for 40 Gbps optical receiver", in *ISSCC Dig. Tech. Papers*, pp. 60–61, 2000.
347. F. T. Chien and Y. J. Chan, "Bandwidth enhancement of transimpedance amplifier by a capacitive peaking design", IEEE J. Solid-State Circuits **34**(8), pp. 1167–1170, 1999.
348. C.-H. Lu and W.-Z. Chen, "Bandwidth enhancement techniques for transimpedance amplifier in CMOS technologies", in *Proc. 27th European Solid-State Circuits Conference (ESSCIRC)*, pp. 192–195, 2001.
349. S. S. Mohan, M. Hershenson, S. P. Boyd, and T. H. Lee, "Bandwidth extension in CMOS with optimized on-chip inductors", IEEE J. Solid-State Circuits **35**(3), pp. 346–355, 2000.
350. H. Zimmermann, *Integrated silicon optoelectronics*, p. 129, Springer, Berlin, Heidelberg, 2000.
351. R. Swoboda, J. Knorr, and H. Zimmermann, "Speed-enhanced OEIC with voltage-up-converter", Electron. Lett. **39**(1), pp. 112–113, 2003.
352. J. F. Dickson, "On-chip high-voltage generation in NMOS integrated circuits using an improved voltage multiplier technique", IEEE J. Solid-State Circuits **11**(3), pp. 374–378, 1976.

353. H. Zimmermann, R. Swoboda, K. Schneider, and J. Knorr, "Comparison of CMOS and BiCMOS Optical Receiver SoCs", in *VLSI Circuits and Systems Conference at SPIE's Int. Symp. on Microtechnologies for the New Millenium*, pp. 598–609, SPIE Proc. Vol. 5117, 2003.
354. J. Knorr, R. Swoboda, and H. Zimmermann, "Speed-enhanced OEIC with area-efficient charge pump and shunt regulator", in *IEEE Int. Symposium on High Performance Electron Devices for Microwave & Optoelectronic Applications (EDMO)*, 2003.
355. S. Brigati, F. Francesconi, D. Gardino, M. Poletti, and A. Maglione, "A 20 Mbit/s integrated photoreceiver with digital outputs in 0.6 µm CMOS technology", in *Proc. 28th European Solid-State Circuits Conference (ESSCIRC)*, pp. 503–506, 2002.
356. S. M. Park and C. Papavassiliou, "On the design of low-noise, Giga-Hertz bandwidth preamplifiers for optical receiver applications", in *Proceedings of ICECS*, vol. 2, pp. 785–788, 1999.
357. M. Förtsch, H. Zimmermann, W. Einbrodt, K. Bach, and H. Pless, "Integrated PIN photodiodes in high-performance BiCMOS technology", in *IEDM Digest Technical Papers*, pp. 801–804, 2002.
358. M. Förtsch and H. Zimmermann, "Integrierter optischer Sensor mit geteilter Fotodiode zur Grenzfrequenzsteigerung", in *Informationstagung Mikroelektronik*, Vienna, 2003.
359. K. Mori, T. Akashi, Y. Tochio, H. Nobuhara, K. Tanaka, N. Nagase, Y. Akimoto, H. Kitasagami, Y. Shimizu, T. Yaname, A. Abe, and M. Kawai, "155.52 Mb/s optical tranceiver modules for ONU/OLT on ATM-PON systems", in *Proc. European Conference on Optical Communication (ECOC)*, pp. 363–366, 1997.
360. S. Yamashita, S. Ide, K. Mori, A. Hayakawa, N. Ueno, and K. Tanaka, "Novel cell-AGC technique for burst-mode CMOS preamplifier with wide dynamic range and high sensitivity for ATM-PON system", in *Proc. 27th European Solid-State Circuits Conference (ESSCIRC)*, pp. 200–203, 2001.
361. Y. Ota and R. G. Swartz, "Burst-mode compatible optical receiver with a large dynamic range", J. Lightwave Technology **8**(12), pp. 1897–1902, 1990.
362. D. Yamazaki, N. Nagase, H. Nobuhara, T. Funaki, and K. Wakao, "156 Mbit/s preamplifier IC with wide dynamic range for ATM-PON application", Electronics Letters **33**(15), pp. 1308–1309, 1997.
363. M. Nakamura, N. Ishihara, Y. Akazawa, and H. Kimura, "An instantaneous response CMOS optical receiver IC with wide dynamic range and evtremely high sensitivity using feed-forward auto-bias adjustment", IEEE J. Solid-State Circuits **30**(9), pp. 991–997, 1995.
364. M. Nakamura, N. Ishihara, and Y. Akazawa, "A 156-Mb/s CMOS optical receiver burst-mode transmission", IEEE J. Solid-State Circuits **33**(8), pp. 1179–1187, 1998.
365. M. Nakamura and N. Ishihara, "1.2 V, 35 mW CMOS optical transceiver ICs for 50 Mbit/s burst-mode communication", Electron. Lett. **35**(5), pp. 394–395, 1999.
366. K. Phang and D. A. Johns, "A 1 V 1 mW front-end with on-chip dynamic gate biasing for a 75 Mb/s optical receiver", in *ISSCC*, pp. 218–219, 2001.
367. V. Peluso, P. Vancorenland, M. Steyaert, and W. Sansen, "900 mV differentail class AB OTA for switched opamp applications", Electron. Lett. **33**(17), pp. 1455–1456, 1997.

368. J. Wilson and J. F. B. Hawkes, *Optoelectronics*, p. 310, Prentice Hall, New York, 1989.
369. A. L. Lentine, K. W. Goossen, J. A. Walker, J. E. Cunningham, W. Y. Jan, T. K. Woodward, A. V. Krishnamoorthy, B. J. Tseng, S. P. Hui, R. E. Leibenguth, L. M. F. Chirovsky, R. A. Novotny, D. B. Buchholz, and R. L. Morrison, "Optoelectronic VLSI switching chip with greater than 1 Tbit/s potential optical I/O bandwidth", Electronics Letters **33**(10), pp. 894–895, 1997.
370. A. V. Krishnamoorthy, R. G. Rozier, T. K. Woodward, P. Chandramani, K. W. Goossen, B. J. Tseng, J. A. Walker, W. Y. Jan, and J. E. Cunningham, "Triggered receivers for optoelectronic VLSI", Electronics Letters **36**(3), pp. 249–250, 2000.
371. K. G. Moerschel, T. Y. Chiu, W. A. Possanza, K. S. Lau, R. G. Swartz, R. A. Mantz, T. Y. M. Liu, K. F. Lee, V. D. Archer, G. R. Hower, G. T. Mazsa, R. E. Carsia, J. A. Pavlo, M. P. Ling, J. L. Dolcin, F. M. Erceg, J. J. Egan, C. J. Fassl, J. T. Glick, and M. A. Prozonic, "BEST: a BiCMOS-compatible super-self-aligned ECL technology", in *IEEE Custom Integrated Circuits Conference*, pp. 18.3.1–18.3.4, 1990.
372. T. Ringbeck, R. Schwarte, and B. Buxbaum, "Introduction of a new opto-electrical phase locked loop in CMOS technology, the PMD-PLL", in *Proc. SPIE*, vol. 3850, pp. 108–115, 1999.

Index

Q-factor, 145
π equivalent circuit, 126
g_m enhancement technique, 293, 295
1/f-noise, 170, 206

a-Si:H, 66
a-Si:H imaging device, 66
a-Si:H photodetector, 66
absorption coefficient, 3, 17, 66
ACCUSIM, 102
active inductors, 294, 295
active-pixel sensor, 62
adaptive image sensor, 165
AIDA, 162
ambient light, 283
ambient photocurrent rejection circuit, 285
amorphous silicon image sensor, 66
analog image detector array, 162
analog-digital system, 211
antireflection coating, 17, 109
application specific integrated circuit, 67, 102
artificial vision, 68
ASIC, 67, 102, 163
ATM-PON, 310
autodoping, 33
avalanche effect, 91, 264

back-gate effect, 103, 248
background light, 208
bandgap, 66
bandwidth, 12, 34, 43, 79, 83, 105, 107, 113, 122, 155, 224
base push-out effect, 33
base-collector diode, 30
basic machine vision, 181
BER, 142
bias current cancellation, 224, 238
BiCMOS
  high-performance BiCMOS, 74
  triple-diffused BiCMOS, 73
BiCMOS OEICs, 222
bipolar transistor, 103
bit error analyzer, 143
bit error rate, 41, 142
bit rate, 29, 41, 79
bit-AGC, 311
blooming, 61
body effect, 205
bond wire inductance, 21
bottom level detector, 311
BSIM3.3, 102
buried collector, 26, 28
buried-channel CCD, 60, 62
burst mode, 310

CADENCE design framework, 102
camera-on-a-chip, 62, 166
capacitance, 22, 23, 107, 113, 245
carrier diffusion, 15, 36
carrier lifetime, 14
carrier mobilities, 9
cascode stage, 161
cascode transistor, 268
CCD image sensor, 58
CD-OEIC, 161
CD-ROM, 212
CDS, 166, 170, 175, 176
cell-AGC, 311
channel length modulation effect, 105
charge pump, 315
charge sensitive amplifier, 37
chemoluminant diagnostics, 192
circuit diagram, 106
circuit simulation, 101

CMOS image sensor, 171
CMOS photonic mixer device, 173
CMOS PMD, 173, 320
CMOS process, 35
CMOS-OEIC, 213
CoC, 62
color detector, 40
color image sensor, 61
combustion monitoring, 31
communication analyzer, 143
compact disk, 160
compensation capacitor, 118
complementary bipolar process, 159
COMPOSER, 106
conductance, 102
correlated double sampling, 166, 170, 175, 176
critical thickness, 95
cross-talk, 162, 252
current comparator, 154, 171
current mirror, 155, 223, 238
current-injection technique, 295
current-starved VCO, 322
current-to-frequency conversion, 193
current/voltage converter, 110

DASP, 92
data rate, 11, 29–31, 267
DBR, 94
decision threshold, 142
DESIGN ARCHITECT, 106
design flowchart, 107
design rule check, 106
design rule file, 106
device simulation, 107
Dickson charge pump, 297, 301
die-as-package, 92
dielectric function, 17
difference amplifier, 222
differential transimpedance amplifier, 284
diffusion coefficient, 13
diffusion length, 13
digital calibration, 166
digital sampling oscilloscope, 143
digital signal processor, 182
digital-versatile-disk, 212
digital-video-disk, 212
digital-video-recording, 212
displacement current, 12
disposable medical probe, 192
distance sensor, 177
distributed RC line, 124
distributed-Bragg-reflector, 94
double photodiode, 38, 81, 84, 113, 222, 225, 237, 245, 274
double-pin photodiode, 274
DRC, 106
drift time, 10
drift velocity, 10, 11
drift zone, 10
drift-diffusion model, 6
driver capability, 106
DSP, 182
DVD, 212
DVD-OEIC, 212
DVR, 212
dynamic gate biasing, 315
dynamic range, 162, 283

Early effect, 105
effective transimpedance, 276
Einstein relation, 13
electromagnetic interference, 212
electron–hole pair, 3, 5
electronic noise, 139
emitter-base photodiode, 30
equivalent circuit, 21
extinction coefficient, 17
extinction ratio, 145, 252
extraction, 106

Fabry–Perot cavity, 89
fall time, 11, 28, 34, 40, 52, 83, 271
fast case, 105
feedback capacitor, 129
feedback network, 133
FET photoreceiver preamplifier, 139
fiber-in-board, 253
fiber-to-the-home, 314
Fibre Channel, 91
figure of merit, 214
fingerprint detector, 171
first-order sigma-delta ADC, 193
fixed pattern noise, 166, 170, 207
flame detection, 31, 159
flicker noise, 206
flip-chip bonding, 156, 158, 317

flip-flop, 148
flux density, 2
focal plane, 198
focal point position detection, 200
FOM, 214
four-quarter photodiode, 307
four-stage current amplifier, 239
Fourier optics, 200
FPN, 166, 168, 170, 207
free space optical interconnects, 246
free-space infrared data link, 283
frequency response, 16, 40, 223, 225
FTTH, 314
full custom design, 107
fundamental absorption, 3

gain-bandwidth-product, 106
gain-peaking, 112
generation rate, 5
Gigabit Ethernet, 91
global sensitivity control, 163
group delay, 115

Hartmann-Shack sensor, 198
HBT, 95, 97
heterojunction bipolar transistor, 95, 97, 317
high-bandwidth OS-BiCMOS-OEIC, 225
high-dynamic-range imager, 166
high-performance photo receivers, 247
high-speed imaging, 164
histogram, 143
HSPICE, 102
HSS, 198
hybrid CMOS VCSEL integration, 155
hybrid integration, 92
hydrogenated amorphous silicon, 66

inductive peaking technique, 293
Infrared Data Association standards, 283
infrared wireless links, 314
input capacitance, 21
input impedance, 125
input resistance, 21, 137
input-node capacitance, 140
integration capacitance, 162
integration time, 162
interference, 88, 89
interline transfer CCD, 60
internal photoeffect, 5
interpolating function, 183
interrupted-SP/N-epi photodiode, 86
intrinsic carrier density, 14
IR wireless receiver, 283
IrDA standards, 283

junction field-effect transistor, 139

Kirk effect, 33

large-area photodetectors, 310
large-signal model, 103
laser driver, 208
latch-up, 53
lateral photodiode, 37
lateral phototransistor, 171
lateral PIN photodiode, 43, 87, 90, 91, 242
layout, 106
layout-versus-schematic, 106
LBPT, 68
LDD MOSFET, 35
lens array, 198
local area network, 91
locally adaptive sensor, 163
LOCOS, 35
logarithmic CMOS camera, 166
low-area CMOS OEIC, 213
low-noise amplifier, 208
low-noise laser driver, 210
LPD, 83
LVS, 106

matching, 215
MDSI, 176
MENTOR design framework, 102
microlens, 62
Miller capacitance, 161
Miller effect, 295
minority carrier diffusion, 13
minority carrier lifetime, 14
mismatch, 207, 236
mixed-signal chip, 211
modified y-parameters, 125
MOSFET, 102
movement detection system, 208

multiple double short time integration, 176

N^+ collector plug, 28
N^+/N-well/P-sub. photodiode, 84
N^+P photodiode, 28, 80
N-well/P-substrate diode, 37
NEP, 178
netlist, 106
noise, 139
noise bandwidth, 208
noise distance, 143
noise-equivalent-power, 178
normalized complex frequency, 126

OCT, 181
OE-PLL, 320
OEIC, 22, 28, 30, 50, 83, 97, 101, 102, 107, 160, 224
offset compensation, 236
OLT, 310
ONU, 310
OPD, 199
open-circuit time-constant, 258
open-circuit time-constant method, 270
open-loop gain, 224, 225
operating point, 101
operating temperature range, 105
operational amplifier, 110, 224, 236
operational transconductance amplifier, 261
optical backplane, 246, 253
optical clock distribution, 31
optical data transmission, 240
optical distance measurement, 172
optical inputs, 153
optical interconnects, 92, 246, 252
optical line terminal, 310
optical low-coherence tomography, 181
optical memories, 150
optical network unit, 310
optical path difference, 199
optical power, 2
optical sensor, 306
optical storage systems, 212
optical wavefront measurements, 198
opto-electrical phase-locked-loop, 320
optoelectronic integrated circuits, 22
OS-BiCMOS-OEIC, 222
OS-OEIC, 212
OTA, 261
output resistance, 137
oversampling ratio, 193
oxide isolation, 25

P^+N photodiode, 39, 149
page-oriented optical memories, 150
parallel optical interconnects, 147
parasitic capacitance, 103, 121
parasitic resistor, 103
passive optical network, 310
penetration depth, 3
phase, 115
phase margin, 118
photo-gate active-pixel sensor, 63
photodiode protection mask, 48
photodiode-type active-pixel sensor, 63
photon, 1
photon energy, 2
photonic mixer device, 71, 320
photoreceiver, 21
phototransistor, 30, 31, 68
  lateral bipolar phototransistor, 68
  vertical bipolar phototransistor, 68
PIN FET preamplifier, 140
PIN photodiode, 29, 34, 50, 79, 97, 107, 113, 213, 244, 253
plastic optical fibers, 306
PMD, 71, 173
PN photodiode, 36, 213
POF, 306
polyimide bonding, 93
polysilicon emitter, 25
polysilicon resistor, 121, 161
POM, 150
postlayout simulation, 107
process complexity, 86
process definition file, 106
process flow, 25, 44, 74
process simulation, 107
punch-through effect, 56

quantum efficiency, 16, 52, 88, 89, 107, 267
  dynamical quantum efficiency, 19, 20
  external quantum efficiency, 17
  internal quantum efficiency, 18, 19
  optical quantum efficiency, 17

stationary quantum efficiency, 19
quartered photodiode, 309

RC interconnect, 124
RC line, 124
reach-through effect, 56
recessed oxide isolation, 25
recombination center, 14
recombination rate, 14
reference photodiode, 154
reflectivity, 17
regions of interest, 170
regulated cascode configuration, 203
replica offset cancellation, 214
replica transimpedance amplifier, 213
resonant photodetector, 89
responsivity, 18, 34, 37, 41, 89
rise time, 11, 28, 34, 40, 52, 83, 271
ROC, 214
ROI, 170

saturation velocity, 10
schematic entry, 106
Schmitt trigger, 172
segmented fourier phase detector, 179
self-biasing cascode amplifier, 174
self-calibration, 166
sense-amplifier, 148
sense-amplifier flip-flop, 150
sensitivity, 145, 213, 224, 225, 270
sensitivity control, 163
sensitivity enhancement, 309
series resistance, 22, 36
SFPD, 179
sheet resistance, 122
shot noise, 141, 197, 208
shunt regulator, 301
shunt-shunt feedback, 133
SiGe HBT receiver, 241
sigma-delta modulator, 193
signal-to-noise ratio, 164, 174, 177, 193
silicon-on-insulator, 87
silicon-on-sapphire, 92, 265
single-bit first-order Σ-Δ modulator, 193
single-chip CMOS camera, 166
slow case, 105
small-signal equivalent circuit, 102
small-signal model, 102
SML detector, 41, 43
SNR, 177, 193
SOI, 87, 264
BESOI, 88
bonded and etched-back SOI, 88
laser-recrystallized SOI, 88
SIMOX, 91
SOS, 92, 265
space-charge capacitance, 21
space-charge region, 7
spatially-modulated-light detector, 41
SPECTRE-S, 102
speed enhancement, 297, 309
speed-responsivity product, 320
spiral inductors, 294
stability, 129
standard buried collector, 25
steering functions, 182
storage mode, 37
sub-micrometer PIN-CMOS-OEIC, 282
substrate removal, 94
subthreshold operation, 195
switched-capacitor amplifier, 170
switched-capacitor circuits, 175
synchronous receiver, 148
systems-on-a-chip, 314

T-type feedback, 160
TDMA, 310
technological flowchart, 110
test of digital circuits, 153
TFA, 67, 162, 164
TFA sensor, 67
thermal noise, 139
thin film on ASIC, 67, 162
time constant, 126
time division multiple access, 310
time-of-flight, 172, 174, 176
time-of-flight method, 71
TOF, 172, 176
total transimpedance, 276
tracking ADC, 208, 209
transconductance, 101, 105
transfer function, 111
transient behavior, 28
transient response, 16, 40
transimpedance amplifier, 110, 222, 236, 264

transimpedance-bandwidth product, 251, 281
transistor models, 102
transistor parameters, 52, 102
transistor transimpedance amplifier, 118, 227
transit frequency, 75, 224, 225
transit time, 12
transmission-line, 123
transresistance amplifier, 133, 137
trench isolation, 25, 88
twin-well CMOS process, 36, 44
two-phase CCD, 61
two-stage amplifier, 236

universally applicable OEIC, 85
UPD, 83
UV sensor, 32
UV-sensitive OEIC, 159

varactor analog image detector, 68
variable-gain transimpedance amplifiers, 284
VBPT, 68
vehicle guidance system, 163
vehicle imaging systems, 164
vertical PIN photodiode, 44
voltage doubler, 315
voltage follower, 113
voltage-up-converter, 297, 301
volume holographic storage systems, 150
VUC, 297

waveguide-in-board, 253
wavelength, 2
wide dynamic range imaging, 164
winner-take-all algorithm, 200
wireless infrared data communication, 283
worst case, 105
WTA algorithm, 200

y-parameters, 124, 133

Zernike polynomials, 199

www.ingramcontent.com/pod-product-compliance
Ingram Content Group UK Ltd.
Pitfield, Milton Keynes, MK11 3LW, UK
UKHW021857190726
13853UKWH00003B/1308
* 9 7 8 3 6 6 2 0 9 9 0 5 6 *